Rethinking the Foundations of Statistics

This important collection of essays is a synthesis of foundational studies in Bayesian decision theory and statistics. An overarching topic of the collection is how the norms for Bayesian decision making should apply in settings with more than one rational decision maker. The essays then trace out some of the consequences of this turn for Bayesian statistics.

There are four principal themes to the collection: cooperative, non-sequential decisions; the representation and measurement of "partially ordered" preferences; non-cooperative, sequential decisions; and pooling rules and Bayesian dynamics for sets of probabilities.

The volume will be particularly valuable to philosophers concerned with decision theory, probability, and statistics, and to statisticians, mathematicians, and economists.

Cambridge Studies in Probability, Induction, and Decision Theory

General editor: Brian Skyrms

Ellery Eells, *Probabilistic Causality*
Richard Jeffrey, *Probability and the Art of Judgment*
Robert C. Koons, *Paradoxes of Belief and Strategic Rationality*
Cristina Bicchieri and Maria Luisa Dalla Chiara (eds.), *Knowledge, Belief, and Strategic Interactions*
Patrick Maher, *Betting on Theories*
Cristina Bicchieri, *Rationality and Coordination*
J. Howard Sobel, *Taking Chances*
Jan von Plato, *Creating Modern Probability: Its Mathematics, Physics, and Philosophy in Historical Perspective*
Ellery Eells and Brian Skyrms (eds.), *Probability and Conditionals*
Cristina Bicchieri, Richard Jeffrey, and Brian Skyrms (eds.), *The Dynamics of Norms*
Patrick Suppes and Mario Zanotti, *Foundations of Probability with Applications*
Paul Weirich, *Equilibrium and Rationality*
Daniel Hausman, *Causal Asymmetries*
William Dembski, *The Design Inference*
James M. Joyce, *The Foundations of Causal Decision Theory*

Rethinking the Foundations of Statistics

JOSEPH B. KADANE

MARK J. SCHERVISH

TEDDY SEIDENFELD

CAMBRIDGE
UNIVERSITY PRESS

32 Avenue of the Americas, New York NY 10013-2473, USA

Cambridge University Press is part of the University of Cambridge.

It furthers the University's mission by disseminating knowledge in the pursuit of education, learning and research at the highest international levels of excellence.

www.cambridge.org
Information on this title: www.cambridge.org/9780521640114

First published 1999

A catalogue record for this publication is available from the British Library

Library of Congress Cataloguing in Publication data

Kadane, Joseph B.
Rethinking the foundations of statistics / Joseph B. Kadane, Mark J. Schervish, Teddy Seidenfeld.
p. cm. – (Cambridge studies in probability, induction, and decision theory)
Includes bibliographical references.
ISBN 0-521-64011-3 (hbk.). – ISBN 0-521-64975-7 (pbk.)
1. Bayesian statistical decision theory. I. Schervish, Mark J. II. Seidenfeld, Teddy. III. Title. IV. Series.
QA279.5.K36 1999
519.5'42-dc21 98-48322
CIP

ISBN 978-0-521-64011-4 Hardback
ISBN 978-0-521-64975-9 Paperback

To our families:
Caroline
Nancy, Margaret, and Meredith
Janice, David, and Adina

and to Jimmie Savage and Morrie DeGroot

Contents

Introduction

In the seventeen years between the first (1954) and second (1971) editions of his book *The Foundations of Statistics*, Savage reported a change in the "climate of opinion" about foundations. That change, he said, "would obliterate rather than restore" his earlier thinking about the relationship between the two major schools of statistics that were the subject of his inquiry. What in the early 1950s started out for Savage as a task of building Bayesian expected utility foundations for common, so-called Frequentist statistics – which for Savage included the Fisher-Neyman-Pearson-Wald program that was dominant in the British-American school from the 1930s – revealed itself, instead, to be an impossibility. Contrary principles separated Bayesian decision theory from what practicing statisticians of the day were taught to do. Significance tests, tests of hypotheses, and confidence intervals give quantitative indices, such as confidence levels, that only accidentally and approximately cohere with the Bayesian theory that Savage hoped might elucidate and justify them.

Since the second edition of *The Foundations of Statistics* and Savage's premature death (both in 1971), we have come to understand much better the extent of the conflict between Bayesian and Frequentist statistical principles of evidence. Some limitations in Frequentist methods, which existed only in the lore of practicing statisticians, gained theoretical footing through the Bayesian point of view. Consider the very general problem of how, within the Frequentist program, to deal with conditional inference, e.g., whether or not to condition on an ancillary statistic. From a Bayesian point of view, conditioning on any ancillary does not affect the resulting posterior distribution. Not so within the Frequentist program, where it is (now) widely recognized as an open challenge.

One version of this problem is optional stopping – whether or not to condition on the experimenter's stopping rule and the resulting changes in Frequentist significance levels that makes. This matter was brought to the fore only in the late 1950s and early 1960s, with much attendant surprise. But by now it has lost its air of "paradox" and is just an accepted part of the difference between the two schools. For example, for the second (1986) edition of *Testing Statistical Hypotheses*, Lehmann added a chapter (10) on "Conditional Inference," noting there that the "discussion will be more tentative than in earlier chapters, and will focus on conceptual aspects more than on technical ones" (p. 539). In our concluding essays (3.7 and 3.8) we revisit the topic of optional stopping. We reach a conclusion only slightly modified from the one Savage did: that the line between Bayesian and Frequentist inference regarding optional stopping is sharp, with the sole exception of reasoning from "improper" priors and finitely additive probability.

Why, then, should anyone continue worrying about foundations? Our view is that foundational studies not only provide for clarification across rival methodologies, but also help point the way to improvements. With that as our banner, this volume collects together sixteen of our essays that deal with what we think are six open issues for Bayesian decision theory and statistics, described below. The reader is alerted that we are not interested, primarily, in issues of classification. Trying to draw a fixed line between what is Bayesian and what is not is futile when that boundary is shifting, as we mean for it to be. Instead, we seek to understand better the scope and limitations of current Bayesian theory with the goal of contributing to its positive growth.

Much of our work on decision theory builds upon one or another of the several foundations laid by Savage (1954), deFinetti (1974), and von Neumann and Morgenstern (1947), together with Anscombe and Aumann (1963). All of these authors arrived at a normative theory of decision making, but by somewhat different means. Fishburn (1970) gives an excellent overview of these and other attempts to lay foundations for decision theory. Here we describe these four theories briefly in order to relate them to the work that we present in our essays.

deFinetti assumed that a statement such as "The probability of event E is 0.6" could be interpreted as a willingness to offer (accept) 0.6 units of some currency for the chance to win (lose) 1.0 units if the event E occurs. That is, the agent who makes such a probability statement believes that the gamble is fair that trades away a sure 0.6 units for the chance to win back 1.0 units (for a net gain of 0.4 units) in case E occurs. If the agent wishes to avoid having a finite number of such fair bets add up to a sure loss, then he/she will have to ensure that the probabilities agree with a finitely additive probability measure. However, deFinetti's theory assumes, in effect, that the units for betting are additive: he develops subjective probability from a theory of additive utility. deFinetti's approach is the subject of our essays 2.1, 2.2, and 2.5.

In their famous theory of cardinal utility, von Neumann and Morgenstern used a decision framework in which an agent holds preferences between *lotteries*: where a lottery is specified by a probability distribution over a set of prizes. When preferences satisfy the axioms of this theory (as summarized in essay 1.2), there exists a unique (cardinal) utility according to which the preferences accord with a ranking of lotteries by their expected utility. In this sense, their theory of utility presupposes probability theory to define the very objects of preference. About 15 years later (and coming after Savage's book), Anscombe and Aumann generalized this framework to one where the agent has preferences over *uncertain* options (called *horse lotteries*), options that consist of mappings from a set of states of nature to a set of von Neumann-Morgenstern lotteries. If an agent's preferences among horse lotteries satisfy several axioms (including those from the von Neumann-Morgenstern theory), then there exists a unique pair consisting of a subjective probability over the states of nature and a (cardinal) utility function over the prizes such that the agent's preferences over horse lotteries agree with a ranking by subjective expected utility. The essays in Part 1 together with essay 2.3 stem from our studies of this approach to decision theory.

Savage (1954) started with a set of options that were merely mappings from states of nature to consequences, rather than mappings to von Neumann-Morgenstern lotteries. He did so to avoid assuming that there existed any agreed-upon probabilities, such as those used (in horse lotteries) to define the von Neumann-Morgenstern lotteries. Thus, Savage's theory takes preference as its primary primitive concept and proceeds without supposing existence of an extraneous concept of probability. Thus, his axioms are somewhat different in kind from those of von Neumann and Morgenstern, and Anscombe and Aumann. But he too concludes that preferences among options should coincide with a ranking by expected utility. Essay 2.4 is particularly concerned with Savage's theory of decision making.

Anscombe and Aumann, Savage, and before them von Neumann and Morgenstern, Ramsey, and deFinetti used normative decision theory as the bedrock for building their theories of rationality. But theirs was decision theory for the single decision maker. An overarching topic in our work is understanding how the norms for decision making should apply in settings with more than one (rational) decision maker and then tracing out some of the consequences of this turn for Bayesian statistics. In connection with multiple agent decision making, we write about these four themes:

1. cooperative, non-sequential decisions;
2. the representation and measurement of "partially ordered" preferences – a relaxed version of traditional, Bayesian expected utility that we use with more than one decision maker;
3. non-cooperative, sequential decisions, i.e., game theory;

and

4. pooling rules and Bayesian dynamics for sets of probabilities.

In our work on representation and measurement of Bayesian preference specifically, we explore in detail the following two themes:

5. the significance of state-dependent utility for, e.g., Savage's foundational program of separating belief from value based on an agent's preferences for acts;

and

6. the significance of additivity assumptions for probability, relating also to dominance principles for preference, in connection with statistical decisions involving infinitely many states.

Next, we explain how these six themes relate to the selection of essays in this volume.

THEMES 1 AND 2. Can the norms of expected utility theory be extended from the standard domain of a coherent individual decision maker to apply also to a cooperative, decision-making group of coherent agents, subject to the following (weak) Pareto rule?

Weak Pareto Rule. The group collectively strictly prefers option h_1 to h_2 provided that each individual does.

The overly simple answer to the question is no. The answer we give is better summarized in these words: "No, but the Bayesian theory can be relaxed by weakening the assumption that all acts are comparable in an agent's preference, the so-called Ordering assumption. This weakening of the canonical Bayesian position allows for Pareto-efficient, cooperative decision making, and it builds on the traditional Bayesian justifications of expected utility."

In the opening essay (1.1), "On the Shared Preferences of Two Bayesian Decision Makers," we present our argument for why the answer to the question is no. We show that Pareto-efficient, cooperative decision making cannot be coherent by the usual (Anscombe-Aumann) standards of Bayesian, subjective expected utility.

In essay 1.2, "Decisions Without Ordering." we summarize why attempts to relax expected utility by removing the so-called Independence assumption for preference (in effect, removing Savage's postulate P2) will not work – that leads to unacceptable decisions in sequential settings. In this same essay we show how several well-known justifications for expected utility theory, e.g., the Dutch Book argument, can be applied to a theory which, instead, relaxes "ordering" (Savage's postulate P1). The upshot is a representation of belief (given utility) by a convex set of probabilities, and a representation of value (given belief) by a convex set of utilities. The essay concludes with the observation that simultaneous representation of belief and value, in the absence of "ordering," will not lead to a convex set of probability-utility pairs.

For a proper comparison with, for example, Savage's foundational work on Bayesian statistics, the alternative theory requires a representation of preference in terms of subjective probabilities and utilities. This we provide in the lengthy third essay (1.3), "A Representation of Partially Ordered Preferences," which concludes Part 1 of the volume. The absence of convexity of the representing set (of probability-utility pairs) relates to the need we find for using mathematical induction, rather than the familiar and more elegant methods that separating hyperplanes offer when convexity obtains. Our approach in this essay raises several serious issues about the project of measuring and separating personal probability and utility based solely on an agent's preferences for acts. This issue we take up in Part 2.

THEMES 2, 5, AND 6. That coherent subjective probability can be reduced to rational preferences over acts has been the thesis of many illustrious authors over the years. Notable among these are deFinetti, Ramsey, and Savage. The common thesis of much of this work is that preferences are rational if and only if the preferred option has higher expected utility. Each author demonstrates this conclusion in a different way, but they all rely on mathematical devices for separating the probability from the utility. Due partly to repeated warnings by Herman Rubin (1987) that "you cannot separate probability from utility," a number of authors have recently begun to examine the various attempts to perform the separation. Two of our essays, "Separating Probability Elicitation from Utilities" (2.1) and "State-dependent Utilities" (2.2), show that the reductions of rational preference to state-independent expected utility performed by Savage, deFinetti, and Anscombe and Aumann all fail to take into account the relationship between the value (utility) for a consequence of an act and the state in which the act awards that consequence.

For example, suppose that an agent is asked to compare two acts h_1 and h_2 whose consequences differ on two events A and A^c. The agent may already own assets whose values also differ on A and A^c. The usual elicitation methodology relates the relative valuations of h_1 and h_2 to the supposed probability of A. Consider a very simple case in which $h_1(s) = r$ for each state s in A and $h_1(s) = 0$ for each state s in A^c. On the other hand let h_2 be a lottery based on some device that the agent agrees has certain probabilities of producing certain random occurrences. In this case let $h_2(s) = r$ with probability p and $h_2(s) = 0$ with probability $1 - p$, for all states s. By varying the value of p, the preference between h_1 and h_2 can be made to change. The value of p that makes the agent indifferent between h_1 and h_2 is the elicited probability of A. However, this value will equal the agent's probability of A if and only if the conditional expected utility of gaining r (in addition to the current fortune) given A is the same as the conditional expected utility of gaining r given A^c. If the agent has a nonlinear utility function and if the agent's current fortune contains assets that take different values depending on whether A or A^c occurs, then those two conditional expected utilities might not be equal. Essay 2.1, "Separating Probability Elicitation from Utilities," considers this example and others like it in more detail.

A somewhat simpler situation occurs when the consequences themselves vary in value from one state to the next. In "State-dependent Utilities" (2.2), we consider examples like the following. Suppose that the different states of nature correspond to different currency exchange rates. It is possible for an agent to express preferences between acts with payoffs in each of several single currencies that (for a single currency) satisfy the "state-independence" axiom of Anscombe and Aumann. This axiom allows the simultaneous determination of a unique probability over states and state-independent utility function over consequences in each currency, separately. However, the "unique" probabilities elicited using different currencies will differ from each other. This is caused by the fact that the utility function is not state-independent in all currencies simultaneously due to the clear dependence of relative values of currencies on the state of nature.

In "Shared Preferences and State-dependent Utilities" (2.3), we return to the central question of Part 1 of this volume, namely, when does there exist a "coherent" Pareto compromise between two "coherent" agents? This time, however, we examine whether a Pareto consensus exists using a generalized version of "coherence": a version of Anscombe-Aumann's subjective expected utility theory. The generalization is obtained by dropping their axiom that is intended to secure a state-independent utility for consequences. That is, in essay 2.3 we investigate whether or not relaxing the assumption that coherent preference over horse lotteries admits a state-independent utility affects the negative result (essay 1.1) on the absence of Pareto-efficient compromises.

The other two essays in Part 2 (2.4 and 2.5) focus on the added complications caused by extending decision theory from cases with finitely many states to those with infinitely many states. This extension is intimately tied to the definition of consequences. For example, an agent might crudely partition the sure-event into finitely many states to do a rough first analysis of the decision problem. Each act then assigns a consequence (or a lottery over consequences) in each state. Later, when the agent realizes that a finer partition is needed, the various "consequences" in each state need to be split up over the finer partition.

For instance, if a state s in the crude analysis is partitioned into $\{(s, i): i = 1, \ldots\}$, the consequence $h(s)$ that an act h awards in state s of the crude analysis might actually not be of constant value on $\{(s, 1), (s, 2), \ldots\}$, the elements of the finer partition. This phenomenon is discussed by Savage under the heading of "Small Worlds." The crude analysis is carried out in a *small world*, and the finer analysis in a *grand world*. If an act h_1 appears to be preferred to act h_2 in every state in a small world, will h_1 necessarily be preferred to h_2 in the grand world analysis? The answer is tied to the degree of additivity of probability. Both Savage and deFinetti wish to avoid mandating that probability be countably additive, thereby allowing the possibility of finitely additive probability. Our essay 2.4, "A Conflict Between Finite Additivity and Avoiding Dutch Book," explores the connection between how the dominance of h_1 over h_2 in small worlds carries over to grand worlds and the additivity of probability.

The final essay in Part 2, "Statistical Implications of Finitely Additive Probability" (2.5), discusses where the distinction between finitely and countably additive probability surfaces in statistical inference. In particular, considerations such as dominance, discussed above, or admissibility are undermined by allowing probability to be finitely additive. On the other hand, finitely additive probability is necessary in order to be able to incorporate many classical (non-Bayesian) methods within the Bayesian framework. The decision of whether or not to mandate countable additivity must weigh the consequences of each decision against each other, which comes as no surprise to anyone familiar with decision theory.

THEMES 3, 4, AND 6. These themes are the subject of Part 3. Here we examine *dynamic* aspects of non-cooperative, multiagent decision making when agents hold shared evidence. The section begins with two papers discussing relations between subjective expected utility theory and game theory. Their motivation is this. When starting on his book project, *The Foundations of Statistics*, Savage speculated that expected utility theory might provide a sound decision theoretic foundation for Wald's theory of statistical decisions, which he based on a minimax choice principle. (Wald's theory was the cutting edge in statistics at the time, in the early 1950s.) It proved otherwise. As Savage discovered, by contrast with subjective expected utility, minimax-loss undervalues observations whereas minimax-regret violates the ordering postulate P1.

A related question arises in connection with simultaneous move game theory. According to the standard von Neumann-Morgenstern analysis, a player in a single-play, two-person zero-sum game "should" play a minimax mixed strategy because only such a strategy has the property that if the opponent knew the player was using that strategy, the opponent would not change from her own minimax mixed strategy. "Subjective Probability and the Theory of Games" (3.1), the first essay in this group, challenges this line of reasoning using Bayesian principles. The argument is that the principal uncertainty a Bayesian faces in a game situation is what the opponent will do. With a subjective opinion about this, it is easy to calculate an optimal strategy for the player. Only when the Bayesian player's belief happens to coincide with the opponent's minimax strategy will his own minimax strategy be optimal. In the typical case of mixed minimax strategies, every mixture of the strategies in his minimax strategy is equally utility maximizing, so there is no particular reason to play a minimax strategy.

The theme that Bayesian analysis of sequential decision problems does not mandate equilibrium behavior is taken up in "Equilibrium, Common Knowledge, and Optimal Sequential Decisions" (3.2). In this context we show, contrary to Aumann, that correlated equilibrium is not a requirement of certain kinds of common knowledge; also, contrary to Bicchieri, that backwards induction is a valid technique for finding optimal strategies.

The third essay in this group, "A Fair Minimax Theorem for Two-Person (Zero-Sum) Games Involving Finitely Additive Strategies," considers the classical von Neumann-Morgenstern game setup in which there are infinitely many strategies available to each player. Using Wald's "bigger integer" game as an example, the essay shows that if one player is limited to countably additive mixed strategies while the other is permitted finitely additive strategies, the latter wins. When both can play finitely additive strategies, the winner depends on the order in which the integrals are taken. Solutions to this conundrum are discussed.

Essay 3.4, "Randomization in a Bayesian Perspective," addresses a persistent puzzle for Bayesians – the proper role of randomization in the theory. The use of randomization as a justification for inference, as was suggested by Fisher, violates (ironically) the very likelihood principle that Fisher introduced and hence is hopeless from a Bayesian viewpoint. But randomization as a design is often used and useful, and yet is hard to understand theoretically. The analysis in this essay points to multiple decision makers as the key ingredient that leads to the attractiveness of randomization. A subsequent paper (Berry and Kadane, 1997) shows an explicit model with multiple decision makers in which it is optimal for a Bayesian experimental designer to randomize.

Another approach to multiple decision makers is to examine when their opinions can be pooled into a single group opinion. One desirable property a pooling method might have is that it be "externally Bayesian," i.e., the same updated pooled group opinion is obtained whether the pooling takes place before or after updating on the basis of a common likelihood function. Essay 3.5, "Characterization of Externally Bayesian Pooling Operators," gives a necessary and sufficient condition for a pooling operator to be externally Bayesian.

Essay 3.6, "An Approach to Consensus and Certainty with Increasing Evidence," examines the long-run dynamics of iterative Bayesian updating for sets of probabilities described in terms of their extreme members. The primary purpose of this work is to examine what happens to a result of Blackwell and Dubins (1962) about the merging of two Bayesian posterior probabilities (under increasing shared evidence) when the "community" of opinions is characterized by different kinds of extreme views. The Blackwell-Dubins paper is, itself, an important generalization of Savage's result about merging with simple *i.i.d.* data. They avoid all but one of Savage's assumptions: that the rival opinions agree on null events.

As a general methodological point, our essay 3.6 notes that there are two distinct roles that topology plays in such results on asymptotic merging: (1) topology is used to determine merging of posterior opinions in terms of the conditions for convergence of distributions, and (2) topology is used to determine the size of the "community" of rival opinions in terms of the conditions for closure of the set of distributions. We show that it is important to mix and match these distinct roles with different topologies, e.g., using the weak-star topology for both does not support merging in the Blackwell-Dubins setting. Contrast this with the important negative finding of Diaconis and Freedman (1986), who, in a (non-parametric) *i.i.d.* setting, use the weak-star topology for both purposes, versus the positive results in the same setting by Barron, Schervish, and Wasserman (1999), who split the jobs between two topologies.

The role of improper prior distributions, or finitely additive prior distributions, is controversial in Bayesian statistics. Some statisticians use improper distributions, especially uniform distributions, as a representation for ignorance (see Kass and Wasserman, 1996, for a review). Others regard this as wasting the opportunity provided by prior distributions to model the opinion of the client. In the context of this subjective view, a restriction to countably additive distributions can be regarded as unnecessarily restrictive (deFinetti, 1974; Kadane and O'Hagan, 1995).

An important consequence of the restriction to countably additive prior distributions is that new (cost-free) information always has non-negative worth, and the expectation of the posterior distribution is the prior distribution. Essay 3.7, "Reasoning to a Foregone Conclusion," shows by example that neither of these is true for finitely additive distributions. In these examples you know what opinion you will hold tomorrow, whatever the data value you observe, and it is different from your prior today. So which is your opinion? And decision problems can be constructed in which it would be profitable to pay not to see data. The essay discusses whether the price for finite additivity is too high, and whether there are remedies for these apparent paradoxes.

In a related essay (3.8), we take up the question of foregone conclusions from the perspective of an observer, watching the experimenter at work. Even with countably additive probability, when will an onlooker agree that a Bayesian experimenter (who updates her opinion by Bayes's rule) cannot lead herself down the garden path to a foregone conclusion through some clever design of an experiment? Our answer applies to non-Bayesian onlookers too!

With this selection of our papers we hope to stir readers, especially those who feel that the "foundations" for Bayesian statistics were set in place years ago, to rethink a host of interrelated, important problems – many of which are still open, we believe. For example, regarding the modified theory of expected utility developed in essay 1.3, there remains hard work to be done creating a useful implementation for the representation theorem given there, using sets of pairs of probabilities and utilities. We are convinced that such an implementation will be valuable, e.g., with incomplete Bayesian elicitations, and to guide the order of questions in an efficient elicitation. Regarding the deep and difficult challenges created by the possibility of state-dependent utility (essay 2.2), the field is turning, slowly we think, to face them. (See Mongin, 1998.) And regarding problems of finitely additive probability (essay 2.5), there remain important open questions about which formal Bayes calculations are valid with "improper" priors. Jeffreys (1961) introduced "improper" priors into Bayesian theory in the late 1930s as a way of justifying some "frequentist" inferences. However, an "improper" prior, e.g., Lebesgue measure on the real line, corresponds to a merely finitely additive probability, not a countably additive one, as it gives equal weight to each unit interval. We know that where non-conglomerability arises, averaging a likelihood against a prior will fail to identify the posterior. But exactly where does non-conglomerability enter through an "improper" prior? We know of only fragmentary answers, e.g., Heath and Sudderth (1989). We expect the study of the foundations of statistics to continue to be fruitful in the years ahead.

We thank Brian Skyrms, the General Editor for the series, and Terence Moore, Executive Editor at Cambridge University Press, for their valuable advice in our selection of papers for this volume and their assistance with its organization. We appreciate the cooperation of the publishers and editors of the books and journals where our papers originally appeared, who have kindly granted permission for these reprintings. They are acknowledged separately at the foot of the first page of each essay. We also thank Margie Smykla for her diligent and patient proofreading, and Heidi Sestrich for her invaluable help indexing. Collectively, we appreciate NSF's financial support of our efforts in creating this volume (DMS-9801401).

REFERENCES

Anscombe, F. J., and Aumann, R. J. (1963). "A definition of subjective probability." *Annals of Math. Stat.* **34**, 199–205.

Barron, A., Schervish, M. J., and Wasserman, L. (1999). "The consistency of posterior distributions in nonparametric problems." *Ann. Stat.* **27**, forthcoming.

Berry, S., and Kadane, J. B. (1997). "Optimal Bayesian randomization." *J. Roy. Stat. Soc.* **Ser B**, **59**, 813–819.

Blackwell, D., and Dubins, L. (1962). "Merging of opinions with increasing information." *Annals of Math. Stat.* **33**, 882–887.

deFinetti, B. (1974). *The Theory of Probability* (2 vols.). New York: Wiley.

Diaconis, P., and Freedman, D. (1986). "On the consistency of Bayes estimates." *Ann. Stat.* **14**, 1–26.

Fishburn, P. C. (1970). *Utility Theory for Decision Making*. New York: Kriefer Publishing Co.

Heath, D., and Sudderth, W. (1989). "Coherent inference from improper priors and from finitely additive priors." *Ann. Stat.* **17**, 907–919.

Jeffreys, H. (1961). *Theory of Probability* (3d ed.; 1st ed., 1939). Oxford: Oxford University Press.

Kadane, J. B., and O'Hagan, A. (1995). "Using finitely additive probability: Uniform distributions on the natural numbers." *J.A.S.A.* **90**, 626–631.

Kass, R., and Wasserman, L. (1996). "The selection of prior distributions by formal rules." *J.A.S.A.* **91**, 1343–1370.

Lehmann, E. (1986). *Testing Statistical Hypotheses* (2d ed.). New York: Wiley.

Mongin, P. (1998). "The paradox of the Bayesian experts and state-dependent utility theory." *J. Math. Econ.* **29**, 331–361.

Rubin, H. (1987). "A Weak System of Axioms for 'Rational' Behavior and the Nonseparability of Utility from Prior," *Statistics and Decisions* **5**, 47–58.
Savage, L. J. (1954). *The Foundations of Statistics*. New York: Wiley.
Savage, L. J., et al. (1962). *The Foundations of Statistical Inference*. London: Methuen.
von Neumann, J., and Morgenstern, O. (1947). *Theory of Games and Economic Behavior* (2d ed.). Princeton: Princeton University Press.

PART 1

Decision Theory for Cooperative Decision Making

1.1

On the Shared Preferences of Two Bayesian Decision Makers

TEDDY SEIDENFELD, JOSEPH B. KADANE, AND MARK J. SCHERVISH

An outstanding challenge for "Bayesian" decision theory is to extend its norms of rationality from individuals to groups. Specifically, can the beliefs and values of several Bayesian decision makers be amalgamated into a single Bayesian profile that respects their common preferences over options? If rational parties to a negotiation can agree on collective actions merely by considering mutual gains, is it not possible to find a consensus Bayes model for their choices? In other words, can their shared strict preferences over acts be reproduced with a Bayesian rationale (maximizing expected utility) from beliefs (probabilities) and desires (utilities) that signify a rational compromise between their rival positions?

Whatever else is to be required of a compromise, we suppose that a consensus Bayes model for the group preserves those strict preferences which the individuals already share. That is, we impose a *weak Pareto condition* on compromises. Whenever all parties to a decision have a common strict preference for one option over another, then any proposed Bayesian group model for their choice – any "neutral" position – must reflect this preference and assign higher expected utility to the Pareto dominating option.

We would like to thank John Broome, Robyn Dawes, Jay Goodman, Mark Kamlet, and Isaac Levi. This work was supported by the Buhl Foundation, ONR Contracts N00014-85-K-0539 and N00014-88-K-0013, and NSF Grant DMS-8805676. Reprinted with permission from *The Journal of Philosophy*, 86, no. 5 (May 1989): 225–244.

Of course, the probabilities and utilities of any one of the agents satisfies this weak Pareto condition. That is, each agent on her own meets this condition – whatever strict preferences they all have, each has. But it is hardly a compromise always to make the group decide all questions based on the preferences of a single member. We call such Pareto solutions *autocratic*. What (nonautocratic) Bayesian compromises are there for the group decisions?

Imagine the dilemma that arises when two Bayesian agents agree on what to do by appeal to their unanimous preferences, but they find no neutral Bayes position (other than their own separate views) to endorse the rationality of their choices. It is our primary purpose in this essay to underscore the ubiquitous nature of this dilemma. Our central result, then, is this. When two Bayesian decision makers differ both in their beliefs (probabilities) and values (utilities), and subject to an assumption that they agree on the preference ranking of some two "constant" acts, the only candidates for a Bayes compromise are the two autocratic solutions. That is, there is no room for a Bayesian compromise. We develop this argument in section I.

In section II we contrast our results with several related theorems in social welfare theory. The question of finding a neutral Bayes model for the group's choices is seen at once to be a variety of Kenneth Arrow's[1] problem for social welfare rules. The Bayesian agents have individual preferences over social acts which are to be amalgamated into a single, Bayesian group preference over social acts. In order to dodge that "impossibility" theorem, some concession to Arrow's result is required.

The alternatives that have been investigated in connection with Arrow's problem follow several directions. One approach is to restrict the domain of applicability of a welfare rule to communities where the individual preferences are not too discrepant, e.g., where the individual preferences conform to Duncan Black's[2] "single-peakedness" condition. (This violates the "unrestricted domain" condition of Arrow's argument.) Our negative finding is unaffected by this consideration, since we show that the dilemma arises for all pairs of Bayesian agents who have even the slightest differences in both their beliefs and their values. There is no gain made here by trying to bracket cases where the agents' probabilities or utilities reflect large, rather than small, discrepancies.

A second approach to avoiding Arrow's impossibility theorem is to add structure to the representation of individual preferences, to add interpersonal utility comparisons not allowed by Arrow's multipartite "independence of irrelevant alternatives" condition. The excellent paper of Kevin Roberts[3] summarizes how different classes of social welfare rules can be achieved by introducing alternative versions of positive interpersonal utility comparisons. Our negative finding about Bayesian social welfare rules applies, however, whether or not interpersonal utility comparisons are made. Thus, another familiar way around Arrow's result does not work when the problem is finding group compromises that are Bayes.

A third approach is to relax the "ordering" requirement, to liberalize the condition that the social choice rule induces a complete ordering of group options where any two social acts may be compared by the compromise social preference relation. In his seminal book on Bayesian decision theory and statistics, Leonard Savage[4] defends the minimax-regret rule for group deliberation by suggesting that the "ordering" postulate (*P*1) does not apply to group preference, though it does apply to individual preferences.

In this connection, Isaac Levi[5] offers an intriguing account, we think, of why the norms on rational choice should be uniform between groups and individuals. The key assumption for a group decision problem is that there is only one agent, the (cooperative) group. Otherwise, there is not one decision problem but, instead, there are the several (noncooperative) deliberations of the separate individuals who, for prudential reasons, take an interest in each other's actions.

In section II.2 we explain how our result impacts on Levi's proposal for achieving a unified decision theory, unified across individual and group decisions. Levi's theory relaxes the "ordering" postulate of Bayesian expected utility. A rational agent (either an individual or a group) need not have a complete preference relation for comparing every two options. The explanation for this departure from Bayesian theory is his view that a rational agent (either an individual or a group) may experience *unresolved* conflicts.

For an individual, the conflicted preferences arise from uncertain beliefs (sets of probabilities) or from indeterminate values (sets of utilities). For a group, the conflicts among the individuals' beliefs and values generate the same uncertainties and indeterminacies when the agent is the cooperative group.

In Levi's theory, however, the compromise position on rational choice is achieved (at the expense of "ordering" for the preference relation) by constructing two *independent* "neutral" positions, one for conflicted beliefs (a set of probabilities) and one for conflicted values (a set of utilities). An incomplete preference relation is formed using inequalities in expected utilities with all pairs from these two sets. Based on our negative result, we show this rule leads to violations of the weak Pareto condition. That is, Levi's theory allows a group to choose a weak Pareto dominated option.

Nonetheless, we are in agreement with Levi on the desirability of relaxing the ordering postulate while preserving a respect for expected utility. Hence, we explore a theory in which preference need not be a complete relation, but also a theory in which preference does not compromise conflicts in beliefs and in values independently.

I. BAYESIAN COMPROMISES BETWEEN TWO BAYESIANS

Here we discuss in detail the case of a group composed of just two Bayesian decision makers. That is, we respond to the question posed in the opening sentences of this essay when only two Bayesian "experts" are involved and, for each, preference is a complete relation. We assume that each agent can compare every two acts, either by a (transitive, antisymmetric) relation of strict preference, or else by an equivalence relation of indifference.

The group decision involves amalgamating the preferences of two Bayesians, whom we shall call Dick and Jane. According to Bayesian decision theory, for simple problems, an agent chooses from among a set of feasible (state-independent[6]) acts according to the principle of maximizing subjective expected utility.

To be more precise, consider the familiar decision matrix in Figure 1. The columns denote a partition into (n) states of nature, $S_1, \ldots, S_n$, about which the agent is uncertain. The rows designate feasible acts, $A_1, \ldots, A_m$, whose outcomes in each state are denoted by the O_{ij}: the outcome of act A_i under state S_j. The agent's uncertainties about the states are given by a probability distribution, $P(S_j)$. The agent's values for outcomes are given by a (von Neumann-Morgenstern) utility function, $U(O_{ij})$. Then, according to the principle of maximizing expected utility, act A_1 is (strictly) preferred to act A_2 whenever

$$\sum_j P(S_j) \cdot U(O_{1j}) > \sum_j P(S_j) \cdot U(O_{2j}).$$

Suppose Dick's preferences over such acts are summarized by the pair (P_1, U_1) of his (personal) probability and utility, while Jane's pref-

	S1	S2			Sj			Sn
A_1	O_{11}	O_{12}			O_{1j}			O_{1n}
A_2	O_{21}	O_{22}			O_{2j}			O_{2n}
A_m	O_{m1}	O_{m2}			O_{mj}			O_{mn}

Figure 1. Decision matrix: acts × states

erences are depicted by the pair (P_2, U_2). What are the alternative Bayesian preference schemes that agree with the (strict) preferences shared by Dick and Jane? That is, for which pairs (P, U) is it the case that:

$$\sum_j P(S_j)\cdot U(O_{1j}) > \sum_j P(S_j)\cdot U(O_{2j}).$$

whenever

$$\sum_j P_k(S_j)\cdot U_k(O_{1j}) > \sum_j P_k(S_j)\cdot U_k(O_{2j}) \quad (k=1,2)?$$

In terms of social welfare rules, our question is this. In group choices where all agents receive the same outcomes (the social acts offer the same prospects to each agent), subject to a weak Pareto condition, what are the Bayesian social welfare rules? (Recall, a social choice rule satisfies the weak Pareto condition provided that an option is inadmissible whenever there is another feasible alternative that everyone strictly prefers. If everyone strictly prefers option B to option A, A is inadmissible whenever B is available.)

The answer is surprising. Stated informally, our result in the case of two agents is that no attractive compromises exist. Only autocratic solutions conform to the weak Pareto condition. That is, a Bayesian

Table 1. *"Horse lotteries" used in fixing the upper and lower probabilities and utilities*

	E	$\neg E$
A_1	r^*	r_*
$A_{2\epsilon}$	$(.9+\epsilon)(r_*)+(.1-\epsilon)(r^*)$	$(.9+\epsilon)(r_*)+(.1-\epsilon)(r^*)$
$A_{3\epsilon}$	$(.7-\epsilon)(r_*)+(.3+\epsilon)(r^*)$	$(.7-\epsilon)(r_*)+(.3+\epsilon)(r^*)$
A_4	r	r
$A_{5\epsilon}$	$(.6-\epsilon)(r_*)+(.4+\epsilon)(r^*)$	$(.6-\epsilon)(r_*)+(.4+\epsilon)(r^*)$

model for the group's collective decisions must use the beliefs and values of a single agent, thereby ignoring the preferences of everyone else whenever the weak Pareto condition does not apply.[7]

It is interesting to contrast the difference between the weak and strong Pareto conditions for group compromises. The strong Pareto condition for social choice rules requires that option A is socially inadmissible whenever there is a feasible option B, which everyone finds either strictly preferable or indifferent to A, and which someone strictly prefers to A. We find that with the imposition of a strong Pareto condition, there are *no* Bayes social welfare rules.

The proofs of these claims are contained in the following simple example. The constructions in the example generalize to every pair, like Dick and Jane, that differs in both beliefs and values.

Example 1: Suppose, as before, Dick and Jane are Bayesian agents with preferences summarized by the probability and utility pairs (P_k, U_k), $k = 1, 2$. Assume that they have different beliefs, $P_1 \neq P_2$. That is, there is some event E with $P_1(E) \neq P_2(E)$. For instance, let Dick assign E a (personal) probability 0.1 while Jane assigns E a (personal) probability 0.3, i.e., $P_1(E) = 0.1$ and $P_2(E) = 0.3$. Also, suppose they have different values, $U_1 \neq U_2$.[8] For simplicity, suppose Dick and Jane have different (cardinal) utilities, in the following sense. They agree on the ranking of two particular rewards: each prefers r^* to r_*, though they differ in their valuation of a third reward r. Let r be a reward with $U_1(r) = 0.1$ and $U_2(r) = 0.4$, while there exist rewards r_* and r^* with $U_1(r_*) = U_2(r_*) = 0$ and $U_1(r^*) = U_2(r^*) = 1$.

Next, we consider pairs of options where Dick and Jane hold common preferences. Examine the acts defined in Table 1, above, on the binary partition formed by the event E.

These acts are "horse lotteries," in the language of F. J. Anscombe and R. J. Aumann.[9] We distinguish a von Neumann-Morgenstern (v.N-M) lottery from a horse lottery. A v.N-M lottery, L, is a specified probability distribution over a set of rewards. For example, a v.N-M lottery that yields the reward r_* with probability 0.6 and the reward r^* with probability 0.4 is denoted by: $L = 0.4r_* + 0.6r^*$. With horse lotteries as acts, the outcome O_{ij} of act A_1 in state S_j (see the decision matrix on page 21) is a v.N-M lottery L_{ij}.

For example, the v.N-M lottery $(0.9 + \varepsilon)(r_*) + (0.1 - \varepsilon)(r^*)$ under event E for act $A_{2\varepsilon}$ corresponds to the outcome: receive reward r_* with probability $(0.9 + \varepsilon)$ and reward r^* with probability $(0.1 - \varepsilon)$. When "horse$_j$" wins the hypothetical race, i.e., when state S_j obtains, the act A_i pays out the v.N-M lottery L_{ij}. Thus, horse lotteries are functions from states to v.N-M lotteries.

Since which state obtains (which "horse" wins) is uncertain, and how rewards are to be valued is not stipulated in advance, "horse lotteries" accommodate both uncertainty over states and (cardinal) utility over rewards.

First, as Dick and Jane agree that

$$0.1 \le P_k(E) \le 0.3 \quad (k = 1, 2)$$

they agree that act A_1 is preferred to each act $A_{2\varepsilon}$, and they agree that each act $A_{3\varepsilon}$ is preferred to A_1 $(0.1 \ge \varepsilon > 0)$. See Table 1.

Likewise, they agree that

$$0.1 \le U_k(r) \le 0.4 \quad (k = 1, 2).$$

Then, also they are unanimous in their preference for A_4 over each act $A_{2\varepsilon}$, while each act $A_{5\varepsilon}$ is preferred to A_4. (All these preferences are "strict.")

In Figure 2 we see the set of pairs of probabilities and utilities agreeing with these shared preferences. That is, Figure 2 is based on the (strict) preferences involving the upper and lower probabilities of event E and the upper and lower utilities of reward r. These bounds for $P(E)$ and $U(r)$ box the family of Bayesian compromises between Dick and Jane with respect to their shared agreements for these (strict) preferences. Subject to the weak Pareto condition, it shows that the set of "neutral" Bayes's models (with respect to *all* their choices – not just for these few comparisons) is some subset of the cross product of weighted averages of their probabilities and weighted averages of their utilities.

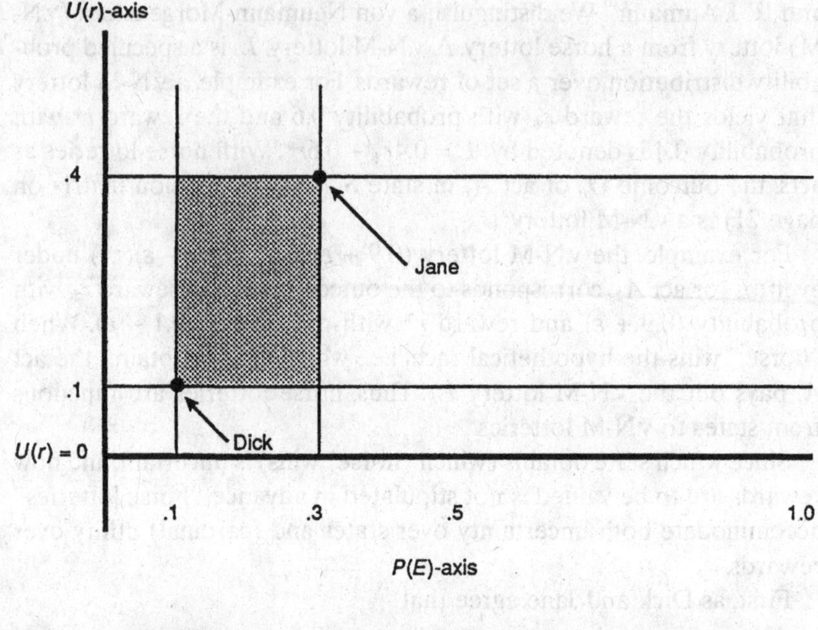

designates the set of probability/utility pairs agreeing with the common preferences of Dick and Jane for the comparisons, above, in table 1.

Figure 2

Next, consider the set of pairs of acts defined in Table 2, over which Dick and Jane also hold common preferences.[10]

For each value, $0 < \varepsilon \leq 0.2$, Dick and Jane agree in their (strict) preference for $A_{7\varepsilon}$ over $A_{6\varepsilon}$. The set of Bayesian compromises for these preferences is *not* connected, however. See Figure 3.

For $0 < \varepsilon < 0.015$ the set of probability/utility pairs $(P(E), U(r))$ for which $A_{7\varepsilon}$ has greater expected utility than $A_{6\varepsilon}$, is bounded by a hyperbola centered at (0.2, 0.25), which satisfies:

$$[P(E) - 0.2][U(r) - 0.25] = 0.015 - \varepsilon.$$

As $\varepsilon \Rightarrow 0$, the hyperbola approaches the pair of points corresponding to Dick and Jane's preferences. When $\varepsilon = 0$, the hyperbola intersects these two points.

If we superimpose the two figures, we obtain Figure 4.

Figure 4 shows that the family of Bayesian agents who agree with these two, even for the few preferences already considered, consists

Table 2. *"Horse lotteries" used in separating the set of compromises between Dick and Jane*

	E	$\neg E$
$A_{6\epsilon}$	$.785(r_*) + .215(r^*)$	$\epsilon(r_*) + .2(r) + (.8 - \epsilon)(r^*)$
$A_{7\epsilon}$	$(.2 - \epsilon)(r_*) + .8(r) + \epsilon(r^*)$	$.165(r_*) + .835(r^*)$

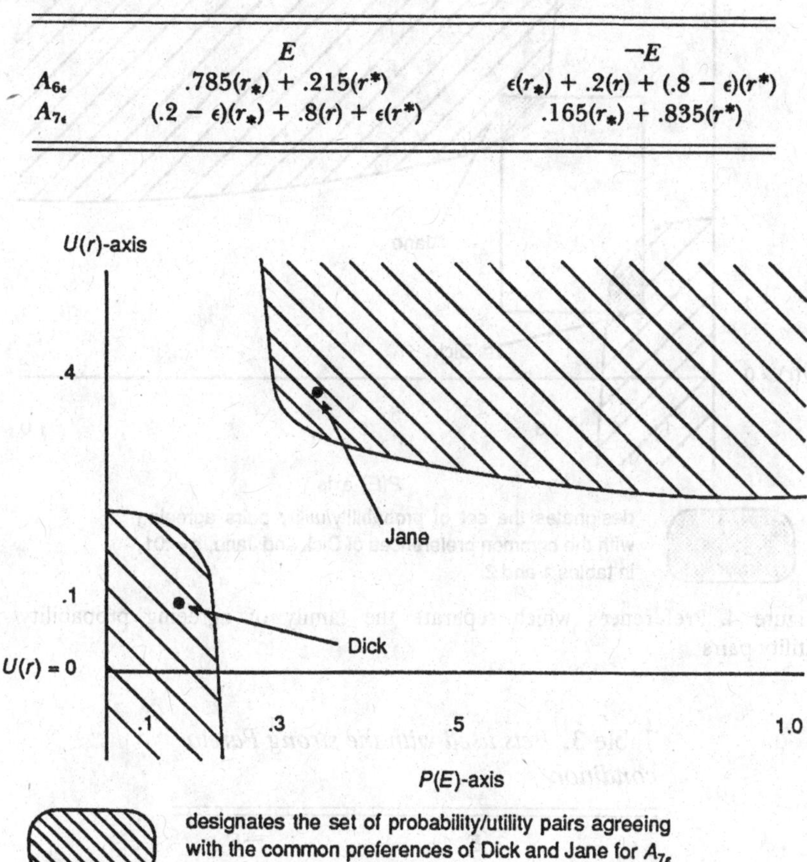

Figure 3. Preferences which separate the family of agreeing probability/utility pairs

exactly of the two themselves: (P_1, U_1) and (P_2, U_2). That is, since the hyperbolas all have negative slopes at the points interior to the box, as $\varepsilon \Rightarrow 0$ the regions of overlap between the two figures collapse onto the two corner points: the points corresponding to Dick and Jane. We may express the lesson of the example as follows. The only Bayes models that preserve Dick and Jane's common strict preferences (a weak Pareto condition) are the two autocratic solutions: choose one of Dick and Jane.

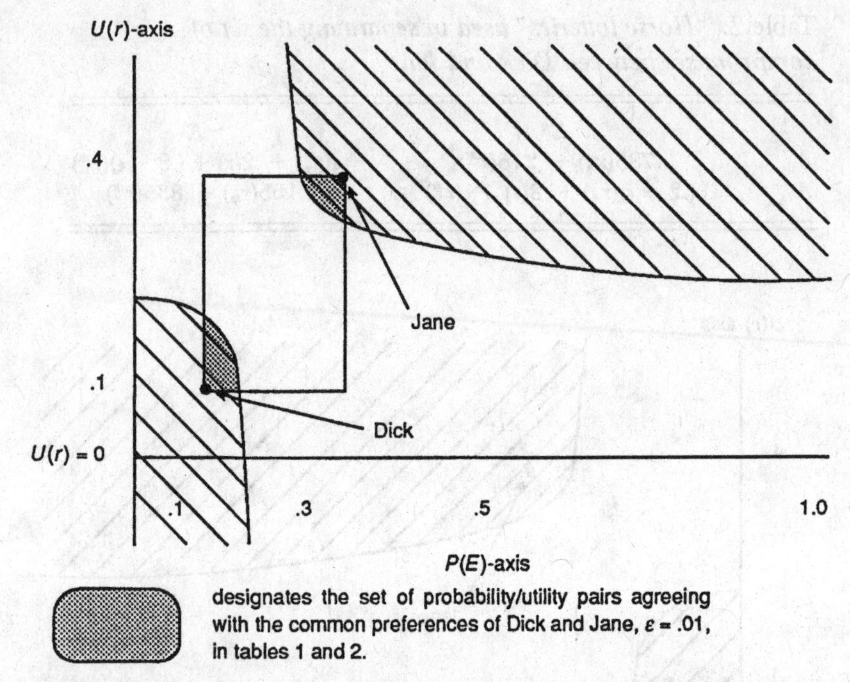

Figure 4. Preferences which separate the family of agreeing probability/utility pairs

Table 3. *Bets used with the strong Pareto condition*

	E	$\neg E$
B_1	r^*	r_*
B_2	$.9(r_*) + .1(r^*)$	$.9(r_*) + .1(r^*)$
B_3	$.7(r_*) + .3(r^*)$	$.7(r_*) + .3(r^*)$

Moreover, if we inquire about Bayes models that satisfy a strong Pareto condition ("When everyone is either indifferent or prefers act A_1 to A_2 and someone prefers A_1, the compromise is for the strict preference."), then there are *no* acceptable solutions. This follows since, e.g., Dick is indifferent between the bets B_1 and B_2 though Jane prefers B_1, and Jane is indifferent between the bets B_1 and B_3 though Dick prefers B_3 (Table 3). Neither autocrat respects both of these strict preferences.

The various constructions in this example generalize easily to any pair of agents with different beliefs and values, subject to the common ranking for some pair of v.N-M lotteries, L^* and L_*.[11] (We choose $U_k(L_*) = 0$ and $U_k(L^*) = 1, k = 1, 2$.)

Theorem 1. Let (P_1, U_1) and (P_2, U_2) be the probability/utility pairs representing two decision makers with different beliefs and preferences over horse lotteries, then:

(i) The set of probability/utility pairs which agree with the strict preferences shared by these two decision makers (weak Pareto) consists exactly of the two pairs themselves – there are autocratic solutions only – no other Bayesian compromises exist: and

(ii) Subject to the strong Pareto condition, the set of probability/utility pairs that agree with these common preferences is empty – there are no Bayes models at all.

Proof. Since $P_1 \neq P_2$ and $U_1 \neq U_2$, there is an event E and a v.N-M lottery L such that $P_1(E) = p_1 < P_2(E) = p_2$, and $U_1(L) = u_1 < U_2(L) = u_2$. Then the agents agree that:

$$p_1 \leq P_k(E) \leq p_2$$
$$u_1 \leq U_k(L) \leq u_2 \quad (k = 1, 2).$$

(*i*) Table 1 is modified by these bounds, creating a "box of compromises," as in Figure 2. The pairs of acts used to "separate" the set of compromises, analogous to those of Table 2, are defined by the hyperbolic equations (provided by Jay Goodman):

$$(P(E) - [p_1 + p_2]/2) \cdot (U(E) - (u_1 + u_2)/2) = (p_2 - p_1)(u_2 - u_1)/4 - \varepsilon$$

for $0 < \varepsilon < (p_2 - p_1)(u_2 - u_1)/4$. Combining these sets of preferences, letting $\varepsilon \Rightarrow 0$, we discover that only the original pairs (P_1, U_1) and (P_2, U_2) remain. By holding fixed the binary partition $\{E, \neg E\}$ and varying the lottery L, we see that the group utility function must be the utility for one of the two agents. Then, by varying the partition, since utility is state-independent, we see that the group probability also must be taken from the same agent.[12] Hence, only autocratic solutions agree with the weak Pareto condition on strict preferences.

(*ii*) Upon adding a strong Pareto requirement for compromises, we may modify the bets from Table 3 to read as in Modified Table 3.

Then, as before, the first agent is indifferent between B_1 and B_2

Modified Table 3. *Bets used with the strong Pareto condition*

	E	$\neg E$
B_1	L^*	L_*
B_2	$(1 - p_1)(L_*) + p_1(L^*)$	$(1 - p_1)(L_*) + p_1(L^*)$
B_3	$(1 - p_2)(L_*) + p_2(L^*)$	$(1 - p_2)(L_*) + p_2(L^*)$

though the second prefers B_1, and the second agent is indifferent between B_1 and B_3 though the first prefers B_3. Neither "autocrat" respects both these strict preferences. Hence, under a strong Pareto condition on compromises, there are none.

II. GROUP DECISION MAKING AND A SEPARATION OF BELIEF AND UTILITY

II.1. SOCIAL WELFARE THEORY. An immediate consequence of theorem 1 is the impossibility of a general, nonautocratic Bayesian social welfare function (subject to the weak Pareto condition). That is, even when interpersonal utility comparisons are admitted, and provided the domain of social acts includes the simple varieties discussed in the previous section, there is no interesting Bayesian solution to the social welfare problem.

That corollary to theorem 1 is obvious when the population consists of two agents. For larger communities, of size n, consider populations where the beliefs and preferences of different agents fall into one of two camps, i.e., where the n-agents are clones of two agents.[13] Thus, the requirement that group welfare satisfies the strong Pareto condition (or else a conjunction of "weak Pareto" and "no autocrats") is inconsistent with the rationality postulates of subjective expected utility theory. Moreover, this result is resistant to the standard cures for Arrow's impossibility theorem. Neither considerations of interpersonal utility nor restrictions on the extent of the discrepancies between the beliefs and values of two Bayesians avoid the dilemma.

The impossibility of Bayes solutions to group decisions involving uncertainty is also discussed in two important papers, one by A. Hylland and R. Zeckhauser, and one by P. Hammond.[14] We detail a contrast between theorem 1 and these earlier results in the Appendix. Briefly stated, those accounts are limited in two ways that leave theorem 1 unaffected.

1. The earlier findings establish the impossibility of a general, Bayesian welfare rule subject to the condition that group preference amalgamates probability and utility independently. In other words, they require that the Bayes model for the group has the group probability defined solely in terms of the individuals' probabilities, and has the group utility defined solely in terms of the individuals' utilities. Theorem 1 applies without this limitation.
2. The earlier findings show that with each (Bayesian) social welfare rule for amalgamation of individual preferences into a group preference (and subject to the independent amalgamation of belief and desire discussed in the previous paragraph), *there is some* profile of individual preferences leading to a failure of the (strong) Pareto condition. Theorem 1, however, shows a failure of the (strong) Pareto condition with each rule and for *every* pair of agents (who differ in both beliefs and desires). Thus, theorem 1 derives a universal statement where the other derives an existential.

II.2. A COMPARISON WITH LEVI'S QUASI-BAYESIAN DECISION THEORY AND A PROBLEM OF INDEPENDENT COMPROMISES OF BELIEFS AND DESIRES. In important papers and books, Levi advocates a unified theory of rational decision making under unresolved conflict, unified between individual and group decisions.[15] Both for individuals and for groups, there is no requirement that an agent's preferences induce an ordering of options – not all acts need be compared by preference.

Levi's decision theory is a liberalization of (Bayesian) expected utility theory. In his theory, an agent's beliefs are represented by a convex set of (personal) probabilities, $\mathcal{P}$, and preferences for outcomes are represented by a convex set of (cardinal) utilities, $\mathcal{U}$. In brief, an option o is an admissible choice from a set of feasible options $\mathcal{O}$, provided that it satisfies a (lexicographically ordered) sequence of maximizations. The first of these is E-admissibility, which requires that an option maximize expected utility for some probability/utility pair (P, U), where $P \in \mathcal{P}$ and $U \in \mathcal{U}$.

Definition: o is *E-admissible* if and only if

$$\exists(P \in \mathcal{P}, U \in \mathcal{U})\forall(o' \in \mathcal{O})\ E_{P,U}(o) \geq E_{P,U}(o').$$

A secondary decision test for narrowing the set of E-admissible options is maximizing a "security" index among those options which are E-admissible. Illustrations of "security" include (a) a vacuous standard – all options have equal security; (b) security indexed by worst ($\inf[\mathcal{U}]$)

payoff; and (c) security indexed by least ($\inf[\mathcal{P} \times \mathcal{U}]$) expected utility. In addition to these two, Levi[16] entertains ternary, etc., maximization requirements, reflecting added structure in the agent's system of values.

Because of the first condition, E-admissibility, an admissible option is "Bayes," i.e., maximizes expected utility for some probability/utility pair. Hence, admissibility takes expected utility theory as a special case, when both $\mathcal{P}$ and $\mathcal{U}$ are unit sets and, e.g., all other value considerations are vacuous. Even when other value considerations are vacuous, however, if either $\mathcal{P}$ or $\mathcal{U}$ is not a singleton, admissibility fails to induce an ordering.[17]

A central theme in Levi's account of choice under unresolved conflict is that a "neutral" position among conflicting beliefs and desires is a position that preserves the shared agreements between the rivals, yet introduces no judgments over which there is disagreement. If an agent experiences a value conflict between, e.g., two (cardinal) utilities U_1 and U_2, then, according to Levi's theory, the convex combinations of these two,

$$\mathcal{U} = \{\alpha U_1 + (1-\alpha)U_2 : 0 \leq \alpha \leq 1\},$$

represents the consensus position of unresolved value conflict from which he makes rational decisions. For example, if John is conflicted between his value assessments of options to acquire art objects – the options rank differently under comparisons of their economic outcomes (U_1) and their aesthetic outcomes (U_2) – nonetheless, by representing his conflicted values with the set $\mathcal{U}$, he may proceed to make rational decisions without first having to resolve the value conflict. The analysis for a consensus position with conflicted beliefs is similar. The "neutral" position for belief is the (convex) set $\mathcal{P}$ of conflicted probabilities.

As we remarked in our introduction, Levi[18] offers arguments that rational deliberation should follow the same standards regardless whether the decision is by an individual or by a cooperative group. The intrapersonal conflicts of values and beliefs for individual decision making should be treated in a like fashion with the parallel interpersonal conflicts for group decisions. Our understanding of his theory is that the "neutral" position for decision under unresolved conflict is found by analyzing desires and beliefs *independently*. That is, the consensus for conflicted preferences over outcomes is a (convex) set of utilities, the consensus for conflicted beliefs is a (convex) set of probabilities, and the independence between them is built into

E-admissibility with the appeal to the Cartesian product of these two sets.

If we apply his method to the group decisions faced by Dick and Jane, our example 1, we see that the E-admissible social acts are those which maximize expected utility for some probability/utility pair in the rectangle pictured in Figure 1. Then, both acts $A_{6\varepsilon}$ and $A_{7\varepsilon}$ are E-admissible in a pairwise choice between them. (The expected utility of act $A_{6\varepsilon}$ is greater for each probability/utility pair within the unshaded region of the rectangle, depicted in Figure 3.) This challenges the claim to the "neutrality" of the E-admissible options, we believe, since E-admissibility fails to respect the shared strict preference for $A_{7\varepsilon}$ over $A_{6\varepsilon}$ E-admissibility conflicts with the weak Pareto condition.

To emphasize this point, Levi's method of group decision making makes identical the E-admissible options in the feasible sets $\mathcal{O}_\varepsilon = \{A_{6\varepsilon}, A_{7\varepsilon}\}$ for the following three groups of Bayesians. Group$_1$ consists of Dick and Jane. Group$_2$ consists of Tom and Mary; where Tom's preferences are summarized by the probability/utility pair (P_1, U_2) and Mary's by (P_2, U_1). Group$_3$ is composed of all four agents: Tom, Dick, Jane, and Mary. For each of the three groups, both options in the sets $\mathcal{O}_\varepsilon$ are E-admissible. Group$_1$ declares a unanimous strict preference for $A_{7\varepsilon}$ over $A_{6\varepsilon}$, however. Group$_2$ declares the opposite. And the members of Group$_3$, of course, find no common preferences over these choices.

Levi's E-admissibility is a normative theory of consensus in *reasons* for preference, the rational causes of decisions. It treats beliefs and values for outcomes as the independent springs for our rational actions. But we discover that this account of consensus in reasons is at odds with the conservation of shared (strict) preferences among options. Then Pareto agreements are taken to be superficial unless they are supported from below by consensus in reasons. Levi's theory makes it a serious question between two exclusive strategies facing Group$_1$: either (merely) appeal to the existing agreements on what to do – choose $A_{7\varepsilon}$ without giving reasons; or, instead, agree first on the "neutral" reasons for the group's choices and let that consensus dictate admissibility – in which case both options are admissible. How is this higher-order decision problem resolved?

At the expense of denying an independent consensus for conflicts of beliefs and values, we avoid this dilemma. In "Decisions Without Ordering,"[19] we propose a theory of choice in which preference over pairs of horse lotteries is a strict partial order. In our theory, preference is represented by a set S of pairs of probabilities and utilities, as follows. Option o_1 is preferred to option o_2 according to the partial order if and only if the expected utility of o_1 is greater than that of o_2 for each probability/utility pair in S.[20] When the strict partial order characterizes a consensus among different (Bayesian) agents, we take the elements of S to be all the probability/utility pairs that agree with the shared (weak Pareto) preferences of those agents. Trivially, the weak Pareto condition is respected with this consensus set. Each element of S, each potential compromise among the agents, preserves their shared agreements. Levi's theory does not satisfy this condition for potential compromises.

The sets S are "conditionally convex": If pairs (P_1, U_1) and (P_1, U_2) belong to S, so too do all pairs of the form (P_1, U_3), where $U_3 = \beta U_1 + (1 - \beta)U_2$ $(0 \leq \beta \leq 1)$. Likewise S is closed under mixtures of pairs that share a common utility. S need not be a (convex) cross product of a set of probabilities and a set of utilities, however, as required in Levi's theory. Nor need S be convex, or even connected – as illustrated in Figure 4. According to this account, the shared preferences for the three groups (above) are represented by three different consensus sets:

$$S_1 = \{(P_1, U_1), (P_2, U_2)\}$$

$$S_2 = \{(P_1, U_2), (P_2, U_1)\}$$

$$S_3 = \{(P, U): P = \alpha P_1 + (1-\alpha)P_2, U = \beta U_1 + (1-\beta)U_2, 0 \leq \alpha, \beta \leq 1\}.$$

Only the third group has (partially ordered) strict preferences agreeing with Levi's E-admissible choices, since S_3 is the cross product of two (convex) sets of probabilities and utilities. In general, the consensus set S is *not* arrived at by analyzing individual beliefs and values independently.

In another study,[21] we explore a representation for strict partial orders over pairs of horse lotteries in terms of such sets S. It remains for us an open question, in general, what set S is generated this way by the preferences of several (quasi-) Bayesian agents – agents each of whose preferences is given by a strict partial order of this very sort. With two Bayesian agents, the quasi-Bayesian group preference generated by the weak Pareto condition is reported in theorem 1. Then, S is the set consisting of the two points. What strict partial order corresponds in this way to the shared preferences of n-Bayesians? To know the answer is to know the Pareto decisions of a panel of Bayesian experts.[22]

APPENDIX

In their 1979 article, Hylland and Zeckhauser pursue a heuristic argument [due to Zeckhauser, "Group Decision and Allocation," Discussion Paper #51 (1968), Harvard Institute of Economic Research] to establish the impossibility of a rule for amalgamating a set of Bayesian preferences into a single Bayesian preference, provided that the rule: (1) applies to all potential sets of Bayesian agents: (2) respects the weak Pareto condition; (3) avoids "Dictators" for the group-probability; (4) combines probabilities and utilities separately; and (5) preserves unanimity of the agents' probabilities in case they all have the same degrees of belief. Also, they show that (2) and (3) may be replaced by the single condition (6): the strong Pareto requirement.

We note, first, that regarding "impossibility," the two results share an equivalence between the conjunction of conditions (2) weak Pareto and (3) no (probability) dictators/no autocrats, and condition (6) the strong Pareto requirement. (The weaker condition, "no (probability) dictators," suffices in their argument because they have condition (4). Instead, we use "no autocrats.") Unlike their result, however, theorem 1 does not place restrictions on the form of the amalgamation, i.e., neither condition (4) nor (5) is required for our analysis. Thus, it is insufficient for finding a Bayes compromise merely to abandon independent amalgamations of probability and utility.

Also, theorem 1 shows that there are none but autocratic, weak Pareto "compromises" for *every* pair of agents with differing beliefs and values. This is in marked contrast with their result which states that, for each Bayesian social welfare rule satisfying 3–5, there is some pair of Bayesian agents whose preferences are amalgamated in violation of (2): the weak Pareto condition.

A somewhat different treatment of the Bayesian group decision problem is found in Hammond's 1981 paper. Hammond's interesting work is concerned with issues of welfare optimality in dynamic (intertemporal) social decisions. Abstracting away from the dynamic features of his analysis, we find the following result about Bayesian amalgamations:

"*Theorem 2*" (Hammond, *op. cit.*, p. 241). Suppose there are n Bayesian agents, whose preferences over their own gains from social acts are represented by the probability/utility pairs (P_k, U_k), $k = 1, \ldots, n$. Then, a Bayesian (Bergson) social welfare function, W, satisfying the strong Pareto condition exists, provided it is of the form (P_w, U_w), where: (i) $P_w = P_k$ $(k = 1, \ldots, n)$, i.e., the n agents all have the same personal probabilities, and (ii) $U_w = \Sigma_k \gamma_k U_k$ $(k = 1, \ldots, n$ and $0 < \gamma_k)$, i.e., social utility is a convex combination of the n individual utilities. Thus, according to this result, there can be no Bayesian amalgamation (subject to the strong Pareto condition) whenever different agents hold different personal probabilities, regardless of the nature of their personal values for outcomes.

Theorem 1 is stronger than this result because, as in the previous comparison, Hammond's conclusion depends upon a restricted form of group amalgamation – a Bergson social welfare rule makes the group utility a function of the individual (interpersonal) utilities, independent of their personal probabilities. A Bergson amalgamation treats beliefs and values separately. (There is, in addition, an assumption that the amalgamation of individual utilities is differentiable, i.e., Hammond's argument requires a smoothly changing Bergson social welfare.) Also paralleling the comparison with Hylland and Zeckhauser's result, Hammond's argument shows that, for every Bergson social welfare rule (unless the agents share a common personal probability), there is some configuration of personal utilities which leads to a violation of the strong Pareto condition. Theorem 1, by contrast, establishes the violation of the strong Pareto condition for each Bayes model and for every pair of agents with different beliefs and values.

Finally, it is worth explaining away an apparent conflict between Hammond's theorem 2 and other findings we have made, concerning Bayesian compromises when values are shared (there is a common utility) and beliefs differ (cf. "Decisions Without Ordering," §4). "Theorem 2" asserts that no Bergson social welfare rule is possible in this case. Our discussion indicates that there exist weak Pareto compromises created by the following pairs: Let the Bayes model use the (assumed) common utility and any convex combination of the personal probability distributions. In the case of two agents, each nonextreme convex combination of the individual probabilities (no one individual is autocrat) creates a compromise that satisfies the strong Pareto condition. [This generalizes to n agents. For conditions under which the strong Pareto compromises require all positive weights, see, e.g., P. C. Fishburn, "On Harsanyi's Utilitarian Cardinal Welfare Theorem," *Theory and Decision*, XVII (1984): 21–28.]

The two claims are not contradictory: they deal with different domains of social acts. In Hammond's presentation, the class of social acts includes the (smaller) domain of our analysis. For Hammond's "theorem 2," the domain of social acts arises by feasible market transactions. In the restricted domain of our analysis, each agent receives the same outcome as every other agent, in every state. Thus, when we suppose that agents share a common utility for individual rewards, perforce it is also a common utility over the "constant" social acts in this limited variety of social choices.

Instead, the social acts used in Hammond's analysis include the commonplace situation where different individuals receive different rewards. The personal utilities of Hammond's analysis reflect the agent's preferences for her individual rewards only. Thus, for Hammond's argument, two agents with the same individual utility will differ in their preferences for some "constant" social acts. If they receive different rewards, one from another, their preferences between two social acts may be in direct opposition despite the common utility for individual rewards. For example, they can have the same (cardinal) utility for money but differ in their ranking of two social options, depending upon which of them receives the greater monetary reward under which of the two acts.

We can extend theorem 1 to the larger domain of Hammond's social choices by recasting our notion of a social option. Let us mean, rather, that an act is an n-tuple of horse lotteries, one for each agent. This is inclusive of the restricted class of social acts in which each agent receives the same horse lottery. Then, as we suppose when arguing that a convex combination of personal probabilities creates a weak Pareto compromise, in order for two agents to share a common utility over outcomes – over the "constant" acts in this extended sense of "option" – they must agree on the ranking of all social acts which award distinct (von Neumann-Morgenstern) lotteries to different individuals. If they have the same utility for these "constant" social acts then [in J. C. Harsanyi's sense; cf. *Rational Behavior and Bargaining Equilibrium in Games and Social Situations* (New York: Cambridge University Press, 1977), ch. 4] they share one "moral" utility over such options. In that case, an agent's preference over "constant" social acts does not depend upon his identity; it does not depend upon which (lottery) reward is his. But when agents have a common utility for personal rewards and evaluate social acts solely in terms of their own gains (as in Hammond's analysis), their preferences do *not* correspond to a "moral" utility. Thus, the appearance of a conflict between the two claims is illusory.

NOTES

1 *Social Choice and Individual Values* (New York: Wiley, 1951). Stated briefly, Arrow's *impossibility theorem* shows that it is inconsistent to posit a social choice rule for amalgamation of individual preferences into a group preference, subject to four conditions: (1) the rule has unrestricted domain – it applies with all varieties of individual preferences and all sets of feasible acts; (2) the rule satisfies the weak Pareto condition – if everyone (strictly) prefers option *A* to option *B*, then the social ordering makes *A* strictly better than *B*; (3) the rule precludes *dictator* solutions – it cannot be that the social ordering is determined by one individual's preferences regardless of the preferences of all others; (4) the rule conforms to an *independence of irrelevant alternatives* condition – the social ordering of a set of feasible options depends solely on the individuals' orderings for these options. (The fourth condition combines a prohibition on interpersonal utility comparisons with an assumption that the social choice rule induces an ordering of acts by revealed preferences.)

2 "The decision of a committee using a special majority," *Econometrica*, XVI (1948): 245–261.

3 "Possibility Theorems with Interpersonally Comparable Welfare Levels," *Review of Economic Studies*, XLVII (1980): 409–420.

4 *The Foundations of Statistics* (New York: Wiley, 1954), §13.5.

5 "Conflict and Social Agency," *Journal of Philosophy*, LXXIX, 5 (May 1982): 231–247.

6 We assume acts and states are probabilistically independent. Later on where we use horse lotteries for acts, as explained in fn. 9, we assume also that states are value-neutral with respect to the lottery prizes.

Interesting discussions of the measurement problem for expected utility theory without either of these assumptions are found in J. Drèze, "Decision Theory with Moral Hazard and State-Dependent Preference," paper #8545, *C.O.R.E.* (1985), Université Catholique de Louvain, Voie du Roman Pays, 34, B-1345, Louvain-la-Neuve, Belgium.

For additional, important commentary about the effects of state-dependent utilities on the measurement of probabilities, see E. Karni, D. Schmeidler, and K. Vind, "On State Dependent Preferences and Subjective Probabilities," *Econometrica*, LI, 4 (July 1983): 1021–1031; H. Rubin, "A Weak System of Axioms for 'Rational' Behavior and the Non-Separability of Utility from Prior," *Statistics and Decisions*, V (1987): 47–58; and J. B. Kadane and R. L. Winkler, "Separating Probability Elicitation from Utilities," *Journal of the American Statistical Association*, LXXXIII, 402 (1988): 357–363.

7 This is not the same as Arrow's concept of a "dictator," since the determination of who is the autocrat may be a function of everyone's preferences, contrary to the requirements for an Arrovian dictator.

8 This assumption is complicated by controversies over interpersonal utility

comparisons. With or without concern for another's preferences, an individual's utility function is defined up to an affine transformation. Regarding this, we suppose that for any two agents there is some pair of v.N-M lotteries, L_* and L^*, which are ranked the same by both – each (strictly) prefers L^* to L_*. (This is mild, as we may introduce two new rewards with this property. We discuss the case of completely opposed preferences in fn. 11.) In case interpersonal utilities are recognized, consider the (perhaps, two) representation(s) of the agents' joint utilities:

$$U_a = \{U_{1a}, U_{2a}\} \text{ and } U_b = \{U_{1b}, U_{2b}\},$$

where

$$U_{1a}(L_*) = U_{2b}(L_*) = 0 \text{ and } U_{1a}(L^*) = U_{2b}(L^*) = 1.$$

Then, the assumption that Dick and Jane have different values is expressed by saying that there is a lottery, L, with $U_{1a}(L) \neq U_{2b}(L)$. That is, their (separate) utility functions do not coincide under affine transformations. Since we are concerned with the set of Bayes compromises that preserve unanimous (strict) preferences, we may use the pair (U_{1a}, U_{2b}) to represent the agents' individual utilities while also respecting interpersonal utility comparisons. For convenience, abbreviate U_{1a} by U_1 and U_{2b} by U_2.

9 "A definition of subjective probability," *Annals of Mathematical Statistics*, XXXIV (1963): 199–205. Their theory of "horse lotteries" is summarized by four axioms on preference. Informally, these four axioms require that: (A1) preference is a weak order; (A2) preference satisfies the "independence" condition (related to Savage's "sure-thing" postulate *P*2); (A3) preference obeys an Archimedean condition – to insure utilities are real-valued; and (A4) preference for v.N-M lotteries is state-independent.

Specifically, the final axiom requires this. Let H_1 be the "constant" horse lottery that awards the same v.N-M lottery L_1 in each state and let H_2 be the "constant" horse lottery that awards L_2 in each state. Let horse lotteries $H_{1'}$ and $H_{2'}$ differ solely in that, for some state s, $H_{1'}(s) = L_1$ and $H_{2'}(s) = L2$. $H_{1'}$ and $H_{2'}$ have the same outcomes in every other state. Then axiom A4 says: H_1 is preferred to H_2 if and only if $H_{1'}$ is preferred to $H_{2'}$.

10 The particularly convenient form of these horse lotteries is due to Jay Goodman, formerly of the Statistics Dept. at Carnegie Mellon University.

11 This assumption cannot be removed. A result due to J. Kadane ["Opposition of Interest in Subjective Bayesian Theory," *Management Science*, XXXI (1985): 1586–1588, Theorem 1] says that two agents hold opposing strict preferences over all pairs of acts if, and only if, they share a common personal probability for the states and have opposite (cardinal) utilities for the rewards. Then, as a simple corollary, if two agents hold different probabilities and diametrically opposed (cardinal) utilities, there will be some pairs of horse lotteries which, by strict preference, they rank in common.

That is, suppose Susan and John have different personal probabilities (denoted P_S and P_J, with $P_S \neq P_J$) and strictly opposed preferences for all pairs of v.N-M lotteries, the "constant" horse-lottery acts. Thus, whenever John strictly prefers lottery L_1 to lottery L_2 Susan's preference goes the other way, and vice versa. There is no pair of lotteries which they rank order the same.

Then, after standardizing their separate utility units with a single pair of v.N-M lotteries (as in fn. 8), where Susan prefers the first and, hence, where John prefers the second of the two lotteries, we have: $U_S = 1 - U_J$.

Nonetheless, as $P_S \neq P_J$, by Kadane's result, there are pairs of horse lotteries which they rank order in the same way. Let E be an event with

$$P_S(E) = p_S < p_J = P_J(E).$$

Consider two horse lotteries $H_1 = \{L_{11}, L_{12}\}$ and $H_2 = \{L_{21}, L_{22}\}$, defined on the binary partition $\{E, \neg E\}$, where the L_{ij} $(i, j = 1, 2)$ are v.N-M lotteries. John prefers H_1 to H_2 if, and only if, $cp_J + d > 0$, where $c = [U_J(L_{11}) + U_J(L_{22}) - U_J(L_{12}) - U_J(L_{21})]$ and $d = [U_J(L_{12}) - U_J(L_{22})]$. Because $U_S = 1 - U_J$, Susan also prefers H_1 to H_2 if, and only if, $cp_s + d < 0$. Hence, with $d < 0 < c$ and $-p_Jc < d < -p_Sc$, John and Susan have common strict preferences for H_1 over H_2.

Other Bayesians will agree with John and Susan in preferring H_1 to H_2 provided they have preferences of the form: (P^*, U_J) or (P_*, U_S) where $P^*(E) > P_J(E)$ and $P_S(E) > P_*(E)$ whenever $P_J(E) > P_S(E)$. That is, there exist nonautocratic, Bayes models for the set of weak Pareto agreements between Susan and John. These models correspond to Bayesian agents with *more extreme* degrees of belief (combined with the respective utility) – models that exaggerate the expected utility differences between the two of them. None of these more extreme Bayes models strikes us as a serious compromise between Susan and John. Thus, even in this exceptional case of diametrically opposed utilities, we cannot locate a viable Bayes compromise for their unanimous preferences.

12 The proof depends upon the "state independence" of utilities. Without it, i.e., if we require only axioms A1–A3 (see fn. 9), there is a continuum of (weak) Pareto, Bayesian compromises using state-dependent utilities. This follows from two other results: (1) Theorem 13.1, p. 176, of P. Fishburn's *Utility Theory for Decision Making* (New York: Krieger, 1979); and (2) the existence of a (convex) set of preferences, satisfying axioms A1, A2, and A3, each of which extends the strict partial order formed by the (weak) Pareto condition. Result (2) is proven in our "A Representation of Partially Ordered Preferences," *Ann. Stat.*, XXIII (1995): 2168–2217 [Chapter 1.3, this volume].

13 This mimics the technique used in the two papers we discuss, by A. Hylland and R. Zeckhauser, "The Impossibility of Bayesian Group Decisions with Separate Aggregation of Beliefs and Values," *Econometrica*, XLVII (1979): 1321–1336, p. 1330, equation 7; and by P. Hammond, "Ex-*ante* and Ex-*post* Welfare Optimality under Uncertainty," *Economica*, XLVIII (1981): 235–250, p. 241, for larger communities of n-many decision makers.

14 Cf. fn. 13 and see John Broome, "Utilitarianism and Expected Utility," *Journal of Philosophy*, LXXXIV, 8 (August 1987): 405–422, who comments on the relationship of Hammond's theorem with utilitartianism. Also, he discusses the problem in two unpublished essays, "Bolker-Jeffrey Decision Theory and Axiomatic Utilitarianism" and "Should Social Preferences Be Consistent?" For a survey of some related issues, see C. Genest and J. Zidek, "Combining

Probability Distributions: A Critique and an Annotated Bibliography," with discussion, *Statistical Science*, I (1986): 114–148.

15 A selection of these works includes his "On Indeterminate Probabilities," *Journal of Philosophy*, LXXI, 13 (July 18, 1974): 391–418; *The Enterprise of Knowledge* (Cambridge: MIT, 1980); "Conflict and Social Agency," *op. cit.*; several entries in *Decisions and Revisions* (New York: Cambridge University Press, 1984); and *Hard Choices* (New York: Cambridge University Press, 1986).

16 *Hard Choices*, §5.7.

17 Then admissible choices fail to satisfy A. K. Sen's property γ for choice rules, "Social Choice Theory: A Re-examination," *Econometrica*, XLV (1977): 53–89. Hence, it is not even a *normal* choice rule in Sen's terminology. For details, see Seidenfeld's Discussion of A. P. Dempster, "Probability, Evidence, and Judgment," in *Bayesian Statistics 2*, Bernardo, DeGroot, Lindley, and Smith, eds. (Amsterdam: North-Holland, 1985), pp. 127–129.

18 "Conflict and Social Agency," *op. cit.*

19 See Chapter 1.2.

20 If the weak Pareto condition fails to specify action, i.e., where there are conflicted recommendations based on the expectation inequalities taken from *S*, then one possibility is Levi's proposal to deploy second-tier considerations in order to choose among those options admissible at the first tier. For example, one might use "security" to compromise choice among the "Pareto" admissible alternatives.

21 See Chapter 1.3, "A Representation of Partially Ordered Preferences."

22 In his doctoral thesis, "Existence of Compromises in Simple Group Decisions" (Dept. of Statistics, Carnegie Mellon University, 1988), Goodman extends the negative finding of theorem 1 to groups of three agents ($n = 3$). Also, he gives sufficient conditions that only autocratic "compromises" exist with groups of *n*-many Bayesian agents.

1.2

Decisions Without Ordering

TEDDY SEIDENFELD, MARK J. SCHERVISH,
AND JOSEPH B. KADANE

ABSTRACT

We review the axiomatic foundations of subjective utility theory with a view toward understanding the implications of each axiom. We consider three different approaches, namely, the construction of utilities in the presence of canonical probabilities, the construction of probabilities in the presence of utilities, and the simultaneous construction of both probabilities and utilities. We focus attention on the axioms of independence and weak ordering. The independence axiom is seen to be necessary to prevent a form of Dutch Book in sequential problems.

Our main focus is to examine the implications of not requiring the weak order axiom. We assume that gambles are partially ordered. We consider both the construction of probabilities when utilities are given and the construction of utilities in the presence of canonical probabilities. In the first case we find that a partially ordered set of gambles leads to a set of probabilities with respect to which the expected utility of a preferred gamble is higher than that of a dispreferred gamble. We illustrate some comparisons with theories of upper and lower probabilities. In the second case, we find that a partially ordered set of gambles leads to a set of lexicographic utilities, each of which ranks preferred gambles higher than dispreferred gambles.

I. INTRODUCTION: SUBJECTIVE EXPECTED UTILITY [SEU] THEORY

The theory of (subjective) expected utility is a normative account of rational decision making under uncertainty. Its well-known tenets are

Reprinted from W. Sieg (ed.), *Acting and Reflecting: The Interdisciplinary Turn in Philosophy* (Dordrecht: Kluwer Academic Publishers, 1990), 143–170.

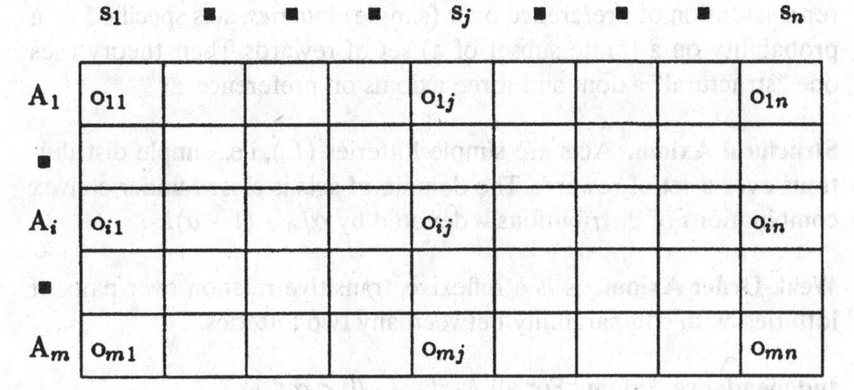

Figure 1. Canonical decision matrix

spotlighted by the familiar, canonical decision problem in which s_j: $j = 1, \ldots, n$ is a partition, and o_{ij} is the outcome of option$_i$ (act$_i$) in state$_j$. That is, acts are functions from states to outcomes. This problem is illustrated in Figure 1.

In the canonical decision problem, states are *value-neutral* and *act independent*. The value of an outcome does not depend upon the state in which it is rewarded, and the choice of an act does not alter the agent's opinion (uncertainty) about the states. In insurance terms, there are no "moral hazards."

General Assumption. Acts are weakly ordered by (weak) preference, $\preceq$, a reflexive, transitive relation with full comparability between any two acts.

Subjective Expected Utility [SEU] Thesis. There is a real-valued utility $U(\ldots)$, defined over outcomes, and a personal probability $p(\ldots)$, defined over states, such that

$$A_1 \preceq A_2 \quad \text{if and only if} \quad \sum_j p(s_j)U(o_{1j}) \leq \sum_j p(s_j)U(o_{2j}).$$

There are several well-trodden approaches to the normative justification of the SEU thesis, which we discuss in the remainder of this section.

I.1. *Utility Given Probability*

The seminal efforts of J. von Neumann and O. Morgenstern (1947) provide necessary and sufficient conditions for an expected utility representation of preference over (simple) *lotteries*: acts specified by a probability on a (finite subset of a) set of rewards. Their theory uses one "structural" axiom and three axioms on preference $\preceq$.

Structural Axiom. Acts are simple lotteries (L_i), i.e., simple distributions over a set of rewards. The domain of acts is closed under convex combinations of distributions – denoted by $\alpha L_1 + (1 - \alpha)L_2$.

Weak-Order Axiom. $\preceq$ is a reflexive, transitive relation over pairs of lotteries, with comparability between any two lotteries.

Independence Axiom. For all L_1, L_2, L_3 ($0 < \alpha \leq 1$),

$$L_1 \preceq L_2 \text{ if and only if } \alpha L_1 + (1-\alpha)L_3 \preceq \alpha L_2 + (1-\alpha)L_3.$$

Archimedean Axiom. For all ($L_1 \prec L_2 \prec L_3$) $\exists(0 < \alpha, \beta < 1)$,

$$\beta L_1 + (1-\beta)L_3 \prec L_2 \prec \alpha L_1 + (1-\alpha)L_3.$$

A particularly simple illustration of this theory involves lotteries over three rewards ($r_1 \prec r_2 \prec r_3$), where the reward r_i is identified with the degenerate lottery having point-mass $P(r_i) = 1$ ($i = 1, 2, 3$). Following the excellent presentation by Machina (1982), we have a simple geometric model for what is permitted by expected utility theory. Figure 2 depicts the consequences of the axioms.

According to the axioms, indifference curves ($\sim$) over lotteries are parallel, straight lines of (finite) positive slope. L_i is (strictly) preferred to L_j, $L_j \prec L_i$, just in case the indifference curve for L_i is to the left of the indifference curve for L_j. Hence, in this setting, expected utility theory permits one degree of freedom for preferences, corresponding to the choice of a slope for the lines of indifference.

Another version of this example occurs with the decision theoretic reconstruction of "most powerful" Neyman-Pearson tests of a simple "null" hypothesis (h_0) versus a simple rival alternative (h_1). We face the binary decision given by the matrix:

	h_0	h_1
accept h_0	a	b
reject h_0	c	d

where we suppose that outcomes b and c are each dispreferred to either outcomes a and d. In the usual jargon, c is the outcome of a type$_1$

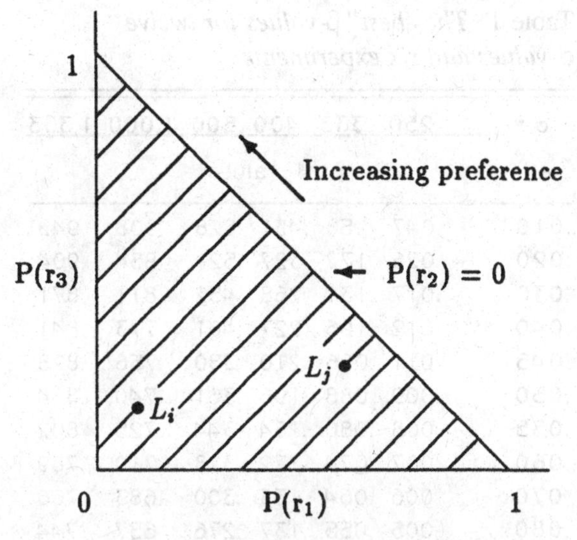

Figure 2. Curves of indifference with three rewards

error and b is the outcome of a type$_2$ error. By the assumption that states are "act independent," without loss of generality, we may rewrite the matrix with utility outcomes:

	h_0	h_1
accept h_0	0	$-(1-x)$
reject h_0	$-x$	0

where $0 < x < 1$. The expected utility hypothesis requires that accepting h_0 is not preferred to ($\preceq$) rejecting h_0 just in case $(1 - p_0)/p_0 \geq x/(1 - x)$, where p_0 is the "prior" probability of h_0.

Suppose we have the option of conducting an experiment E (with a sample space of possible experimental outcomes denoted by Ω), where the conditional probabilities $p(\cdot|h_0)$ and $p(\cdot|h_1)$ over Ω are specified by the description of E. A (Neyman-Pearson) statistical test of h_0 against h_1, based on E, is defined by a critical region $\mathcal{R} \subset \Omega$; with the understanding that h_0 is rejected iff $\mathcal{R}$ occurs. Associated with each statistical test are two quantities: (α, β), where $\alpha = p(\mathcal{R}|h_0)$ is the probability of a type$_1$ error, and $\beta = p(\mathcal{R}^c|h_1)$ is the probability of a type$_2$ error.

According to the N-P theory, two tests may be compared by their (α, β) numbers. Say that T_2 dominates T_1 if $(\alpha_2 \leq \alpha_1)$, $(\beta_2 \leq \beta_1)$ and at least one of these inequalities is strict. This agrees with the ranking of

Table 1. *The "best" β-values for twelve α-values and six experiments*

σ =	.250	.333	.400	.500	1.000	1.333
α	β-values					
.010	.047	.250	.431	.628	.908	.942
.020	.026	.172	.327	.521	.854	.904
.030	.017	.131	.268	.452	.811	.871
.040	.012	.106	.227	.401	.773	.841
.045	.011	.096	.210	.380	.756	.828
.050	.009	.088	.196	.361	.740	.814
.055	.008	.080	.184	.344	.725	.802
.060	.007	.074	.172	.328	.710	.789
.070	.006	.064	.153	.300	.683	.766
.080	.005	.055	.137	.276	.657	.744
.090	.004	.049	.123	.255	.633	.722
.100	.003	.043	.111	.236	.611	.702

tests by their expected utility since (prior to observing the outcome of the experiment) the expected utility of test T, having errors (α, β), is given by:

$$-[x \cdot p(\mathcal{R}\&h_0)+(1-x)\cdot p(\mathcal{R}^c\&h_1)] = -[x\cdot\alpha\cdot p_0+(1-x)\cdot\beta\cdot(1-p_0)],$$

so that $T_1 < T_2$ if T_2 dominates T_1 (except for the trivial cases of certainty: $p_0 = 0$ or $p_0 = 1$, when $T_2 \sim T_1$ is possible still – but then there hardly is need for a "test" of h_0).

Given an experiment E, there are numerous, mutually undominated tests based on E. For example, consider the family of undominated tests of h_0: $\mu = 0$ versus h_1: $\mu = 1$ from the observation of a normally distributed random variable $X \sim N[\mu, \sigma^2]$, with specified variance σ^2. These are just the family of "best," i.e., most powerful tests of h_0 versus h_1 – which, by the Neyman-Pearson lemma, is the family of likelihood ratio tests for the datum x. Table 1 lists some (α, β) values for undominated tests from six such experiments: $\sigma = 1/4$; $= 1/3$; $= 2/5$; $= 1/2$; $= 1$; and $= 4/3$.

Three of these families, corresponding to $\sigma = 1/3$; $\sigma = 1/2$; and $\sigma = 4/3$, are depicted by the curves in Figure 3. The graph shows the tangents to these three curves at $\alpha = 0.05$. The "0.05-α-level" tangents are

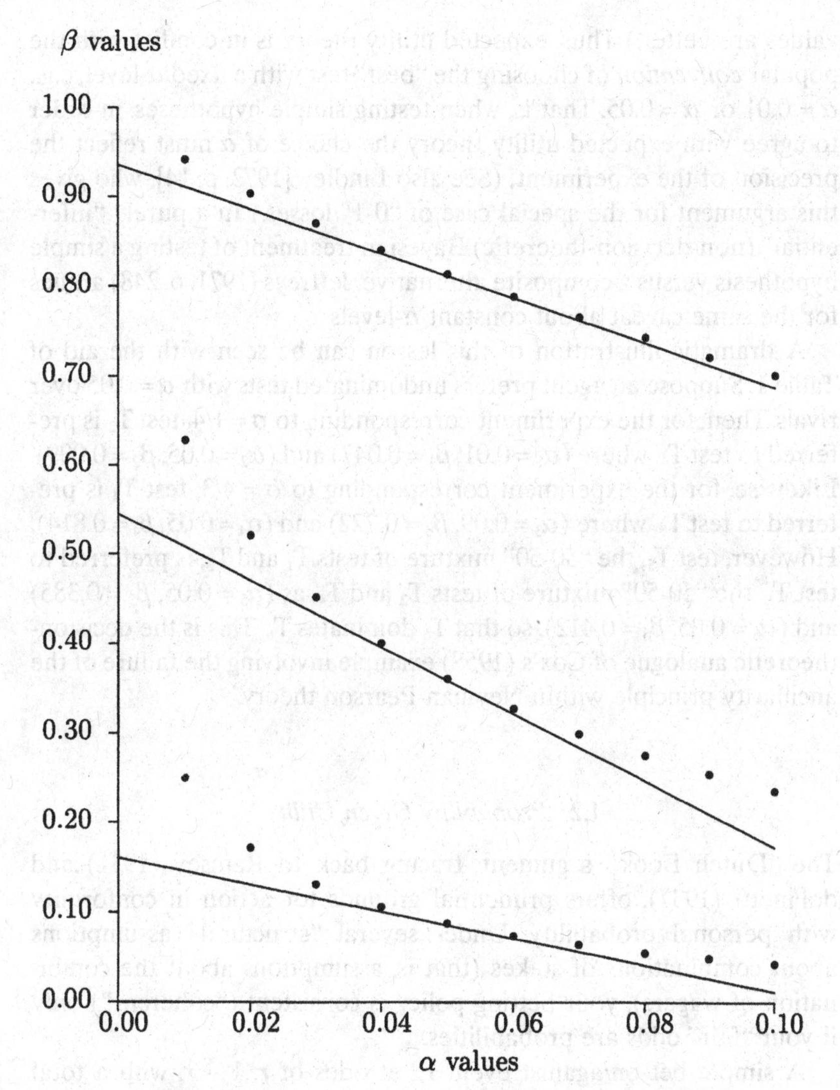

Figure 3. Families of (α, β) pairs for undominated tests

not parallel. A statistical test of h_0 versus h_1 is a lottery involving the three prizes $-x$, $-(1 - x)$, 0. As before, if the preferences among such tests satisfy the expected utility hypothesis, then the indifference curves (of equally desirable tests) are parallel straight lines.

In Figure 3, these indifference curves have negative slopes equal to $-xp_0/(1 - x)(1 - p_0)$. (The slopes are negative because *smaller* (α, β)

values are better.) Thus, expected utility theory is in conflict with the popular *convention* of choosing the "best" test with a fixed α-level, e.g., $\alpha = 0.01$ or $\alpha = 0.05$. That is, when testing simple hypotheses, in order to agree with expected utility theory the choice of α must reflect the precision of the experiment. (See also Lindley [1972, p. 14], who gives this argument for the special case of "0-1" losses.) In a purely "inferential" (non-decision-theoretic) Bayesian treatment of testing a simple hypothesis versus a composite alternative, Jeffreys (1971, p. 248) argues for the same caveat about constant α-levels.

A dramatic illustration of this lesson can be seen with the aid of Table 1. Suppose an agent prefers undominated tests with $\alpha = 0.05$ over rivals. Then, for the experiment corresponding to $\sigma = 1/4$, test T_2 is preferred to test T_1, where ($\alpha_1 = 0.01, \beta_1 = 0.047$) and ($\alpha_2 = 0.05, \beta_2 = 0.009$). Likewise, for the experiment corresponding to $\sigma = 4/3$, test T_4 is preferred to test T_3, where ($\alpha_3 = 0.09, \beta_3 = 0.722$) and ($\alpha_4 = 0.05, \beta_4 = 0.814$). However, test T_5, the "50-50" mixture of tests T_1 and T_3, is preferred to test T_6, the "50-50" mixture of tests T_2 and T_4, as ($\alpha_5 = 0.05, \beta_5 = 0.385$) and ($\alpha_6 = 0.05, \beta_6 = 0.412$), so that T_5 dominates T_6. This is the decision-theoretic analogue of Cox's (1958) example involving the failure of the ancillarity principle within Neyman-Pearson theory.

I.2. *Probability Given Utility*

The "Dutch Book" argument, tracing back to Ramsey (1931) and deFinetti (1937), offers prudential grounds for action in conformity with personal probability. Under several "structural" assumptions about combinations of stakes (that is, assumptions about the combination of wagers), your betting policy is consistent ("coherent") only if your "fair" odds are probabilities.

A simple bet on/against event E, at odds of $r: 1 - r$, with a total stake $S > 0$ (say, bets are in $ units), is specified by its payoffs, as follows:

	E	¬E
bet on E	win $(1 - r)S$	lose rS
bet against E	lose $(1 - r)S$	win rS

(By writing $S < 0$ we can reverse betting "on" or "against.")

The general assumption (that acts are weakly ordered by $\preceq$) entails that there is a preference among the options betting on, betting against and abstaining from betting (whose consequences are "status quo," or net \$0, regardless of whether E or ¬E). The special ("structural") assumptions about the stakes for bets require, in addition:

a. Given an event E, a betting rate r: $1 - r$ and a stake S, your preferences satisfy exactly one of three profiles. Either:
   betting on $\prec$ abstaining $\prec$ betting against E,
   or betting on $\sim$ abstaining $\sim$ betting against E,
   or betting against $\prec$ abstaining $\prec$ betting on E.
b. The (finite) conjunction of favorable/fair/unfavorable bets is favorable/fair/unfavorable. (A conjunction of bets is favorable in case it is preferred to abstaining, unfavorable if dispreferred to abstaining, and fair if indifferent to abstaining.)
c. Your preference for outcomes is continuous in rates; in particular, each event E carries a unique "fair odds" r_E for betting on E.

Note: It follows from these assumptions that your attitude towards a simple bet is independent of the size of the stake.

Dutch Book Theorem. If your fair betting odds are not probabilities, then your preferences are incoherent, i.e., inconsistent with the preference for sure-gains. Specifically, then there is some "favorable" combination of bets which is dominated by abstaining, i.e., some "favorable" combination where you pay out in each state of a finite (exhaustive) partition. (See Shimony (1955), for an elegant proof using the linear structure of these bets.)

The Dutch Book argument can be extended to include conditional probability, $p(\cdot|\cdot)$, through the device of called-off bets. A called-off bet on (against) H given E, at odds of r: $(1 - r)$ with total stake S (>0), is specified by its payoffs, as follows.

	$H \cap E$	$\neg H \cap E$	$\neg E$
bet on H	win $(1 - r)S$	lose rS	0 (the bet is called off)
bet against H	lose $(1 - r)S$	win rS	0 (the bet is called off)

By including called-off bets within the domain of act to be judged favorable/indifferent/unfavorable against abstaining, and subject to the same structural assumptions (a–c) imposed above, coherence of "fair"

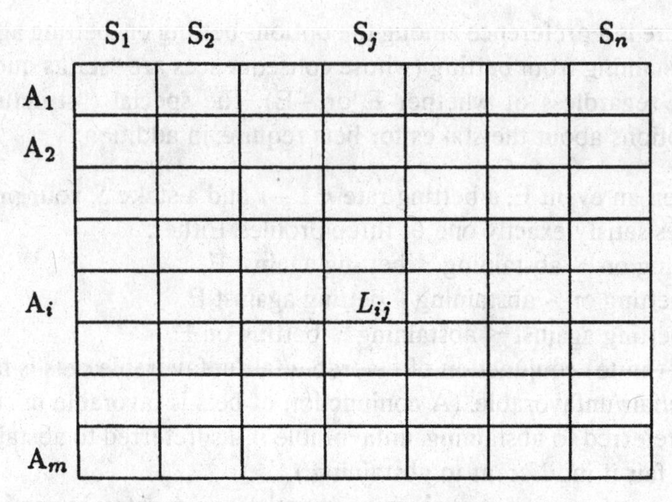

Figure 4. Anscombe-Aumann "horse lotteries"

betting odds entails: $r_{(H|E)} \cdot r_E = r_{(H \cap E)}$, where "$r_{(H|E)}$" is the "fair called-off" odds on H given E. This result gives the basis for interpreting conditional probability, $p(H|E)$, by the fair "called-off:" odds $r_{(H|E)}$, for then we have:

$$p(H|E) \cdot p(E) = p(H \cap E),$$

the axiomatic requirement for conditional probabilities.

I.3. *Simultaneous Axiomatizations of (Personal) Probability and Utility*

We distinguish two varieties:

i. without extraneous "chances," as in Savage's (1954) theory.
ii. with extraneous "chances," a continuation of the von Neumann-Morgenstern approach, as in Anscombe & Aumann's (1963) theory of "horse lotteries." Horse lotteries are a generalization of lotteries, as illustrated in Figure 4.

An outcome of act A_i, when state S_j obtains (when "horse$_j$" wins), is the von Neumann-Morgenstern lottery L_{ij}. The Anscombe-Aumann theory is the result of taking the von Neumann-Morgenstern axiomatization of $\preceq$ (the Weak-order, Independence and Archimedean postulates), and adding an assumption that states are value-neutral.

II. INDEPENDENCE AND CONSISTENCY IN SEQUENTIAL CHOICES

We are interested in relaxing the "ordering" postulate, without abandoning the normative standard of coherence (consistency) and without losing the representation ("measurement") of our modified theory. First, however, let us compare two programs for generalizing expected utility in order to justify the concern for consistency:

*Program ¬**I** – Delete the "Independence" Postulate.* Illustrations: Samuelson (1950); Kahneman & Tversky's "Prospect Theory" (1979); Allais (1979); Fishburn (1981); Chew & Macrimmon (1979); McClennen (1983); and especially Machina (1982, 1983 – which has an extensive bibliography).

*Program ¬**O** – Delete the "Ordering" Postulate.* Illustrations: I. J. Good (1952); C. A. B. Smith (1961) – related to the Dutch Book argument; I. Levi (1974, 1980); Suppes (1974); Walley & Fine (1979); Wolfenson & Fine (1982); Schick (1984).

And in Group Decisions: Savage (1954, §7.2); Kadane & Sedransk (1980); and Kadane (1996) – applied to clinical trials.

Also, "regret" models involve a failure of "ordering" if we define the relation $\preceq$ by their choice functions, which violate (Sen's properties α and β, 1977) "independence of irrelevant alternatives": Savage (1954, §13.5); Bell & Raiffa (1979); Loomes & Sugden (1982); and Fishburn (1983).

A CRITICISM OF PROGRAM ¬I. Consider elementary problems where we apply the modified theory ¬**I** to simple lotteries. Thus, we discuss the case, like the von Neumann-Morgenstern setting, where "probability" is given and we try to quantify (represent) the value of "rewards."

There is a technical difficulty with the theory that results from just the two postulates of "weak-ordering" and the usual "Archimedean" requirement. It is that these two are insufficient to guarantee a real-valued "utility" representation of $\preceq$ (see Fishburn, 1970, §3.1). We can avoid this detail and also simplify our discussion by assuming that lotteries are over (continuous) monetary rewards; we assume that lotteries have $-equivalents and more $ is better.

Under these assumptions and to underscore the normative status of coherence, let us investigate what happens when a particular consequence of "independence" is denied.

Mixture Dominance ("Betweenness"). If lotteries L_1, L_2 are each preferred (dispreferred) to a lottery L_3, so too each convex combination of L_1 and L_2 is preferred (dispreferred) to L_3.

Here is an illustration of sequential inconsistency for a failure of mixture dominance. Let $L_1 \sim L_2 \sim \$5.00$, but $0.5L_1 + 0.5L_2 \sim \$6.00$: the agent prefers the "50-50" mixture of L_1 and L_2 to each of them separately. Then, by continuity of (ordinal) utility over dollar payoffs, there is a fee, $-\$\ \varepsilon$, such that, e.g.,

$$L_1 \sim L_2 \prec 0.5(L_1 - \varepsilon) + 0.5(L_2 - \varepsilon) \sim \$5.75 \prec 0.5L_1 + 0.5L_2,$$

where $L_i - \varepsilon$ denotes the modification of L_i obtained by reducing each payoff in L_i by the fee $\$\ \varepsilon$. Assume $\$4.00 \prec (L_i - \varepsilon)(i = 1, 2)$.

Consider two versions of a sequential decision problem, depicted by the decision trees in Figures 5 and 6. "Choice" nodes are denoted by a □ and "chance" nodes are denoted by •. In the first version (Figure 5), at node **A** the agent may choose between plans **1** and **2**. These lead to *terminal* choices at nodes **B**, depending upon how a "fair" coin lands at the intervening chance nodes. If the agent chooses plan **1** (at **A**) and the coin lands "heads," he faces a (terminal) choice between lottery L_1 and the certain prize of \$5.50. If, instead, the coin lands "tails," he faces a (terminal) choice between L_2 and the certain prize of \$5.50.

The decision tree is known to the agent in advance. He can anticipate (at **A**) how he will choose at subsequent nodes, if only he knows what his preferences will be at those junctures. In the problem at hand, we suppose the agent knows that, at **B**, he will not change his preferences over lotteries. (There is nothing in the flip of the coin to warrant a shift in his valuation of specified, von Neumann-Morgenstern lotteries.) For example, according to our assumptions, at **A** he prefers a certain \$5.50 to the lottery L_1. Thus, we assume that at **B**, too, he prefers the \$5.50 to L_1.

Then, at **A**, the agent knows which terminal options he will choose at nodes **B** and plans accordingly. If he selects plan **1**, he will get \$5.50. If he selects plan **2**, he will get lottery $L_1 - \varepsilon$ with probability 1/2 and he will get lottery $L_2 - \varepsilon$ with probability 1/2. But this he values \$5.75; hence, plan **2** is adopted.

The decision program ¬**I** requires the "ordering" postulate for terminal decisions. Thus, at choice nodes such as **B**, the agent is indifferent between lotteries that are judged equally desirable ($\sim$) according to his preferences ($\preceq$). The second version of the sequential choice problem (Figure 6) results by replacing the lotteries at the (terminal)

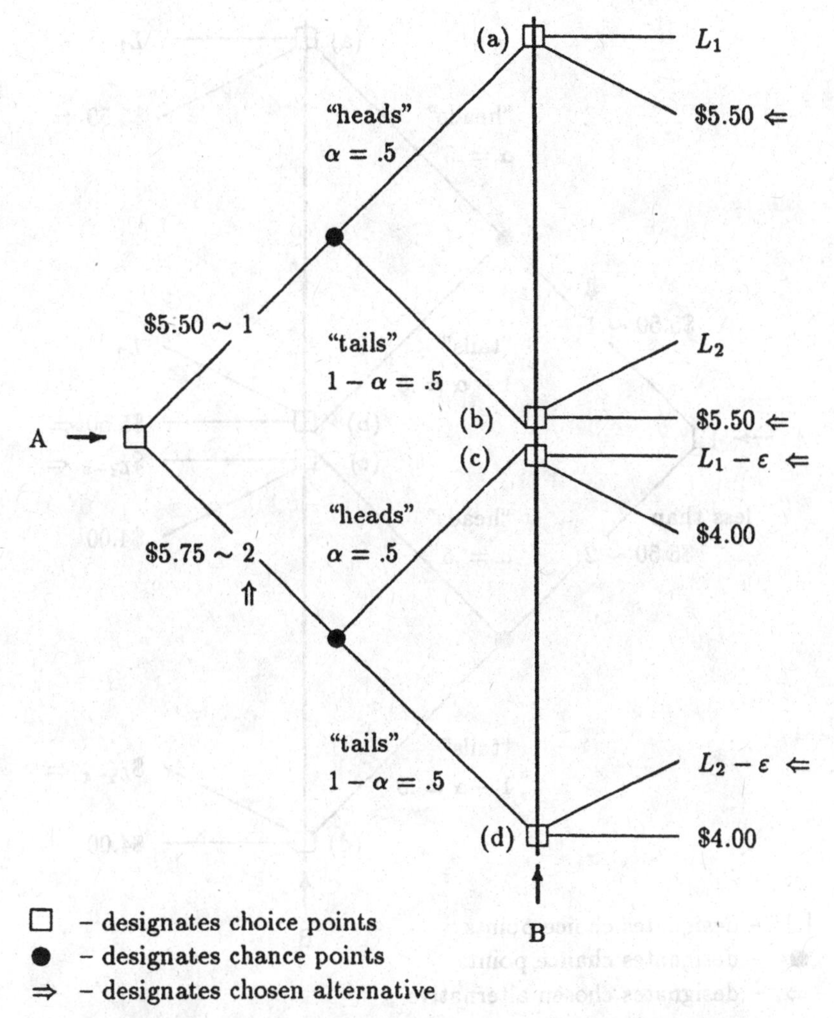

Figure 5. First version of the sequential decision: an illustration of sequential incoherence for a failure of mixture dominance ("betweenness"). At choice node A option 2 is preferred to option 1. At each choice node B this preference is reversed.

nodes **B** by their sure-dollar equivalents under ~. In this version, by the same reasoning, the agent rejects plan **2** and adopts plan **1**. This is an inconsistency within the program since, at nodes **B**, the agent's preferences are given by the weak-ordering, $\preceq$, yet his (sequential) choices do not respect the indifferences, ~, generated by $\preceq$.

Let us call such inconsistency in sequential decisions an episode of

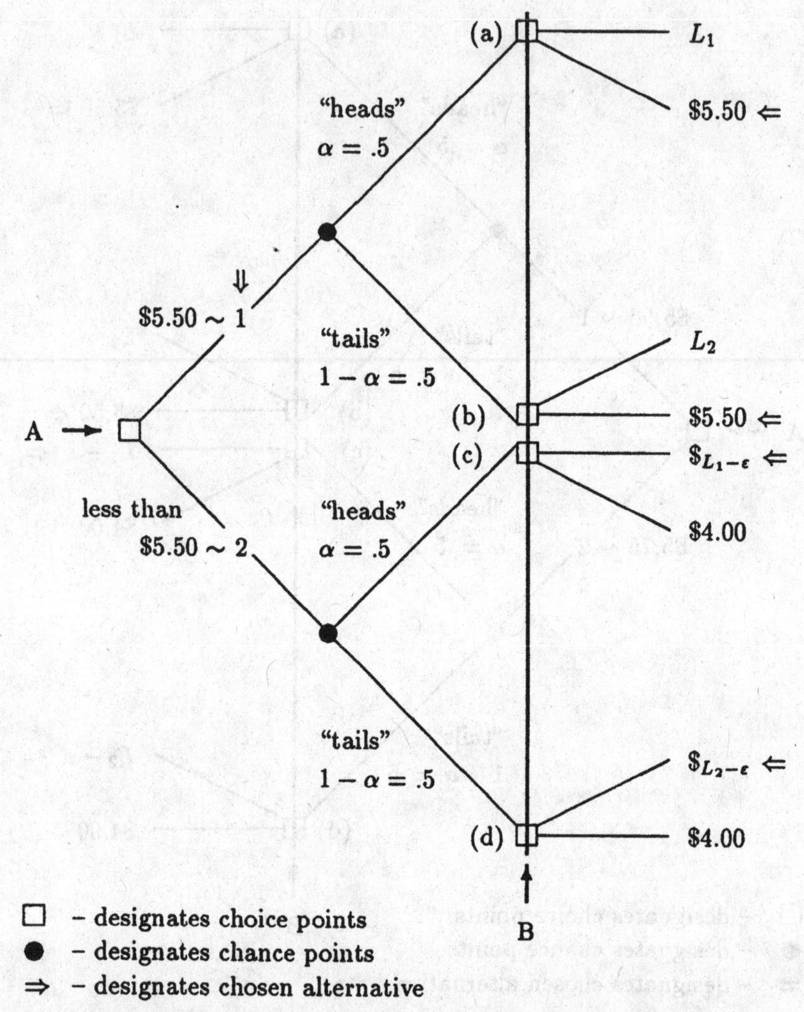

Figure 6. Second version of the sequential decision: an illustration of sequential incoherence for a failure of mixture dominance ("betweenness"). At choice node A option 1 is preferred to option 2. The tree results by replacing $L_i - \varepsilon$ $(i = 1, 2)$ from Figure 6.5 with \$-equivalents under $\preceq$.

"sequential incoherence." Then, we can generalize this example and show:

Theorem. If $\preceq$ is a weak order **(1)** of simple lotteries satisfying the Archimedean postulate **(3)** with sure-dollar equivalents for lotteries, and if $\preceq$ respects stochastic dominance in payoffs (a greater chance at more $ is better), then a failure of "independence" **(2)** entails an episode of sequential incoherence (see Seidenfeld, 1988).

However, as Levi's decision theory – one which relaxes the ordering postulate rather than "independence" – avoids sequential incoherence (Levi, 1986), we see that it is not necessary for decisions to agree with expected utility theory in order that they be sequentially coherent.

III. REPRESENTATION OF PREFERENCES WITHOUT "ORDERING"

Next, we discuss the representation of an alternative theory falling within program **¬O**: to weaken the "ordering" assumption. Again, let us begin with the more elementary problem where we try to quantify values for the rewards when "probability" is given – analogous to the von Neumann-Morgenstern setting.

Let $R = \{r_i: i = 1, \ldots\}$ be a countable set of rewards, and let $L = \{L: L$ is a discrete lottery, a discrete P on $R\}$. As before, define the convex combination of two lotteries $\alpha L_1 + (1 - \alpha)L_2 = L_3$, by $P_3 = \alpha P_1 + (1 - \alpha)P_2$. We consider a theory with three axioms:

Axiom 1. Preference $\prec$ is a strict partial order, being transitive and irreflexive. (This weakens the "weak order" assumption, since non-comparability, ~, need not be transitive.)

Axiom 2. (independence). For all L_1, L_2, and L_3, and for all $1 \geq \alpha > 0$,

$$L_1 \prec L_2 \text{ iff } \alpha L_1 + (1-\alpha)L_3 \prec \alpha L_2 + (1-\alpha)L_3.$$

Axiom 3. A suitable Archimedean requirement. (Difficulties with axiom 3 are discussed below.)

Say that a real-valued utility U *agrees with* the partial order $\prec$ iff

$$\sum_i P_1(r_i)U(r_i) < \sum_i P_2(r_i)U(r_i) \text{ whenever } L_1 \prec L_2.$$

We hope to show that $\prec$ is represented by a (convex) set of agreeing utilities. That is, we seek to show there is a (maximal) set of agreeing utilities, $\mathbf{U}_{\prec}$, where (by the **unanimity** rule)

$$L_1 \prec L_2 \text{ iff for each } U \in \mathbf{U} \prec \sum_i P_1(r_i)U(r_i) < \sum_i P_2(r_i)U(r_i).$$

Aside on Related Results. Aumann (1962) proved that when R is finite, there exists a real-valued utility agreeing with $\prec$, provided axioms like 1–3 hold. A lottery is *simple* if its support is a finite set of rewards. Kannai (1963) extended Aumann's result to simple lotteries on a countable reward set by strengthening the Archimedean axiom 3. (More precisely, these theories deal with a *reflexive* and transitive partial order – which identifies indifferences – not just with the irreflexive part $\prec$.) These two studies, as well as Fishburn's (1970, ch. 9) simplification of Aumann's work, use an embedding of the partial order in a separable, normed linear space. Their proofs have a common theme. Represent a lottery L by a vector of its probability P, with coordinates corresponding to the elements of R. Because a lottery is simple, all but finitely many of its coordinates are zero. Call a vector difference $(P_2 - P_1)$ "favorable" when $L_1 \prec L_2$. The set of "favorable" vectors forms a convex cone, and a Separating Hyperplane Theorem (Klee, 1955) yields a utility. (However, the separability assumption prohibits using this method when, e.g., the reward set R is uncountable.)

There are three observations which help to explain some of the difficulties that arise in carrying out our project for representing preferences given by partial orders.

1. The usual Archimedean axiom won't do; it is too restrictive.

Example 1. $R = \{r_0 \prec r^* \prec r_1\}$ but for no $0 < \alpha < 1$ is it the case that $\alpha r_0 + (1 - \alpha)r_1 \prec r^*$. However, this partial order can be represented by a set of utilities, $\mathbf{U} = \{U_x: 0 < x < 1\}$ with $U_x(r_0) = 0$, $U_x(r_1) = 1$ and $U_x(r^*) = x$. This is illustrated in Figure 7.

Hence, in general, to represent a partial order generated by a set of utilities, a weakening of the usual Archimedean postulate is necessary.

2. Two different convex sets of utilities can generate the same partial order. That is, given convex sets $\mathbf{U}_1$ and $\mathbf{U}_2$, we can difine the partial orders $\prec_1$ and $\prec_2$ according to the "unanimity" rule. However,

Example 2. It may be that $\prec_1 = \prec_2$, though $\mathbf{U}_1 \neq \mathbf{U}_2$. See Figure 8 for an illustration.

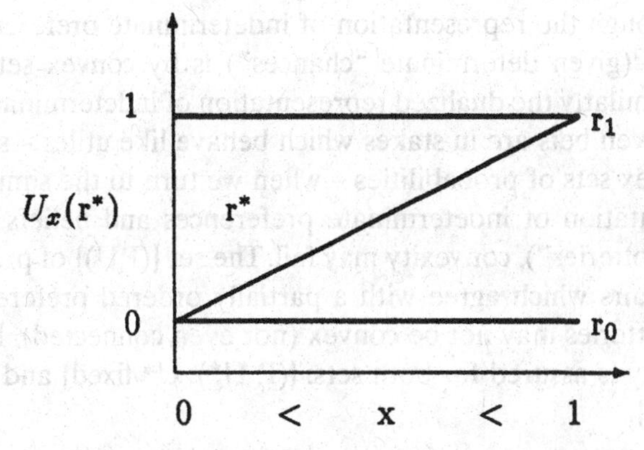

Figure 7. Example of restrictions of the usual Archimedean axiom

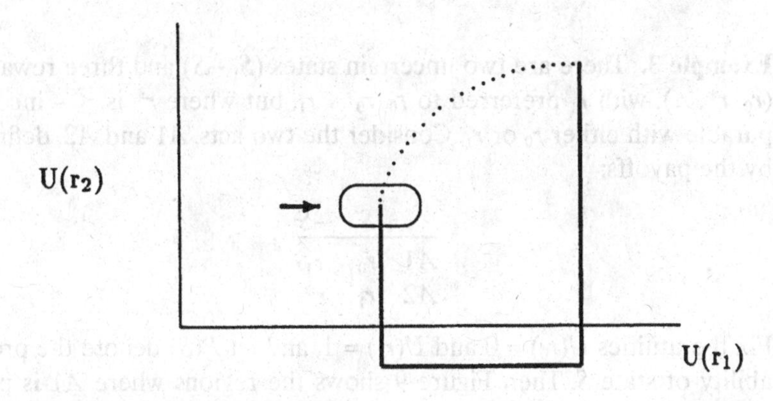

Figure 8. Two convex sets of utilities which generate the same partial order. The two (convex) sets differ by the presence of the point identified by the arrow. The common partial order is generated by the "unanimity" rule

When we shift from representing indeterminate utility (given determinate "chances") to the dual task of representing indeterminate probability (given a determinate utility – by assuming favorable bets combine according to the Dutch Book assumptions – see §IV), this phenomenon causes difficulties for the representation of conditional probabilities. (Also, contrast this with Aumann's example, 1964, p. 210.)

3. Last, though the representation of indeterminate preferences over lotteries (given determinate "chances") is by convex sets of utilities – similarly the dualized representation of indeterminate betting odds (given bets are in stakes which behave like utiles – see §IV) is by convex sets of probabilities – when we turn to the simultaneous representation of indeterminate preferences and beliefs (through "horse lotteries"), convexity may fail. The set {(P,U)} of probability-utility pairs which agree with a partially ordered preference over horse lotteries may not be convex (nor even connected). However, convexity is assured for both sets: {(P, U*): U* fixed} and {(P*, U): P* fixed}.

Here is an example of non-convexity of the set of probability-utility pairs agreeing with a partial order, $\prec$, over "horse lotteries."

Example 3. There are two uncertain states ($S, \neg S$) and three rewards (r_0, r^*, r_1), with r_1 preferred to r_0, $r_0 \prec r_1$, but where r^* is $\prec$ – incomparable with either r_0 or r_1. Consider the two acts, $A1$ and $A2$, defined by the payoffs:

	S	$\neg S$
$A1$	r_0	r_1
$A2$	r_1	r^*

Fix the utilities $U(r_0) = 0$ and $U(r_1) = 1$, and let $P(S)$ denote the probability of state S. Then Figure 9 shows the regions where $A1$ is preferred or $A2$ is preferred.

This example shows why the proof techniques based on the Separating Hyperplane results are inappropriate for identifying the (maximal) set of pairs: $\{(P, U): (P, U)$ agrees with $\prec\}$ for "horse lotteries."

Our proof procedure for giving a representation of a strict preference over horse lotteries is to modify Szpilrajn's (1930) argument that, by transfinite induction, every partial order may be extended to a total order. The modification involves preserving the other axioms: "Independence," "Archimedes," and "value-neutrality" of states. In the Appendix we illustrate this technique for representing strict partial orders of von Neumann-Morgenstern lotteries by convex sets of (lexicographic) utilities.

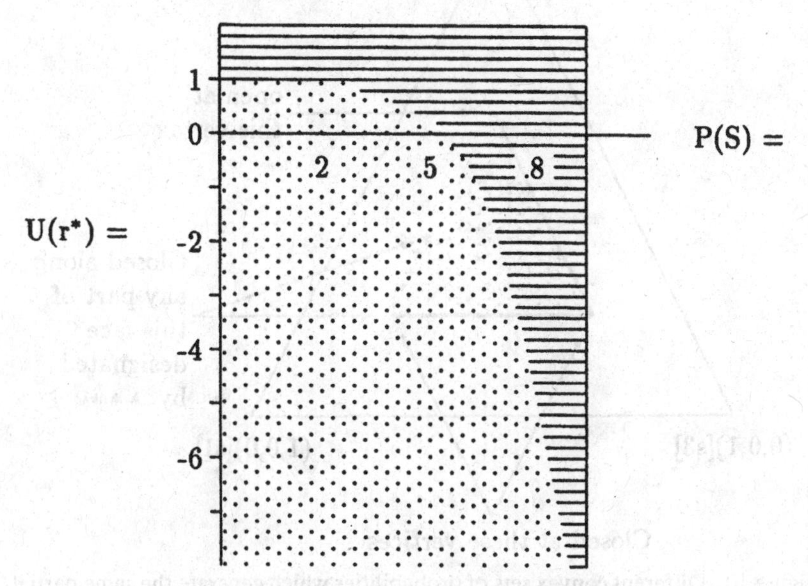

Figure 9. Regions of preference for Example 3

IV. REPRESENTATION OF BELIEFS WITHOUT "ORDERING"

By appeal to the Separating Hyperplanes theorem, we may generalize the Dutch Book argument to establish the coherence of beliefs for partially ordered gambles, including the case (discussed by C. A. B. Smith, 1961) of "medial" odds. Consider the finite partition of states $\{s_j: j = 1, \ldots, n\}$, and define a gamble as a vector of n real-values, $A_i = \langle r_{i1}, \ldots, r_{in} \rangle$, where r_{ij} is the (utility of the) reward generated by A_i when state s_j obtains. Denote the constant gamble $r_j = 0$ (corresponding to "no bet," or "status quo") by **O**, and define the set of *favorable gambles*, $\mathcal{F}$, to be those which are preferred to **O** in pairwise comparisons. As in the Dutch Book argument, we make structural assumptions about the value of the rewards, assuring that the magnitudes of the rewards behave like utilities.

STRUCTURAL ASSUMPTIONS

i. Weak dominance over **O**. If $r_{ij} \geq 0$ (all j) with a strict inequality for some j, then A_i is favorable.

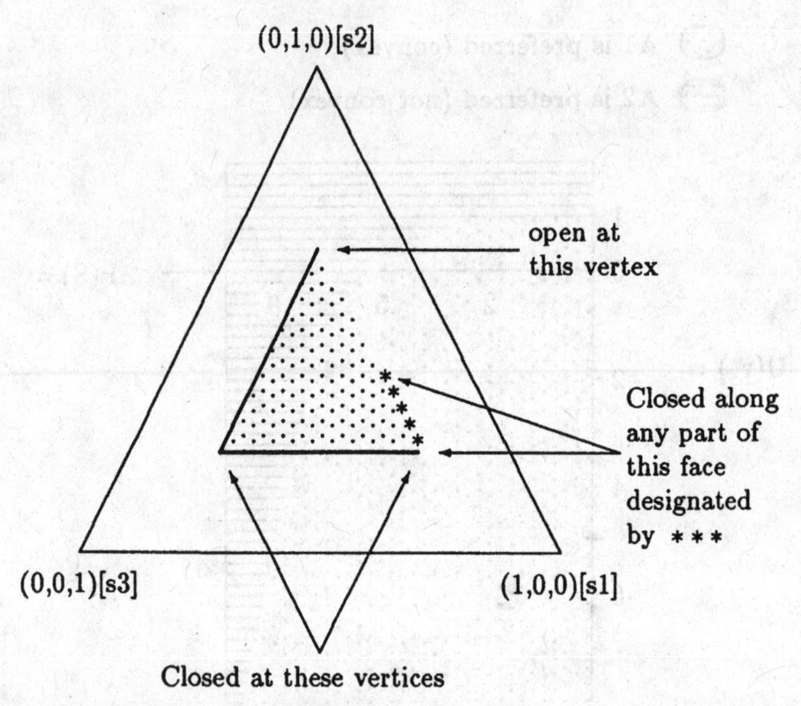

Figure 10. Different convex sets of probabilities which generate the same partial order under the "unanimity" rule

ii. Scalars. If A_i is favorable, so too is $cA_i = \langle \ldots, cr_{ij}, \ldots \rangle$, for $c > 0$.
iii. Convex combinations. If A_h and A_i are favorable, so too is the convex combination $xA_h + (1-x)A_i = \langle \ldots, xr_{hj} + (1-x)r_{ij}, \ldots \rangle$, for $0 \leq x \leq 1$.

REPRESENTATION THEOREMS RELATING TO $\mathcal{F}$

Theorem 1. Coherence of $\mathcal{F}$:

i. $\mathbf{O} \notin \mathcal{F}$ iff there is a maximal, non-empty convex set $\mathcal{P}$ of probabilities with the property that $\forall A_i \in \mathcal{F}, \forall p \in \mathcal{P}, \Sigma_j p(s_j) r_{ij} > 0$.
ii. Moreover, if $\mathcal{F}$ is open, or if $\mathcal{F} \cup \{\mathbf{O}\}$ is closed, then $A_i \in \mathcal{F}$ provided $\forall p \in \mathcal{P}, \Sigma_j p(s_j) r_{ij} > 0$.

We may extend this to include conditional probabilities by paralleling the device of "called-off" bets, used to show coherence of conditional odds in the Dutch Book argument. Then:

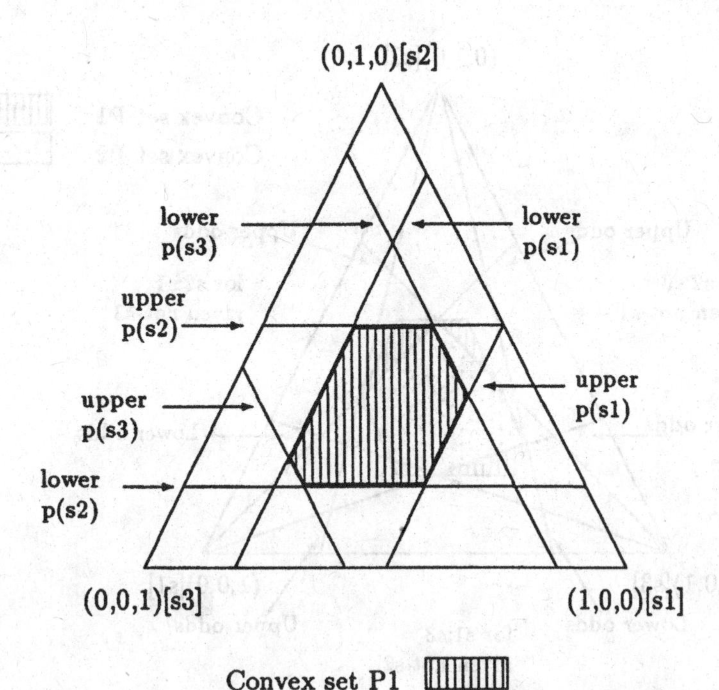

Figure 11. Supporting lines determined by odds alone

Theorem 2. Coherence of conditionally favorable gambles: Let $\mathcal{F}_E(\subset \mathcal{F})$ be the set of favorable gambles, called off in case event E fails to occur, i.e., $\forall A_i \in \mathcal{F}_E$, $r_{ij} = 0$ if $s_j \in E^C$. Assume that $\mathbf{O} \notin \mathcal{F}$.

i. Then $\forall A_i \in \mathcal{F}_E, \forall p \in \mathcal{P}, \Sigma_j p(s_j|E)r_{ij} > 0$.

ii. Moreover, if A_i is called off when E fails and either $\mathcal{F}_E$ is open or $\mathcal{F}_E \cup \{\mathbf{O}\}$ is closed, then $A_i \in \mathcal{F}_E$ provided $\forall p \in \mathcal{P}, \Sigma_j p(s_j|E)r_{ij} > 0$.

In both theorems, the closure conditions imposed in clause (ii) reflect the severity of the problem illustrated in Figure 10, which is dual to the problem illustrated in Example 2, p. 54.

The favorable gambles $\mathcal{F}$ are a subset of those preferred to "no bet" under the partial order ($\prec_{\mathcal{P}}$), generated by the "unanimity" rule adapted to the set $\mathcal{P}$. Denote the closure of $\mathcal{F}$ by cl($\mathcal{F}$), and denote by $\mathcal{F}^-$ the set that results from taking each favorable gamble and changing the sign of its payoffs. It is straightforward to verify that $\mathcal{P}$ is a unit set (expected utility theory) just in case $\prec_{\mathcal{P}}$ is a weak-order. That occurs if and only if cl($\mathcal{F}$) $\cup \mathcal{F}^- = \mathcal{R}^n$ (the space of all gambles on the n states s_j). In other words, when $\mathcal{P}$ is not a unit set, there will be

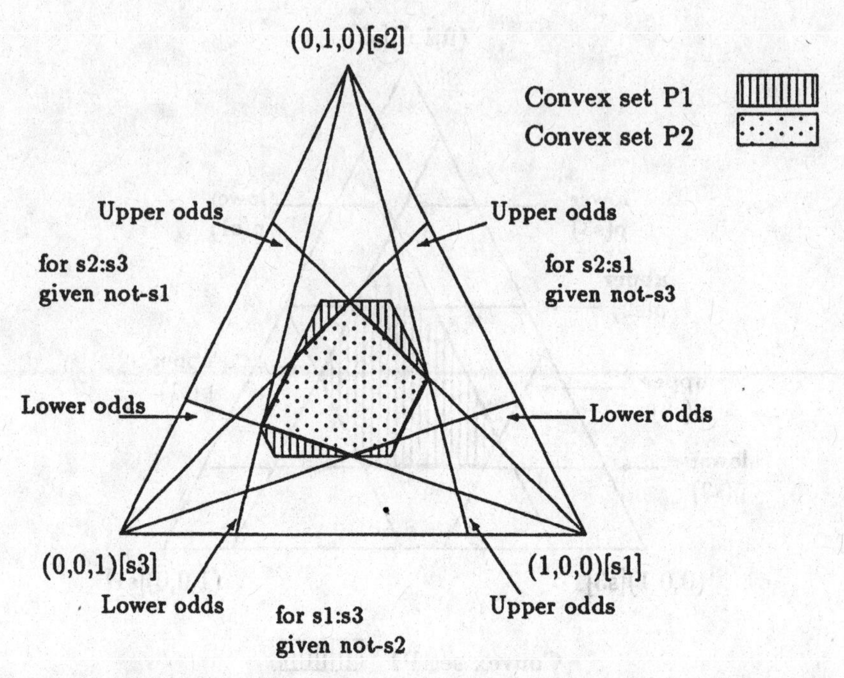

Figure 12. Supporting lines determined by odds and called-off bets

gambles A_1 and A_2 with $A_1 \prec_{\mathcal{P}} A_2$ but where none of A_1, A_1^-, A_2, or A_2^- is favorable.

We illustrate sets $\mathcal{P}$ for the elementary case of three states, $n = 3$ in Figures 11–13. The figures use barycentric coordinates. Each trinomial distribution on $\{s_1,s_2,s_3\}$ is a point in the simplex having vertices: ⟨(100) (010) (001)⟩. Figure 10 shows different convex sets of probabilities that generate the same preferences under the "unanimity" rule. Figure 11 shows the supporting lines for a set $\mathcal{P}_1$ which arises merely by specifying odds at which betting "on" and "against" the (atomic) events s_j become favorable. The set $\mathcal{P}_1$ is the largest one agreeing with these upper and lower probabilities. As noted by Levi (1980, p. 198), typically, infinitely many convex subsets of $\mathcal{P}_1$ carry the same probability intervals.

Figure 12 illustrates the supporting lines for a set $\mathcal{P}_2$ given, in addition, by bounds on the favorable "called-off" bets $\mathcal{F}_{s_j}c$. $\mathcal{P}_2$ is properly included within $\mathcal{P}_1$, has the same upper and lower probabilities, and is the largest set agreeing with all six pairs of unconditional and conditional odds. As Levi (1974, and 1980, p. 202) points out, we can distin-

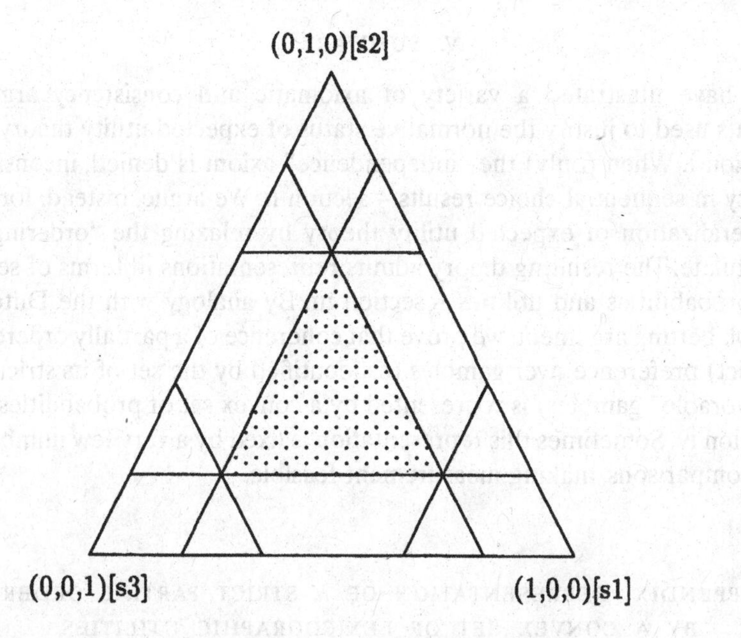

A convex set such that no proper subset has the same upper and lower probabilities for the atoms.

Figure 13. Supporting lines which overdetermine the vertices

guish between two sets having different supporting lines, e.g., $\mathcal{P}_1$ and $\mathcal{P}_2$, with a gamble that is favorable for only one of them.

Figure 13 illustrates how just a few supporting lines can overdetermine the vertices (and thereby all) of a convex set. The simplest case is when the supporting lines corresponding to the upper and lower unconditional odds fix the convex set, $\mathcal{P}_3$, uniquely. That is, there is no proper subset of $\mathcal{P}_3$ with the same upper and lower probabilities. Hence, the set of favorable gambles, $\mathcal{F}$, is fully determined once these upper and lower betting odds are given. (This corrects a minor error in Levi's (1980, p. 202) presentation. There, the set "B_t," has upper and lower unconditional and conditional odds which overdetermine its vertices. Thus, the proper subset "B_t'" does not have the same range of unconditional and conditional odds as "B_t".) We plan to investigate the computational issues relating to the measurement of a convex set, $\mathcal{P}$, using the set of favorable gambles, $\mathcal{F}$. How efficiently can we locate supporting lines which overdetermine the vertices of a set?

V. SUMMARY

We have illustrated a variety of axiomatic and consistency arguments used to justify the normative status of expected utility theory – section I. When (only) the "independence" axiom is denied, inconsistency in sequential choice results – section II. We argue, instead, for a generalization of expected utility theory by relaxing the "ordering" postulate. The resulting theory admits representations in terms of sets of probabilities and utilities – section III. By analogy with the Dutch Book betting argument, we prove that coherence of a partially ordered (strict) preference over gambles (as identified by the set of its strictly "favorable" gambles) is represented by a convex set of probabilities – section IV. Sometimes this representation is fixed by a very few number of comparisons, making measurement feasible.

APPENDIX: REPRESENTATION OF A STRICT PARTIAL ORDER BY A CONVEX SET OF LEXICOGRAPHIC UTILITIES

Defs. Let REW be a set of rewards, REW = $\{r_\alpha : \alpha \le \beta\}$. A *lottery*, L, is a discrete probability distribution over REW, $L = \{p(\cdot) : p(r_\alpha) \ge 0, \Sigma p(r_\alpha) = 1\}$. Let $Supp(L)$ be the support of $p(\cdot)$. (A *simple lottery* is a lottery with finite support.) Denote by L_{REW} the set of simple lotteries over REW. Given two lotteries $L_1 = \{p_1(\cdot)\}$ and $L_2 = \{p_2(\cdot)\}$, define their convex combination by $L_3 = xL_1 + (1-x)L_2 = \{xp_1(\cdot) + (1-x)p_2(\cdot)\}$. Then, L_{REW} is a (Herstein & Milnor, 1953) *mixture set*.

The following two are our axioms for a strict partial order, $\triangleright$, over L_{REW}.

Axiom 1. $\triangleright$ is a transitive and irreflexive relation on $L_{REW} \times L_{REW}$.

Axiom 2 (Independence)**.** For all L_1, L_2 and L_3, and for each $0 < x \le 1$:

$$xL_1(1-x)L_3 \triangleright xL_2 + (1-x)L_3 \text{ iff } L_1 \triangleright L_2.$$

Def. When neither $L_1 \triangleright L_2$ nor $L_2 \triangleright L_1$, we say the two lotteries are *incomparable* (by preference), which we denote by $L_1 \sim L_2$.

Incomparability is not transitive, unless $\triangleright$ is a weak order.

Theorem 1. Let REW be a reward set of arbitrary cardinality and let L_{REW} be the set of simple lotteries over these rewards. Let $\triangleright$ be a strict partial order defined over elements of L_{REW}. Then there is an extension of $\triangleright$ to $\triangleright^*$ which is a total ordering of L_{REW} satisfying axiom 2.

Combining Theorem 1 with Hausner's (1954) important result (since a total order is a "pure" weak order), we arrive at the following consequence.

Corollary 1. There is a lexicographic real-valued utility, $\mho$, which agrees with $\triangleright$, i.e., if $L_1 \triangleright L_2$ then $E_{\mho}[L_1] < E_{\mho}[L_2]$.

(Note: A lexicographic utility $\mho$ is a (well-ordered) sequence of real-valued utilities, $\mho = \{U_\alpha: U_\alpha$ is a real-valued utility, for each $\alpha < \beta\}$. When $\mho$ is a lexicographic utility, then $E_{\mho}[L_1] < E_{\mho}[L_2]$ is said to obtain if $E_{U_\alpha}[L_1] < E_{U_\alpha}[L_2]$ at the first utility U_α in the sequence $\mho$ which gives L_1 and L_2 different expected values, provided one such U_α exists.)

Proof of Theorem 1. Let $\{L_\gamma: \gamma < k$ (γ ranging over ordinals, k a cardinal)$\}$ be a well ordering of L_{REW}. Let $\triangleright$ be a partial order on L_{REW} satisfying axioms 1 and 2. By induction, we define a sequence of extensions of $\triangleright$, $\{\triangleright_\lambda: \lambda \le k\}$, where each $\triangleright_\lambda$ preserves both axioms and where $\triangleright_k$ is a total order on L_{REW}. The partial order $\triangleright_\lambda$, corresponding to stage λ in the k sequence of extensions, is obtained by contrasting lotteries L_α and L_β, where $\Gamma(\alpha,\beta) = \lambda$ under the canonical well ordering Γ of $k \times k \to k$. We define extensions for successor and limit ordinals separately.

Successor Ordinals. Suppose $\triangleright_\lambda$ satisfies axioms 1 and 2. Let $\Gamma(\alpha,\beta) = \lambda + 1$ and (for convenience) suppose $\max[\alpha,\beta] = \beta$. Define $\triangleright_{\lambda+1}$ as follows.

Case 1: If $\alpha = \beta$, then $\triangleright_{\lambda+1} = \triangleright_\lambda$.

Otherwise,

Case 2: $L_\mu \triangleright_{\lambda+1} L_\nu$ iff either

(i) $L_\mu \triangleright_\lambda L_\nu$ (so $\triangleright_{\lambda+1}$ extends $\triangleright_\lambda$), or
(ii) $L_\alpha \sim_\lambda L_\beta$ & $\exists(0 < x < 1)$ with $xL_\mu + (1-x)L_\beta \triangleright_\lambda$ (or $=$) $xL_\nu + (1-x)L_\alpha$.

Limit Ordinals. Let $\Gamma(\alpha,\beta) = \lambda < k$, a limit, and (for convenience) again assume $\max[\alpha,\beta] = \beta$.

Case 1: If $\alpha = \beta$, then take $\triangleright_\lambda = \cup_{\delta<\lambda}(\triangleright_\delta)$. That is, $L_\mu \triangleright_\lambda L_\nu$ obtains just in case $\exists(\delta < \lambda) L_\mu \triangleright_\delta L_\nu$.

Case 2: If $\alpha \neq \beta$, then define $\triangleright_\lambda$ as: $L_\mu \triangleright_\lambda L_\nu$ **iff** either (i) $\exists(\delta < \lambda) L_\mu$

$\rhd_\delta L_\nu$ (so $\rhd_\lambda$ extends all preceding $\rhd_\delta$), or (ii) $\forall(\delta < \lambda)L_\alpha \sim_\delta L_\beta$ & $\exists(\delta < \lambda)\exists(0 < x < 1)$ with $xL_\mu + (1 - x)L_\beta \rhd_\delta$ (or =) $xL_\nu + (1 - x)L_\alpha$. Next, we show (by transfinite induction) that $\rhd_\lambda$ satisfies the two axioms, assuming $\rhd(= \rhd_0)$ does. First, consider successor stages where the extension is of the form $\rhd_{\lambda+1}$.

Axiom 1 – irreflexivity. We argue indirectly. Assume, for some lottery L_μ, $L_\mu \rhd_{\lambda+1} L_\mu$. Since $L_\mu \rhd_\lambda L_\mu$ is precluded, by hypothesis of induction, it must be that (ii): $\exists(0 < x < 1)$ with

$$xL_\mu + (1-x)L_\beta \rhd_\lambda (\text{or} =)\ xL_\mu + (1-x)L_\alpha.$$

Since $\rhd_\lambda$ satisfies axiom 2, $L_\beta \rhd_\lambda$ (or =) L_α. If either $L_\beta \rhd_\lambda L_\alpha$ or $L_\beta = L_\alpha$, then $\rhd_{\lambda+1} = \rhd_\lambda$, contradicting the hypothesis $L_\mu \rhd_{\lambda+1} L_\mu$.

Axiom 1 – transitivity. Assume $L_\mu \rhd_{\lambda+1} L_\nu$ and $L_\nu \rhd_{\lambda+1} L_\psi$. There are four cases to consider, since each $\rhd_{\lambda+1}$ relation may obtain in one of two ways. The combination where clause (ii) is used for both provides the greatest generality (the other cases being analyzed in the same way). Thus, we have: $\exists(0 < x, y < 1)$ with

$$xL_\mu + (1-x)L_\beta \rhd_\lambda (\text{or} =)\ xL_\nu + (1-x)L_\alpha$$

and also

$$yL_\nu + (1-y)L_\beta \rhd_\lambda (\text{or} =)\ yL_\psi + (1-y)L_\alpha.$$

Since $\rhd_\lambda$ satisfies axioms 1 and 2, we may "mix" these to yield

$$w(xL_\mu + (1-x)L_\beta) + (1-w)(yL_\nu + (1-y)L_\beta)$$

$\rhd_\lambda$ (or =)

$$w(xL_\nu + (1-x)L_\beta) + (1-w)(yL_\psi + (1-y)L_\beta).$$

Choose $w \cdot x = (1 - w)y$, cancel the common "L_ν" terms (according to axiom 2), regroup (by "reduction") to arrive at: $\exists(0 < v < 1)$

$$vL_\mu + (1-v)L_\beta \rhd_\lambda (\text{or} =)\ vL_\psi + (1-v)L_\alpha,$$

where $v = wx/(1 - y + wy)$. Hence, $L_\mu \rhd_{\lambda+1} L_\psi$, as desired.

Axiom 2. We are to show $L_\mu \rhd_{\lambda+1} L_\nu$ **iff**

$$xL_\mu + (1-x)L_\psi \rhd_{\lambda+1} xL_\nu + (1-x)L_\psi.$$

There are two cases.

Case 1: $L_\mu \rhd_\lambda L_\nu$ occurs just in case $xL_\mu + (1 - x)L_\psi \rhd_\lambda xL_\nu + (1 - x)L_\psi$ (by axiom 2). By the definition of $\rhd_{\lambda+1}$, we obtain the desired result:

$$xL_\mu + (1-x)L_\psi \rhd_{\lambda+1} xL_\nu + (1-x)L_\psi.$$

Case 2: $\nu L_\mu + (1-\nu)L_\beta \rhd_\lambda$ (or =) $\nu L_\psi + (1-\nu)L_\alpha$ occurs just in case

$$yL_\psi + (1-y)(\nu L_\mu + (1-\nu)L_\beta) \rhd_\lambda (or =)$$
$$yL_\psi + (1-y)(\nu L_\psi + (1-\nu)L_\alpha),$$

according to axiom 2. Choose $y = \nu(1-x)/[\nu(1-x)+x]$, regroup terms to yield: $w(xL_\mu + (1-x)L_\psi) + (1-x)L_\beta \rhd_\lambda$ (or =) $w(xL_\nu + (1-x)L_\psi) + (1-x)L_\alpha$, where $w = \nu/[\nu(1-x)+x]$. By the definition of $\rhd_{\lambda+1}$, we obtain the desired result:

$$xL_\mu + (1-x)L_\psi \rhd_{\lambda+1} xL_\nu + (1-x)L_\psi.$$

This establishes axioms 1 and 2 for successor stages, $\rhd_{\lambda+1}$.

The argument with limit stages is similar.

Axiom 1 – irreflexivity. Again, we argue indirectly. Assume $L_\mu \rhd_\lambda L_\mu$. By hypothesis of induction $\neg\exists(\delta < \lambda)\ L_\mu \rhd_\delta L_\mu$. So we may assume $L_\alpha \neq L_\beta$ and $\forall(\delta < \lambda) L_\alpha \sim_\delta L_\beta$ and $\exists(\delta < \lambda)\exists(0 < x < 1)$ with $xL_\mu + (1-x)L_\beta \rhd_\delta$ (or =) $xL_\mu + (1-x)L_\alpha$. But by the hypothesis of induction $\rhd_\delta$ satisfies axiom 2, hence, $L_\beta \rhd_\delta$ (or =) L_α, a contradiction.

Axiom 1 – transitivity. Assume $L_\mu \rhd_\lambda L_\nu$ and $L_\nu \rhd_\lambda L_\psi$. Again there are four cases, and again we discuss the most general case where clause (ii) is used to obtain these $\rhd_\lambda$ – preferences. Thus, we have: $\exists(0 < x, y < 1)\ \exists(\delta, \delta' < \lambda)$ with

$$xL_\mu + (1-x)L_\beta \rhd_\delta \text{ (or =) } xL_\nu + (1-x)L_\alpha$$

and also

$$yL_\nu + (1-y)L_\beta \rhd_{\delta'} \text{ (or =) } yL_\psi + (1-y)L_\alpha.$$

Without loss of generality, let $\delta = max[\delta,\delta']$. Then

$$yL_\nu + (1-y)L_\beta \rhd_\delta \text{ (or =) } yL_\psi + (1-y)L_\alpha,$$

since $\rhd_\delta$ extends $\rhd_{\delta'}$. Now, repeat the "mixing" and "cancellation" steps used with the parallel case for successor stages. This yields the desired conclusion: $L_\mu \rhd_\lambda L_\psi$.

Axiom 2. For this axiom, the reasoning is the same as used with axiom 2 in the successor case, modified to apply to the appropriate (preceding) stage $\rhd_\delta$.

Last, define $\rhd_k = \cup_{\delta<k} (\rhd_\delta)$. Hence, $\rhd_k$ is a total order of L_{REW} which satisfies axiom 2. Every two (distinct) lotteries are compared under $\rhd_k$, i.e., $\forall(L_\alpha \neq L_\beta \in L_{REW})\ L_\alpha \rhd_k L_\beta$ or $L_\beta \rhd_k L_\alpha$. □

Next, we state without proof a simple lemma.

Lemma 1. If lexicographic utilities $\mho_1$ and $\mho_2$ both agree with the strict partial order $\rhd$, then so too does their convex mixture $x\mho_1 + (1-x)\mho_2$. Also, sets of lexicographic utilities generate a strict partial order according to the "unanimity" rule, as we now show.

Lemma 2. Each set of lexicographic utilities $\mathcal{U} = \{\mho\text{: } \mho \text{ is a lexicographic utility over REW}\}$ induces a strict partial order $\rhd_{\mathcal{U}}$ (satisfying axioms 1 and 2) under the "unanimity" rule:

$$L_\alpha \rhd_{\mathcal{U}} L_\beta \textbf{ iff } \forall(\mho \in \mathcal{U})\, E_\mho[L_\alpha] < E_\mho[L-\beta]$$

Proof. The lemma is evident from the fact that each lexicographic utility induces a weak-ordering $\preceq_\mho$ of L_{REW}, satisfying axiom 2, according to the definition:

$$L_\alpha \prec_\mho L_\beta \textbf{ iff } E_\mho[L_\alpha] < E_\mho[L_\beta].$$

Recall, $E_\mho[L_\alpha] < E_\mho[L_\beta]$ obtains if $E_U[L_\alpha] < E_u\,[L_\beta]$ for the first utility U (if one exists) in the sequence $\mho$ which assigns L_α and L_β different expected utilities. Each utility U (hence, $\preceq_U$), supports axiom 2 as:

$$E_U[L_\alpha] < E_U[L_\beta] \textbf{ iff } E_U[xL_\alpha + (1-x)L_\gamma] < E_U[xL_\beta + (1-x)L_\gamma]. \quad \Box$$

As is evident from the proof of Theorem 1, if $L \sim L'$, i.e., if neither $L \rhd L'$ nor $L' \rhd L$, then there are alternative extensions of $\rhd$ in which $L \rhd_\delta L'$ and in which $L' \rhd_\delta L$. This observation, together with the two lemmas and Corollary 1, establishes the following representation for strict partial orders $\rhd$.

Theorem 2. Each strict partial order $\rhd$ over a set L_{REW} is identified by a maximal, convex set $\mathcal{U}$ of lexicographic utilities that agree with it. In symbols, $\rhd = \rhd_{\mathcal{U}}$, where $\rhd_{\mathcal{U}}$ is the strict partial order induced by $\mathcal{U}$ under the "unanimity" rule.

Of course, in light of problem (2) (p. 54), it can be that there is a proper (convex) subset $\mathcal{U}' \subset \mathcal{U}$ where $\rhd = \rhd_{\mathcal{U}'}$ as well; hence, the maximality of $\mathcal{U}$ is necessary for uniqueness of the representation.

REFERENCES

Allais, M. (1979) "The So-Called Allais Paradox and Rational Decisions Under Uncertainty," in Allais and Hagen (eds.), *Expected Utility Hypotheses and the Allais Paradox*. D. Reidel: Dordrecht.

Anscombe, F. J., and Aumann, R. J. (1963) "A Definition of Subjective Probability," *Annals of Math. Stat.*, **34**, 199–205.
Aumann, R. J. (1962) "Utility Theory Without the Completeness Axiom," *Econometrica*, **30**, 445–462.
Aumann, R. J. (1964) "Utility Theory Without the Completeness Axiom: A Correction," *Econometrica*, **32**, 210–212.
Bell, D., and Raiffa, H. (1979) "Decision Regret: A Component of Risk Aversion," MS., Harvard University.
Chew Soo Hong and MacCrimmon, K. R. (1979) "Alpha-Nu Choice Theory: A Generalization of Expected Utility Theory," working paper, University of British Columbia.
Cox, D. R. (1958) "Some Problems Connected with Statistical Inference," *Annals of Math. Stat.*, **29**, 357–363.
deFinetti, B. (1937) "La prévision: ses lois logiques, ses sources subjectives," *Annals de l'Institut Henri Poincaré*, **7**, 1–68.
Fishburn, P. C. (1970) *Utility Theory for Decision Making*. Kriefer Publishing Co.: N.Y.
Fishburn, P. C. (1981) "An Axiomatic Characterization of Skew-Symmetric Bilinear Functionals, with Applications to Utility Theory," *Economic Letters*, **8**, 311–313.
Fishburn, P. C. (1983) "Nontransitive Measurable Utility," *J. Math. Psych.*, **26**, 31–67.
Good, I. J. (1952) "Rational Decisions," *J. Royal Stat. Soc. B*, **14**, 107–114.
Hausner, M. (1954) "Multidimensional Utilities," in R. M. Thrall, C. H. Coombs, and R. L. Davis (eds.), *Decision Processes*. Wiley: N.Y.
Herstein, I. N. and Milnor, J. (1953) "An Axiomatic Approach to Measurable Utility," *Econometrica*, **21**, 291–297.
Jeffreys, H. (1971) *Theory of Probability*, 3rd ed. Oxford University Press: Oxford.
Kadane, J., and Sedransk, N. (1980) "Toward a More Ethical Clinical Trial," in Bernardo et al. (eds.), *Bayesian Statistics*. University Press: Valencia.
Kadane, J. B. (ed.) (1996) *Bayesian Methods and Ethics in a Clinical Trial Design*. Wiley: N.Y.
Kahneman, D., and Tversky, A. (1979) "Prospect Theory: An Analysis of Decision Under Risk," *Econometrica*, **47**, 263–291.
Kannai, Y. (1963) "Existence of a Utility in Infinite Dimensional Partially Ordered Spaces," *Israel J. of Math.*, **1**, 229–234.
Klee, V. L. (1955) "Separation Properties of Convex Cones," *Proc. Amer. Math. Soc.*, **6**, 313–318.
Levi, I. (1974) "On Indeterminate Probabilities," *J. Phil.*, **71**, 391–418.
Levi, I. (1980) *The Enterprise of Knowledge*. MIT Press: Cambridge.
Levi, I. (1986) "The Paradoxes of Allais and Ellsberg," *Economics and Philosophy*, **2**, 23–53.
Lindley, D. V. (1972) *Bayesian Statistics: A Review*. SIAM: Philadelphia.
Loomes, G., and Sugden, R. (1982) "Regret Theory: An Alternative Theory of Rational Choice Under Uncertainty," *Economic J.*, **92**, 805–824.
McClennen, E. F. (1983) "Sure Thing Doubts," in B. Stigum and F. Wenstop

(eds.), *Foundations of Utility and Risk Theory with Applications*. D. Reidel: Dordrecht.
Machina, M. (1982) "'Expected Utility' Analysis Without the Independence Axiom," *Econometrica*, **50**, 277–323.
Machina, M. (1983) "The Economic Theory of Individual Behavior Toward Risk: Theory, Evidence and New Directions," Dept. of Economics, U.C.S.D.: San Diego, CA 92093. Tech. Report #433.
Ramsey, F. P. (1931) "Truth and Probability," in *The Foundations of Mathematics and Other Essays*. Kegan, Paul, Trench, Trubner, and Co. Ltd.: London.
Samuelson, P. (1950) "Probability and the Attempts to Measure Utility," *Economic Review*, **1**, 167–173.
Savage, L. J. (1954) *The Foundations of Statistics*. Wiley: N.Y.
Schick, F. (1984) *Having Reasons*. Princeton Univ. Press: Princeton.
Seidenfeld, T. (1988) "Decision Theory Without Independence or Without Ordering, What Is the Difference?" with discussion, *Economics and Philosophy*, **4**, 267–315.
Sen, A. K. (1977) "Social Choice Theory: A Re-examination," *Econometrica*, **45**, 53–89.
Shimony, A. (1955) "Coherence and the Axioms of Probability," *J. Symbolic Logic*, **20**, 1–28.
Smith, C. A. B. (1961) "Consistency in Statistical Inference and Decision," *J. Royal Stat. Soc. B*, **23**, 1–25.
Suppes, P. (1974) "The Measurement of Belief," *J. Royal Stat. Soc. B*, **36**, 160–175.
Szpilrajn, E. (1930) "Sur l'extension de l'ordre partiel," *Fundamenta Mathematicae*, **16**, 386–389.
von Neumann, J., and Morgenstern, O. (1947) *Theory of Games and Economic Behavior*, 2nd ed. Princeton Univ. Press: Princeton.
Walley, P., and Fine, T. (1979) "Varieties of Modal (Classificatory) and Comparative Probability," *Synthese*, **41**, 321–374.
Wolfenson, M., and Fine, T. (1982) "Bayes-like Decision Making with Upper and Lower Probabilities," *J. Amer. Stat. Assoc.* **77**, 80–88.

1.3

A Representation of Partially Ordered Preferences

TEDDY SEIDENFELD, MARK J. SCHERVISH, AND JOSEPH B. KADANE

ABSTRACT

This chapter considers decision-theoretic foundations for robust Bayesian statistics. We modify the approach of Ramsey, deFinetti, Savage and Anscombe, and Aumann in giving axioms for a theory of *robust* preferences. We establish that preferences which satisfy axioms for robust preferences can be represented by a set of expected utilities. In the presence of two axioms relating to state-independent utility, robust preferences are represented by a *set* of probability/utility pairs, where the utilities are almost state-independent (in a sense which we make precise). Our goal is to focus on preference alone and to extract whatever probability and/or utility information is contained in the preference relation when that is merely a partial order. This is in contrast with the usual approach to Bayesian robustness that begins with a class of "priors" or "likelihoods," and a single loss function, in order to derive preferences from these probability/utility assumptions.

The research of Teddy Seidenfeld was supported by the Buhl Foundation and NSF Grant SES-92-08942. Mark Schervish's research on this project was supported through ONR Contract N00014-91-J-1024 and NSF Grant DMS-88-05676. Joseph Kadane's research was supported by NSF grants SES-89-0002 and DMS-90-05858, and ONR contract N00014-80-J-1851.

Reprinted from *The Annals of Statistics*, 23, no. 6 (1995): 2168–2217, with permission of the Institute of Mathematical Statistics.

The authors thank Peter Fishburn for his careful reading of an earlier version of this essay and for providing us with insightful and extensive feedback. Also, we thank Robert Nau for stimulating exchanges relating to matters of state-dependent utility. We are exceedingly grateful to two *Annals of Statistics* referees, whose patient and detailed readings of our long manuscript resulted in many helpful ideas and useful suggestions.

I. INTRODUCTION AND OVERVIEW

I.1. *Robust Bayesian Preferences*

This essay is about decision-theoretic foundations for robust Bayesian statistics. The fruitful tradition of Ramsey (1931), deFinetti (1937), Savage (1954), and Anscombe and Aumann (1963) seeks to ground Bayesian inference on a *normative* theory of rational choice. Rather than accept the traditional probability models and loss functions as given, Savage is explicit about the foundations. He axiomatizes a theory of preference using a binary relation over acts, $A_1 \lesssim A_2$, "act A_1 is not preferred to act A_2." Then, he shows that $\lesssim$ is represented by a unique personal probability (state-independent) utility pair according to subjective expected utility. That is, he shows there exists exactly one pair (p, U) such that, for all acts A_1 and A_2, $A_1 \lesssim A_2$ if and only if $E_{p,U}[A_1] \leq E_{p,U}[A_2]$. [More precisely, in Savage's theory what is needed to justify the assertion that p is the agent's personal probability is the added assumption that each consequence has a constant value in each state. Unfortunately this is ineffable in Savage's language of preference over acts. [See Schervish, Seidenfeld and Kadane (1990).] We discuss this in Section IV, below.

In recent years, either under the headings of *Bayesian robustness* [Berger (1985), Section 4.7; Hartigan (1983), Chapter 12; Kadane (1984)] or *sensitivity analysis* [Rios Insua (1990)], it has become an increasingly important issue to show how to arrive at Bayesian conclusions from logically weaker assumptions than are required by the traditional Bayesian theory. Given data and a particular likelihood from a statistical model, for example, how large a class of prior probabilities leads to a class of posterior probabilities that are in agreement about some event of interest? Our work differs from the common trend in Bayesian robustness in much the same way that Savage's work differs from the traditional use of probability models and loss functions in Bayesian decision theory. Our goal is to axiomatize robust preferences directly, rather than to robustify given statistical models. Results in this theory are strikingly different from those obtained in the existing Bayesian robustness literature.

For an illustration of the difference, suppose two Bayesian agents each rank the desirability of Anscombe–Aumann ("horse lottery") acts according to his/her subjective expected utility. ("Horse lotteries" are defined in Section II.) Let (p_1, U_1) and (p_2, U_2) be the probability/utility pairs representing these two decision makers and assume they have different beliefs and values: that is, assume $p_1 \neq p_2$ and $U_1 \neq U_2$. Denote by $\prec_1$ and $\prec_2$ their respective (strict) preference relations, each a weak order over acts. Suppose now our goal is to find those coherent (Anscombe–Aumann) preference relations $\lesssim$ corresponding to probability/utility pairs (p, U) such that the following *Pareto condition* applies:

$$\text{If } E_{p_1,U_1}[A_1] < E_{p_1,U_1}[A_2] \text{ and } E_{p_2,U_2}[A_1] < E_{p_2,U_2}[A_2],$$
$$\text{then } E_{p,U}[A_1] < E_{p,U}[A_2].$$

In words, when both agents strictly prefer act A_2 to act A_1, then this shared preference is robust for all efficient, cooperative Bayesian decisions that the pair make together. [We assume that though the two Bayesian agents may discuss their individual preferences, nonetheless, some differences remain in their beliefs and in their values even after such conversations. See DeGroot (1974) for a rival model.] We have the following theorem.

Theorem 1. [Seidenfeld, Kadane and Schervish (1989)]. *Assume there exists a pair of prizes* $\{r_*, r^*\}$ *which the two agents rank in the same order:* $r_* \prec_i r^*$ $(i = 1, 2)$. *Then the set of probability/utility pairs, each of which satisfies the Anscombe–Aumann theory and each of which agrees with the strict preferences shared by these two decision makers, consists exactly of the two pairs themselves* $\{(p_1, U_1), (p_2, U_2)\}$. *There are no other coherent, Pareto compromises.* [*There is no coherent weak order meeting the strong Pareto condition, which requires that* $A_1 \prec A_2$ *if* $A_1 \precsim_i A_2$ $(i = 1, 2)$ *and at least one of these two preferences is strict.*]

Thus, with respect to Pareto-robust preferences, the set of probability/utility pairs for the problem of two distinct Bayesians is not connected and therefore not convex. Hence, a common method of proof – separating hyperplanes (used to develop expected utility representations) – is not available in our investigation. This is just one way in which our methods differ from the usual robust Bayesian analysis. We want the strict preferences held in common by two Bayesians to be a special case of robust preferences. Applied to a class of weak orders, the Pareto condition creates a strict partial order $\prec$: to wit, the binary relation $\prec$ is irreflexive and transitive.

Our view of robustness is that sometimes a person does not have a (strict) preference for act A_1 over act A_2 nor for A_2 over act A_1 nor are they indifferent options. Assume that strict preference is a transitive relation. Then such a person's preferences are modeled by a partial order. We ask, under what assumptions on this partial order is there a set of probability/utility pairs agreeing with it according to expected utility theory, a set which characterizes that partial order? In this we are exploring the possibility pointed out by Savage [(1954), page 21].

The general form of our inquiry is as follows. Axiomatize coherent preference $\prec$ as a partial order and establish a representation for it in terms of a set of probability/utility pairs. That is, we characterize each coherent, partially ordered preference $\prec$ in terms of the set of coherent weak orders $\{\precsim\}$ that extend it. We rely on the usual expected utility theory to depict each coherent weak order $\precsim$ by one probability/utility pair (p, U). Thus, we model $\prec$ by a set of probability/utility pairs.

In contrast with Savage's theory, which uses only personal probability, our approach is based on Anscombe–Aumann's "horse lottery" theory. Preferences over "horse lotteries" accommodate both personal and extraneous (agent-invariant) probabilities. Also, by characterizing strict preference in terms of a set of probability/utility pairs, we improve so-called one-way representations of, for example, Fishburn [(1982), Section 11], as we show more than existence of an agreeing probability/utility pair.

I.2. *Overview*

In outline, our approach is as follows: In Section II we introduce axioms for a partial order over Anscombe–Aumann horse lotteries (HL). Anscombe–Aumann theory contains three substantive axioms that incorporate the (von Neumann–Morgenstern) theory of cardinal utility for simple acts:

1. A postulate that preference ($\precsim$) is a weak order – analogous to Savage's P1.
2. The independence postulate – analogous to Savage's P2, "sure thing."
3. An Archimedean condition, which plays an analogous role to Savage's P6.

Our replacement axioms for these are:

HL Axiom 1. A postulate that strict preference ($\prec$) is a strict partial order.

HL Axiom 2. The independence postulate.

HL Axiom 3. A modified Archimedean axiom for discrete (not just simple) lotteries.

To avoid triviality, a commonplace assumption of expected utility theory is that not all acts are indifferent; for example, there exist two acts, **W** and **B** that do *not* satisfy $\mathbf{B} \precsim \mathbf{W}$. Let $\prec$ obey our (three) preference axioms on a domain of horse-lottery acts $\mathbf{H_R}$. We show (Theorem 2) how to extend $\prec$ to a preference $\prec'$ over a larger domain that includes two new acts **B** (best) and **W** (worst) where

$$(\forall H_1, H_2 \in \mathbf{H_R}) \quad [(\mathbf{B} \prec' H_1 \prec' \mathbf{W}) \ \& \\ (H_1 \prec' H_2 \text{ if and only if } H_1 \prec' H_2)]$$

Then we establish three related theorems (Theorems 3, 4 and 5): $\prec$ is represented by a nonempty (maximal and convex) set $\mathcal{V} = \{V: \mathbf{H_R} \rightarrow (0, 1)\}$ of bounded, real-valued cardinal utilities $V(\cdot)$ defined for acts. Each $V \in \mathcal{V}$ induces a weak order $\precsim_V$ that agrees with $\prec$ on the domain of simple acts and almost agrees (Definition 10b) with $\prec$ on all acts. Moreover, given a set Z of bounded, real-valued cardinal (so-called) "linear" utilities, $Z(\cdot)$ defined on $\mathbf{H_R}$, the partial order formed using the Pareto condition with the set Z satisfies our three axioms for preference.

In the light of the surprising "shape" that the family of agreeing subjective expected utilities can have (Theorem 1), we employ a modification of Szpilrajn's (1930) transfinite induction for extending a partial order. We show how to extend a partial order while preserving the other preference axioms. The proofs of all results appear in the Appendix. *Also, we number definitions, lemmas, and corollaries to coincide with their logical order in our arguments, regardless of whether they appear for the first time in the body of the text or in the Appendix.*

In Section IV we turn our attention to the representation of $\mathcal{V}$ as a set of subjective expected utilities. We discuss when a linear utility V over acts also is a subjective expected utility for a probability/utility pair (p, U). Corollary 4.1 gives a representation of $\mathcal{V}$ in terms of sets of probability/state-dependent utility pairs $(p, \{U_j: j = 1, \ldots, n\})$, where the utility $U_j(L)$ of a (von Neumann–Morgenstern) lottery L may depend upon the accompanying state s_j. (This follows up the issue raised in the first paragraph in Section I.1.) In Section IV.3 we introduce two axioms (HL Axioms 4 and 5) that parallel the fourth Anscombe–Aumann postulate. That postulate (and our replacements for it) permits a representation of preference using a (nearly) state-independent utility: where (with high personal probability) the value of a lottery L does not depend (by more than amount $\varepsilon > 0$) upon the state in which it is awarded. In Section IV.3 we lean heavily on the proof technique of Section III in order to find a representation for the partial order $\prec$ in terms of a set of agreeing pairs of probabilities and (nearly) state-independent utilities, Lemma 4.3 and Theorem 6.

Section V is about conditional preference. Two theorems (Theorems 7 and 8) relate conditional (called-off) partially ordered preferences and Bayesian updating of the family of unconditional personal probabilities that agree with an unconditional partially ordered preference. We provide an example involving conditional probability that highlights the nonconvexity of the agreeing sets. In Section VI we conclude with a review of several features that distinguish our results.

II. THE FORMAL THEORY

II.1. *The Act Space: A Domain for the Preference Relation*

We provide a representation for a partially ordered strict preference relation over (discrete) Anscombe–Aumann (1963) horse lotteries – acts that generalize von Neumann–Morgenstern (1947) lotteries to allow for uncertainty over states of nature.

Let $\mathbf{R}$ be a set of *rewards*. We develop our theory for countable sets $\mathbf{R}$.

Definition 1. A *simple* (von Neumann–Morgenstern) *lottery* is a simple probability distribution P over $\mathbf{R}$, that is, a distribution with finite support. A *discrete lottery* is a countably additive probability over $\mathbf{R}$ (with a countable support). Denote a lottery by L and its distribution by P.

Horse lotteries are defined with respect to a finite partition of states. Let π be a finite partition of the sure event S into n disjoint, mutually exhaustive nonempty (sets of) states, $\pi = \{s_1, \ldots, s_n: s_i \cap s_j = \varnothing$ iff $i \neq j$ and $\cup_{j \le n}(s_j) = S\}$. Strictly speaking, elements of π are subsets of S. We take this approach rather than supposing S is finite, for example, rather than assuming $S = \{s_1, \ldots, s_n\}$. Then our analysis allows for elaborations of a given preference relation in a larger domain of acts defined over (finite) refinements of the partition π. Having made this point, we allow ourselves the familiar convention of equating the set state s_j with its elements. That is, for notational convenience, often we shall use s_j when we intend "members of s_j."

Definition 2. A *simple* (or *discrete*) *horse lottery* is a function from states to simple (or to discrete) lotteries. Denote a horse lottery by H and denote the space of (discrete) horse lotteries on the reward set $\mathbf{R}$ by $\mathbf{H_R}$.

In the tradition where acts are functions from states to outcomes, a horse lottery is an act with a lottery outcome. For example, the act that yields a 50–50 chance at \$10 and \$20 provided the Republicans win the next Presidential election, and which yields a 0.25 chance at \$5 and a 0.75 chance at \$10 if the Republicans do not win, is a horse lottery over a binary partition with two states: Republicans win and Republicans do not win. Thus, a *constant* horse lottery is just a von Neumann–Morgenstern lottery, and a proper subset of these are the constant von Neumann–Morgenstern lotteries, that is, the acts which yield a specific reward for certain.

Next, we define the operation of convex combination of two horse lotteries, "+", as the state-by-state mixture of their respective v.N–M lottery outcomes. Thus:

Definition 3. $xH_1 + (1 - x)H_2 = H_3 = \{xL_{1j} + (1 - x)L_{2j}: j = 1, \ldots, n;\ 0 \leq x \leq 1\}$.

The mixture of two lotteries is a lottery $xL_1 + (1 - x)L_2 = L_3$, where $P_3(r) = xP_1(r) + (1 - x)P_2(r)$. For the special cases where each H is a "constant" act, that is, if H_1 is the lottery L_1 and H_2 is L_2, Definition 3 coincides with the von Neumann–Morgenstern operation of "+" for lotteries.

II.2. The Axioms for Order and Independence

The von Neumann–Morgenstern theory of preference over (simple) lotteries is encapsulated by three axioms:

1. The assumption that preference $\precsim$ is a weak-order relation.
2. The independence postulate (formulated below).
3. An Archimedean condition (discussed below).

These axioms may be applied to horse lotteries also. Then the three axioms guarantee that (i) preference is represented by a (cardinal) utility V over acts with a property (ii) that utility distributes over convex combinations. To wit, given these three axioms:

i. There exists a real-value V defined on acts, unique up to positive linear transformations, where $V(H_1) \leq V(H_2)$ if and only if $H_1 \precsim H_2$; and
ii. $V[xH_1 + (1 - x)H_2] = xV(H_1) + (1 - x)V(H_2)$.

Definition 4. When a preference relation over acts satisfies (i) and (ii) we say it has the *expected* (or *linear*) *utility* property and we call V an agreeing expected (or, linear) utility for $\precsim$.

Anscombe–Aumann theory requires a fourth postulate ensuring the existence of a unique decomposition of V as a subjective expected, *state-independent* utility. That is, subject to a fourth postulate for preference, there exists a (unique) personal probability p defined on states and a utility U defined on lotteries (independent of states) so that:

iii. $$V(H) = \sum_{j=1}^{n} p(s_j)U(L_j).$$

[Recall the notation $H(s_j) = L_j$.] Key, here, is that U is a state-independent utility, defined on lotteries independent of the state in which they occur. To be precise, let H_L be the constant horse lottery that yields lottery L in each state, $H_L(s_j) = L$.

Definition 5. The utility $\{U_j: j = 1, \ldots, n\}$ is *state-independent* when, for each lottery L and pair of states s_j and $s_{j'}$,

$$U_j(L) = U_{j'}(L) = U(L).$$

[For our purposes, and following the usual practice, the condition of Definition 5 is required only for states s_j and $s_{j'}$ that are not "null," i.e., only when $p(s_j) \neq 0$ is it worth restricting U_j in a decomposition of a linear utility V.] If the utility is state-independent, for convenience, we drop the subscript (for states) and abbreviate it U. When (iii) obtains for a state-independent utility U, $V(H_L) = U(L)$.

Definition 6. When a preference relation over acts satisfies (i)–(iii) we say it has the *subjective expected* (*state-independent*) *utility* property and we say the pair (p, U) agrees with $\precsim$.

In contrast to (iii), a decomposition of V by a subjective (possibly) state-dependent utility allows

iii*. $$V(H) = \sum_{j=1}^{n} p(s_j)U_j(L_j).$$

We examine such state-dependent decompositions in Section IV.1.

Next, we propose versions of the first two Anscombe–Aumann axioms to accommodate our theory of preference as a partial order. We postpone our discussion of the Archimedean axiom to Section II.4 to allow for a timely account of "indifference" in Section II.3.

In this essay, a partial order $\prec$ identifies a *strict* preference relation.

HL Axiom 1. $\prec$ is a strict partial order. It is a transitive and irreflexive relation on $\mathbf{H_R} \times \mathbf{H_R}$.

Definition 7. When neither $H_1 \prec H_2$ nor $H_2 \prec H_1$, we say the two lotteries are *incomparable* by preference, which we denote as $H_1 \sim H_2$. When $\sim$ is transitive – corresponding to a weak order – then the relation $\precsim$ (standing for "$\prec$ or $\sim$") identifies a *weak* preference relation. Hereafter, we shall mean by "preference" the strict preference relation.

HL Axiom 2 (Independence). $\forall$ (H_1, H_2 and H_3) and for each $0 < x \leq 1$,

$$xH_1 + (1-x)H_3 \prec xH_2 + (1-x)H_3 \quad \text{if and only if } H_1 \prec H_2.$$

This version of independence may be used, also, as the second axiom in the Anscombe–Aumann theory or in the von Neumann–Morgenstern theory.

II.3. *Indifference* ($\approx$)

Next, we define a (transitive) relation of indifference, $\approx$, based on $\prec$, which will play a central role in our extension of $\prec$ to a weak order. [See Fishburn (1979), Exercises 9.1 and 9.4, pages 126–127, for additional discussion.]

Definition 8 (Indifference). $H_1 \approx H_2$ iff $\forall\, H_3, H_4$ $(0 < x \leq 1)$,

$$xH_1 + (1-x)H_3 \sim H_4 \quad \text{iff } xH_2 + (1-x)H_3 \sim H_4.$$

We establish several useful corollaries of the HL Axioms 1 and 2 about indifference.

Corollary 2.1. *$\forall\, H_1, H_2$, if $H_1 \approx H_2$, then $H_1 \sim H_2$.*

That is, when two acts are indifferent, neither is preferred to the other.

Corollary 2.2. *$\approx$ is an equivalence relation.*

Corollary 2.3. *$H_1 \approx H_2$ if and only if $\forall\, H_3, H_4$, and $0 < x \leq 1$,*

$$xH_1 + (1-x)H_3 \prec (\succ)\, H_4 \quad \textit{iff } xH_2 + (1-x)H_3 \prec (\succ)\, H_4.$$

Corollary 2.4.

$$\forall\, 0 < x \leq 1, H, H_1 \approx H_2 \quad \textit{iff } xH_1 + (1-x)H \approx xH_2 + (1-x)H.$$

Corollaries 2.3 and 2.4 establish important substitution properties for elements of the same indifference equivalence class.

II.4. *The Archimedean Axiom: Continuity of Preference*

First, define *convergence* for acts. Let $\{H_n\}$ be a denumerable sequence of horse lotteries.

Definition 9. $\{H_n\}$ *converges* to a lottery H^*, denoted by $\{H_n\} \Rightarrow H^*$, just in case the respective discrete lottery distributions $\{P_j^n(\cdot)\}$ converge (point-wise) to the lottery distribution $P_j^*(\cdot)$.

The third (Archimedean) axiom precludes infinitesimal degrees of preference. As we show in Theorem 4, it suffices for representing preferences by sets of agreeing real-valued utilities.

HL Axiom 3. Let $\{H_n\} \Rightarrow H$ and $\{M_n\} \Rightarrow M$.

a. If $\forall\, n(H_n \prec M_n)$ and $(M \prec N)$, then $(H \prec N)$.
b. If $\forall\, n(H_n \prec M_n)$ and $(J \prec H)$, then $(J \prec M)$.

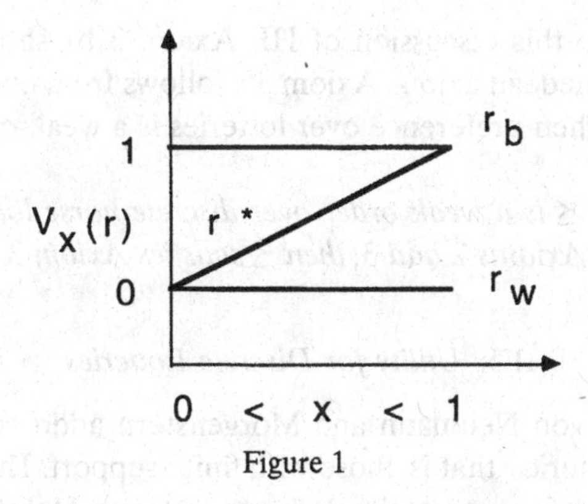

Figure 1

The familiar Archimedean condition from Anscombe–Aumann theory (also from von Neumann–Morgenstern theory), denoted here as Axiom 3*, is this:

Axiom 3*. Whenever $H_1 \prec H_2 \prec H_3$, $\exists\ (0 < x, y < 1)$, $yH_1 + (1-y)H_3 \prec H_2 \prec xH_1 + (1-x)H_3$.

However, Axiom 3* is overly restrictive for our purposes, as a simple example shows.

Example 2.1. Consider a set of three rewards $\mathbf{R} = \{r_w, r^*, r_b\}$ and a minimal, one element partition comprising the single sure state $\pi = \{s\}$. Then $\mathbf{H_R}$ is the set of von Neumann–Morgenstern lotteries on $\mathbf{R}$. (Denote by $\mathbf{r}$ the horse lottery with constant prize r.) Let $\mathcal{V} = \{V_x: 0 < x < 1\}$ be a (convex) set of linear utilities, where $V_x(\mathbf{r}_w) = 0$, $V_x(\mathbf{r}^*) = x$ and $V_x(\mathbf{r}_b) = 1$. Figure 1 graphs these utilities.

Let $\prec_{\mathcal{V}}$ be the partial order on lotteries generated by this set of utilities according to the weak Pareto condition. That is, $L_1 \prec_{\mathcal{V}} L_2$ iff $(\forall\ V \in \mathcal{V})\ E_V[L_1] < E_V[L_2]$. By Theorem 4 (below), $\prec_{\mathcal{V}}$ satisfies HL Axioms 1 and 2. However, it fails Axiom 3*. Specifically, $\mathbf{r}_w \prec_{\mathcal{V}} \mathbf{r}^* \prec_{\mathcal{V}} \mathbf{r}_b$, but $\forall\ (0 < y < 1)$, $y\mathbf{r}_w + (1-y)\mathbf{r}_b \sim_{\mathcal{V}} \mathbf{r}^*$. Of course, $\prec_{\mathcal{V}}$ is represented by the (convex) set of agreeing (real-valued) utilities $\mathcal{V}$.

Next, we provide a connection between the Archimedean condition HL Axiom 3 and ≈-indifference.

Corollary 2.5. *Let $\{H_n\}, \{H'_n\} \Rightarrow H$; $\{M_n\}, \{M'_n\} \Rightarrow M$. If $\forall\ n(H_n \prec M_n$ and $M'_n \prec H'_n)$, then $H \approx M$.*

We conclude this discussion of HL Axiom 3 by showing that the familiar Archimedean axiom, Axiom 3*, follows from our replacement HL Axiom 3 when preference over lotteries is a weak order.

Lemma 2.1. *If* $\precsim$ *is a weak order over discrete horse lotteries meeting conditions HL Axioms* 2 *and* 3, *then* $\precsim$ *satisfies Axiom* 3*.

II.5. *Utility for Discrete Lotteries*

The theory of von Neumann and Morgenstern addresses preference over simple lotteries, that is, those with finite support. These constitute the subdomain of constant, simple horse lotteries. However, there are weakly ordered preferences over lotteries which satisfy the expected utility hypothesis (Definition 4) for simple lotteries, that is, which are represented by a cardinal, linear utility V over the domain of *simple* lotteries, but which fail the expected utility hypothesis over the larger domain of discrete lotteries. [See Fishburn's (1979), Section 10, discussion; also Fishburn (1982), Section 11.3. A similar problem arises in Savage's (1954) theory, as shown by Seidenfeld and Schervish (1983). In a related matter, Aumann's (1962, 1964) argument about a utility for a partially ordered preference does not apply when the set of rewards is denumerable rather than finite, even though all lotteries are simple. Kannai (1963) showed that, and strengthened Aumann's Archimedean condition to remedy the problem.] We address this problem with an extended dominance condition.

Let **r** denote the simple horse lottery with constant prize r. Recall that H_L denotes the constant horse lottery that yields L in each state. Consider the following dominance principle:

Dominance. $\forall$ ($\mathbf{r}$, H_L) if for each $r_n \in \text{supp}(L)$, $\mathbf{r_n} \prec \mathbf{r}$, then it is not the case that ($\mathbf{r} \prec H_L$) [or, alternatively, if universally, $\mathbf{r} \prec \mathbf{r_n}$, then not ($H_L \prec \mathbf{r}$)].

This weak dominance condition contrasts each reward r with the lottery L through the (countably many) constant horse lotteries $\mathbf{r_n}$ taken from L's support. The condition precludes a preference for L over **r** if it occurs for **r** over each $\mathbf{r_n}$.

Our first three axioms yield dominance, as the next lemma establishes.

Lemma 2.2. *HL Axioms* 1–3 *entail dominance.*

Based on Lemma 2.2, we may apply Fishburn's [(1979), page 139] Theorem 10.5 to argue that a weakly ordered preference $\precsim$ over *discrete* (horse) lotteries which satisfies our HL Axioms 2 and 3 has the expected utility property. (See Remark 1.) The import for our representation theorems is given in terms of *agreeing* and *almost agreeing* utilities for a partial order:

Definition 10a. A utility *V agrees with* a partial order $\prec$ iff $V[H_1] < V[H_2]$ whenever $(H_1 \prec H_2)$.

Definition 10b. A utility *V almost agrees* with a partial order $\prec$ iff $V[H_1] \leq V[H_2]$ whenever $(H_1 \prec H_2)$.

Thus, when our strategy for extending a partial order $\prec$ to a weak order $\precsim$ succeeds, it induces a linear utility V that agrees with $\prec$ for *discrete* horse lotteries, not just for simple ones.

Unfortunately, our argument for extending a partial order $\prec$ produces a set of expected utilities $\{V\}$ each of which agrees with $\prec$ for simple acts, and only *almost agrees* with it for discrete acts. Of course, by itself the condition of "almost agreeing" is quite weak. A utility V that makes all options indifferent almost agrees with every partially ordered preference. The point, however, is that we consider an almost agreeing utility for a partial order $\prec$ only in the case in which it agrees with $\prec$ on all simple acts. [This idea parallels a similar distinction between a qualitative probability (a weak order on events) and a quantitative probability that agrees or almost agrees with it. See Savage (1954), Section 3.3.] Through Corollary 3.2, we provide a sufficient condition for the existence of a (convex) set of utilities that agree with $\prec$ on all of $\mathbf{H_R}$.

Remark 1. Fishburn's Theorem 10.5 uses the traditional Archimedean axiom Axiom 3*. However, by Lemma 2.2, Axiom 3* follows from the assumptions that $\precsim$ is a weak order satisfying HL Axioms 2 and 3. Also, dominance is equivalent to Fishburn's [(1979), page 138] Axiom 4c, given the other three axioms and our structural assumptions about the domain of lotteries.

Next in our discussion of the axioms, we settle the question whether a partial order $\prec$ satisfying HL Axioms 1–3 admits an *unbounded* utility that agrees with it or agrees with it on simple lotteries. It is well known that utilities for von Neumann–Morgenstern lotteries are finite. In light of our assumption that all discrete lotteries are acts, utilities that agree with $\prec$ are bounded as well.

Corollary 2.6. *Let $\prec$ satisfy HL Axioms* 1 *and* 2. *If V agrees with $\prec$, it is bounded. Hence, all utilities that agree with $\prec$ are bounded.*

As noted above, we sometimes construct a utility V that agrees with a partial order $\prec$ for all simple lotteries but (merely) almost agrees with $\prec$ for discrete lotteries. Thus, as V may fail to agree with $\prec$ on nonsimple acts, it is worthwhile to show (appealing only to simple acts) that each utility V we construct is bounded. For this purpose we formalize a condition that a partially ordered preference is bounded, and establish it as a corollary of two of our axioms, HL Axioms 1 and 3. From the fact that a partial order $\prec$ is bounded, we show that a utility V agreeing with it on simple acts also is bounded.

Call a countable (finite or denumerably infinite) sequence of lotteries $\{H_n: n = 1, \ldots\}$ an *increasing* (*decreasing*) *chain* if $H_i \prec H_j$ ($H_j \prec H_i$) whenever $i < j$. The following concepts deal with chains of strict preference.

Definition 11a. Say $\prec$ is *bounded above* if, for each increasing chain $\{H_n\}$,

$$\lim_{n\to\infty} \sup\{x: (H_2 \prec xH_1 + (1-x)H_n)\} < 1.$$

Definition 11b. Say $\prec$ is *bounded below* if, for each decreasing chain $\{H_n\}$,

$$\lim_{n\to\infty} \sup\{x: (xH_1 + (1-x)H_n \prec H_2)\} < 1.$$

Definition 11c. Call $\prec$ *bounded* if it is bounded both above and below.

Lemma 2.3. *If $\prec$ satisfies HL Axioms* 1 *and* 3, *then $\prec$ is bounded, that is, all $\prec$-chains are bounded.*

Also, Lemma 2.3 yields the following claim about utilities for rewards:

Corollary 2.7. *Let W be a (real-valued) utility and assume that, in the domain of simple lotteries, its strict order $\prec_w$ satisfies HL Axioms* 1 *and* 3. *Then* $\sup_{\mathbf{R}}|W(r)| < \infty$.

Recall that a linear utility V is defined only up to a positive linear transformation. We use the facts reported by Corollaries 2.6 and 2.7 to standardize the units (0 and 1) for each V in a set of agreeing utilities $\mathcal{V}$ (agreeing with $\prec$ on simple acts, at least).

Definition 11d. A set $\mathcal{V} = \{V\}$ of utilities is *bounded* if, for some standardization of its elements,

$$\sup_{\mathcal{V},\mathbf{H_R}} |V(H)| < \infty.$$

The problem we face is this. Though each $V \in \mathcal{V}$ is bounded, there exist what are for our purposes undesirable standardizations of the V's which fail to satisfy Definition 11d. For an illustration, recall Example 2.1. There, the domain of (simple) lotteries $\mathbf{H_R}$ is generated by three rewards $\mathbf{R} = \{r_w, r^*, r_b\}$ using a partition of one (sure) state. That is, Example 2.1 is about preferences over von Neumann–Morgenstern lotteries. A partially ordered preference $\prec_{\mathcal{V}}$ over $\mathbf{H_R}$ arises (by the Pareto rule) from the convex set of utilities $\mathcal{V} = \{V_x: 0 < x < 1\}$, where $V_x(\mathbf{r}_w) = 0$, $V_x(\mathbf{r}^*) = x$ and $V_x(\mathbf{r}_b) = 1$. That is, $H_{L1} \prec_{\mathcal{V}} H_{L2}$ iff $V(H_{L1}) < V(H_{L2})$ for each $V \in \mathcal{V}$.

Obviously, the two constant acts (the rewards) $\mathbf{r_w}$ and $\mathbf{r_b}$ bound the partial order $\prec_{\mathcal{V}}$, that is, for each act H_L different from $\mathbf{r_w}$ and $\mathbf{r_b}$, $\mathbf{r_w} \prec_{\mathcal{V}} H \prec_{\mathcal{V}} \mathbf{r_b}$. Moreover, in this standardization of $\mathcal{V}$, $\sup_{\mathcal{V},\mathbf{H_R}}|V(H)| = 1$. Hence, it is a bounded set of utilities. However, we may standardize the elements of $\mathcal{V}$ so that it fails the condition in question. Rewrite each V_X, instead, so that $V_X(\mathbf{r}_w) = 0$, $V_X(\mathbf{r}^*) = 1$ and $V_X(\mathbf{r}_b) = 1/X$. Then $\lim_{X\to 0} V_X(\mathbf{r}_b) = \infty$.

To ensure a simple standardization which establishes our $\mathcal{V}$s are, indeed, bounded sets of utilities, we verify that (without loss of generality) we may introduce two rewards W and B (analogous to r_w and r_b in Example 2.1) that serve to bound the preferences for all other acts: Theorem 2. Then, the sets $\mathcal{V}$ are bounded sets of utilities since we standardize all $V \in \mathcal{V}$ with $V(\mathbf{W}) = 0$ and $V(\mathbf{B}) = 1$.

In this section we show how to extend the domain of a partially ordered preference by bounding it with "worst" and "best" acts. First, however, we review two concepts of "null" events.

Definition 12. An *event e* is the set of states in a subset T of π: $(\forall\ e)\ \exists\ (T \subset \pi),\ e = \cup_{s \in T}[s]$.

Definition 13. Call H_1 and H_2 a pair of *e-called-off* acts when $H_1(s) = H_2(s)$ if $s \notin e$.

Distinguish two senses of "null" events.

Definition 14a. An event e is *potentially null* iff for each pair of e-called-off acts H_1 and H_2, $H_1 \sim H_2$.

Definition 14b. Event e is *essentially null* iff for each pair of e-called-off acts H_1 and H_2, $H_1 \approx H_2$.

It is evident that when event e is essentially null, so too is each state that comprises it. Denote by **n** the union of the essentially null states. It follows (as is proven next) that the union of essentially null states is an essentially null event. Hence, **n** is the maximal essentially null event.

Corollary 2.8. *Let $N \subset \pi$ be the subset of all essentially null states $N = \{s_{j_1}, \ldots, s_{j_k}\}$, with $\mathbf{n} = \cup_{s \in N}(s)$. Then* **n** *is essentially null.*

Theorem 2. *Assume $\prec$ is a partially ordered preference (satisfying HL Axioms 1–3) over a set of discrete horse lotteries $\mathbf{H_R}$, defined on the partition $\pi' = \{s_j: j = 1, \ldots, n\}$. Let $\mathbf{R'} = \mathbf{R} \cup \{\mathbf{W}, \mathbf{B}\}$, where neither W nor B is an element of* **R**. *Then we may extend $\prec$ to a partially ordered preference $\prec'$ over $\mathbf{H_R}$, so that:*

1. $\prec'/\mathbf{H_R} = \prec$. *That is, $\prec'$ restricted to $\mathbf{H_R}$ is just $\prec$.*
2. $\forall\ (H \in \mathbf{H_R}),\ \mathbf{W} \prec' H \prec' \mathbf{B}$.
3. $\prec'$ *satisfies HL Axioms 1–3.*

Since $\mathbf{n} = S$ iff $\prec$ is trivial, that is, iff $\forall\ (H_1, H_2),\ H_1 \approx H_2$ [also, iff $\forall\ (H_1, H_2),\ H_1 \sim H_2$], without loss of generality, by Theorem 2, assume preference is not trivial by including rewards W and B. (This proposition, warranted by Theorem 2, is the counterpart in our theory to Savage's P5.)

III. EXTENDING STRICT PARTIAL ORDERS: THE INDUCTIVE ARGUMENT

III.1. *An Overview*

Let $\mathbf{R} = \{r_1, r_2, \ldots\}$ be a countable (finite or denumerable) set of rewards and let $\prec$ be a preference over $\mathbf{H_R}$ satisfying HL Axioms 1–3. Based on Theorem 2, without loss of generality, assume the existence of two distinguished rewards *not* in $\mathbf{R}$: reward W, where $\mathbf{W}$ is the worst act, and reward B, where $\mathbf{B}$ is the best act. Acts $\mathbf{W}$ and $\mathbf{B}$ are to serve as the common 0 and 1 in a (convex) set $\mathcal{V}$ of bounded utility functions V that agree with $\prec$. Hence for all $H \in \mathbf{H_R}$, $\mathbf{W} \prec H \prec \mathbf{B}$.

Let us highlight the major results in this section of our essay.

Our strategy is to use a transfinite induction to extend the preference $\prec$ (a partial order) to a weak order $\precsim$ over *simple* horse lotteries in $\mathbf{H_R}$. Let $\prec$ $(=\prec_0)$ serve as the basis for the induction. The induction at the ith stage extension of $\prec$, $\prec_i$, obtains by assigning a utility v_i to act $\hat{H}_i$, $V(\hat{H}_i) = v_i$, so that $\hat{H}_i \approx_i v_i\mathbf{B} + (1 - v_i)\mathbf{W}$. The quantity v_i is chosen (in accord with Definitions 20 and 25 in the Appendix) from a (convex) set of *target utilities* for $\hat{H}_i$, $\mathcal{T}_i(\hat{H}_i)$. The sequence $\{\hat{H}_i\}$ is chosen (see Definition 26 in the Appendix) so that the limit stage $\prec_w$ is a weak order over $\mathbf{H_R}$. We use $\mathbf{W}$ and $\mathbf{B}$ as the 0 and 1 of our utility, as follows.

Assume $\{H_n\} \Rightarrow H$ and $H_n \in \mathbf{H_R}$. The general target sets $\mathcal{T}_i(H)$ are defined through endpoints that bound the candidate utilities:

Definition 17. Let $v_i^*(H)$ be the lim inf of the quantities x_n for which $H_n \prec_{i-1} x_n\mathbf{B} + (1 - x_n)\mathbf{W}$.

Definition 18. Let $v_{i*}(H)$ be the lim sup of the quantities x_n for which $x_n\mathbf{B} + (1 - x_n)\mathbf{W} \prec_{i-1} H_n$.

[The "utility" bounds $v_*(H)$ and $v^*(H)$ do not depend upon which sequence $\{H_n\} \Rightarrow H$ is used, as explained in the Appendix.] Next, define the (*closed*) *target set of utilities* for an act $H \in \mathbf{H_R}$:

Definition 19. $\mathcal{T}(H) = \{v: v_*(H) \le v \le v^*(H)\}$.

We report two key properties of $\mathcal{T}(H)$ with the following lemma.

Lemma 3.1. *Assume $\prec$ satisfies the three axioms. Then:*

i. $v_*(H) \le v^*(H)$; *and*
ii. $v_*(H) = v^*(H) = v_H$ *iff* $H \approx v_H B + (1 - v_H)W$.

Our plan succeeds because $\prec_i$ extends $\prec_{i-1}$, it satisfies the three HL axioms and it preserves the $\approx_{i-1}$-indifference relations. Each weak order $\precsim_V = \precsim_w$ (corresponding to the limit stage relation "$\prec_w$ or $\approx_w$") is defined by inequalities in expected V-utility, based on the utilities v_i for each act $\tilde{H}_i$ in a (finite or) denumerable class $\mathcal{H} \subset \mathbf{H_R}$. (As explained below, $\mathcal{H}$ is finite or denumerable depending upon whether $\mathbf{R}$ is.) We choose $\mathcal{H}$ to form a basis for $\precsim_V$, that is, each $H \in \mathbf{H_R}$ is a limit point of simple acts and each simple act has its utility fixed by some finite stage of the transfinite induction. Then, $\precsim_V$ extends $\prec$ on simple acts in $\mathbf{H_R}$. Also, the utility V *almost agrees* with $\prec$ over the discrete lotteries $\mathbf{H_R}$. That is, if $H_1 \prec H_2$, then $H_1 \precsim_V H_2$. In Corollary 3.1, we provide sufficient conditions under which $\precsim_V$ extends $\prec$ for all the acts in $\mathbf{H_R}$. (See Remark 2.)

We show in Theorem 4 that each set $\mathcal{Z}$ of (bounded, standardized) real-valued utility functions over $\mathbf{R}$ induces a partial order, $\prec_{\mathcal{Z}}$, according to the Pareto preference relation, and $\prec_{\mathcal{Z}}$ satisfies our axioms. Of course, each utility $Z \in \mathcal{Z}$ agrees with $\prec_{\mathcal{Z}}$. That is, $\mathcal{L}$ is a subset of the set of all utilities agreeing with $\prec_{\mathcal{Z}}$. However [Chapter 1.2 in this volume], distinct convex sets of bounded utilities may induce the same strict partial order. Thus, our representation of the partial order $\prec$ is in terms of the largest convex set of agreeing linear utilities – the union of all sets of utilities where each set induces $\prec$ according to the Pareto condition.

Assume $\prec$ satisfies our axioms and let $\mathcal{Z}^S$ be the nonempty (convex) set of bounded utilities that agree with $\prec$ for simple acts. That is, $\mathcal{Z}^S$ is the set of all bounded utilities with the property that, for simple acts H_1 and H_2, $H_1 \prec H_2$ only if for each utility Z in $\mathcal{Z}^S$, the expected Z-utility of H_2 is greater than that of H_1. Let $\mathcal{V}$ be the nonempty (convex) set of utilities created for $\prec$ by (our method of induction in) Theorem 3. Theorem 5 asserts $\phi \ne \mathcal{V} = \mathcal{Z}^S$. Last, when the conditions of Corollary 3.1 apply, then (Corollary 3.2) $\mathcal{V}$ is the nonempty set of all utilities that agree with $\prec$.

Remark 2. If the Archimedean axiom is ignored and only simple lotteries are considered, the induction for extending $\prec$ to a weak order (in fact, to a total order) $\precsim$ over $\mathbf{H_R}$ is elementary and applies without a cardinality restriction on the reward set $\mathbf{R}$ and without need of the special acts $\mathbf{W}$ and $\mathbf{B}$. See the Appendix to Chapter 1.2. There, we show the following: Let κ be the cardinality of $\mathbf{R}$. Using Hausner's (1954) result, the order $\precsim$ ($=\prec_\kappa$) is a lexicographic expected utility.

III.2. *The Central Theorem*

Theorem 3. *Let $\prec$ be a nontrivial partial order over $\mathbf{H_R}$ satisfying* HL *Axioms* 1–3. *Then:*

i. *For simple lotteries in $\mathbf{H_R}$, $\prec$ can be extended to a weak order $\precsim_w = \precsim$ satisfying* HL *Axioms* 2 *and* 3. *That is, $\precsim$ is uniquely represented by a (bounded) real-valued utility V over $\mathbf{R}$ which agrees with $\prec$ for simple acts. In symbols, $\forall$ (simple $H_1, H_2 \in \mathbf{H_R}$), if $(H_1 \prec H_2)$, then $E_V[H_1] \leq E_V[H_2]$, and if $(H_1 \approx H_2)$, then $E_V[H_1] = E_V[H_2]$.*
ii. *V almost agrees with $\prec$. $\forall$ $(H_1, H_2 \in \mathbf{H_R})$, if $(H_1 \prec H_2)$, then $E_V[H_1] \leq E_V[H_2]$.*

It is instructive to illustrate how $\precsim_w$ may fail to agree with $\prec$ for some nonsimple lotteries in $\mathbf{H_R}$. The example motivates a condition on $\prec$ which proves sufficient for $\prec_w$ to extend $\prec$.

Example 3.1. Let $\mathcal{W} = \{W_j: j = 1, \ldots\}$ be a countable set of utilities on $\mathbf{R} = \{r_i: i = 1, \ldots\}$ with the two properties that $W_j(\mathbf{r}_m) = 0.25$ if $m \neq 2j$, while $W_j(\mathbf{r}_{2j}) = 0.5$. According to Theorem 4 (below), under the (weak) Pareto rule, $\mathcal{W}$ induces a partial order $\prec_{\mathcal{W}}$ which satisfies our three horse lottery axioms. Define the constant, nonsimple acts H_a and H_b by $H_a = \{P(r_i) = 1/2^m$ if $i = 2m - 1$, $P(r_i) = 0$ otherwise$\}$ and $H_b = \{P(r_i) = 1/2^m$ if $i = 2m$, $P(r_i) = 0$ otherwise$\}$. Then, evidently $(H_a \prec_{\mathcal{W}} H_b)$. However, at the kth stage $\prec_k$ in the extension of $\prec_{\mathcal{W}}$, we may arrange our choices of utilities for rewards so that $V(r_k) = 0.25$ $(k = 1, \ldots)$. However, then $\precsim_w$ does not extend $\prec_{\mathcal{W}}$ as $H_a \approx_w H_b$.

Definition 28. Given two subsets of $\prec$-preferences $\mathcal{Q}$ and $\mathcal{R}$, say that $\mathcal{Q}$ *is a basis for* $\mathcal{R}$ if every preference, $(H_1 < H_2) \in \mathcal{R}$, is a consequence (under HL Axioms 1–3) of preferences in $\mathcal{Q}$.

Corollary 3.1. *If there exists a countable basis* $\mathcal{B}$ *for* $\prec$, *then there exists a (bounded) real-valued utility* V *and corresponding weak order* $\precsim$ *that agrees with* $\prec$ *on all of* $\mathbf{H_R}$. *(Thus, a sufficient condition for the existence of an agreeing* $\precsim$ *is that* $\prec$ *is a separable partial order.)*

Next, we show that our axioms are not overly restrictive for representing a partial order by a (convex) set of agreeing utilities. We investigate relationships between a partial order $\prec_Z$ (formed by the Pareto rule with a set $\mathcal{Z}$ of utilities) and the set $\mathcal{V}$ of utilities created by induction on $\prec_Z$. Let $\mathcal{Z}$ be a set of bounded utilities on $\mathbf{R}$, standardized so that for $Z \in \mathcal{Z}$ and $H \in \mathbf{H_R}$, $0 = Z(\mathbf{W}) < Z(H) < Z(\mathbf{B}) = 1$. Define the relation $\prec_Z$ on $\mathbf{H_R}$ by the Pareto condition:

Definition 29. $(H_1 \prec_Z H_2)$ iff $\forall\ Z \in \mathcal{Z}$, $Z(H_1) < Z(H_2)$.

Theorem 4. $\prec_Z$ *satisfies* HL *Axioms* 1–3.

Next, let $\mathcal{V}$ be the set of utilities that can be generated from the partial order $\prec$ according to our induction. Let $\mathcal{Z}^S$ be the set of all bounded utilities Z that agree with $\prec$ on simple lotteries.

Theorem 5. $\phi \neq \mathcal{V} = \mathcal{Z}^S$.

Last, assume $\prec$ satisfies our axioms, let $\mathcal{Z}$ be the set of all utilities that agree with $\prec$ and let $\mathcal{V}$ be the set of utilities created by (our induction in) Theorem 3. We state three immediate corollaries of Theorem 5:

Corollary 3.2. *When* $\prec$ *satisfies the separability condition of Corollary* 3.1, *then* $\phi \neq \mathcal{V} = \mathcal{Z}$.

Corollary 3.3. *The set* $\mathcal{V}$ *does not depend upon the ordering of* $\mathcal{H}$.

Corollary 3.4. *The set* $\mathcal{V}$ *is convex.*

IV. A REPRESENTATION OF $\prec$ IN TERMS OF PROBABILITIES AND STATE-DEPENDENT UTILITIES

IV.1. *The Underdetermination of Personal Probability by HL Axioms 1–3*

Let $\mathcal{V}$ be the set of utilities V, each of which (by Theorem 5) corresponds to a limit stage $\precsim_V$ in our inductive extensions of the partial order $\prec$. According to Theorem 5, $\mathcal{V}$ is the set of all and only utilities that agree with $\prec$ on simple acts. According to Corollary 3.2, when $\prec$ is separable, $\mathcal{V}$ is the set of utilities that agree with $\prec$. We examine decompositions of $V \in \mathcal{V}$ as a subjective expected (state-dependent) utility.

Let $\precsim$ be a weak order over the discrete horse lotteries $\mathbf{H_R}$. Let $p(\cdot)$ be a (personal) probability defined on states in π, with $P(\mathbf{n}) = 0$ for the set of $\precsim$-null states $\mathbf{n}$. Finally, let $U_j(\cdot)$ be a (possibly) state-dependent utility on the discrete v.N-M lotteries $\mathbf{L_R}$, defined for the $\precsim$-nonnull states s_j. That is, for each $\precsim$-nonnull state, $s_j \notin \mathbf{n}$, U_j is a v.N-M utility. (For completeness, we may take U_j to be a constant function when $s_j \in \mathbf{n}$.)

Defintion 30. Say that $\precsim$ represented as a *subjective expected (state-dependent) utility* by the pair $(p, \{U_j: j = 1, \ldots, n\})$, whenever

$$H_1 \precsim H_2 \textit{ iff } \sum_j p(s_j)U_j(L_{1j}) \leq \sum_j p(s_j)U_j(L_{2j}). \qquad (4.1)$$

For convenience, abbreviate the probability/(state-dependent) utility pairs as (p, U_j).

We rely on a result due to Fishburn [(1979), Theorem 13.1] to show that each $\precsim_V$, $V \in \mathcal{V}$, bears the subjective expected utility property for a large class of (p, U_j) pairs. In fact, for each such $\precsim_V$, the (p, U_j) pairs range over all mutually absolutely continuous probabilities defined on the $\precsim_V$-nonnull states. Specifically:

Lemma 4.1. *Let $\precsim$ be a (nontrivial) weak order on $\mathbf{H_R}$ satisfying* HL *Axioms* 2 *and* 3. *For each probability $p(\cdot)$ with support the (nonempty) set of $\precsim$-nonnull states, there is a (possibly) state-dependent utility $U_j(\cdot)$ on discrete lotteries for which $\precsim$ has property* (4.1) *under* (p, U_j).

Putting Theorem 5 and Lemma 4.1 together, we have the following corollary:

Corollary 4.1. *There exists a set of pairs* $\{(p(\cdot), U_j(\cdot))\}$, *with* p *a* personal probability *defined on the set of* $\precsim_V$*-nonnull states in* p *and* U_j *a* state-dependent utility *over the discrete lotteries, where* $\forall$ $(H \in \mathbf{H_R})$, $V(H) = \Sigma_j p(s_j) \times U_j(L_j)$.

[Being linear utilities, the U_j have the expected utility property for lotteries. That is, $U_j(xL_1 + (1 - x)L_2) = xU_j(L_1) + (1 - x)U_jL_2$. Moreover, for each $V \in \mathcal{V}$, the set of personal probabilities $\{p: \exists\ U_j$ with $\precsim_V$ represented by $(p, U_j)]$ is closed under the relation of mutual absolute continuity.]

IV.2. *State-independent Utilities and a Counterexample*

Anscombe and Aumann's theory of horse lotteries introduces a fourth axiom which suffices for a *unique* expected utility representation of a weak order $\precsim$ by a pair (p, U), with p a personal probability over states and U a *state-independent* utility over rewards.

Recall Definition 5, when U is state-independent, a lottery L has the same utility independent of the (nonnull) state s_j. Hence [as in Savage's (1954) theory], U assigns a constant utility across (nonnull) states to each "constant" act. In Anscombe and Aumann's theory, then (4.1) is strengthened to read:

$$H_1 \precsim H_2 \textit{ iff } \sum_j p(s_j)U(L_{1j}) \leq \sum_j p(s_j)U(L_{2j}) \qquad (4.2)$$

and each $\precsim$ is so represented by a *unique* (p, U) pair.

The existence of a state-independent utility for $\precsim$ is assured through a contrast between (unconditional) preferences over constant horse lotteries and preferences over s_j-called-off horse lotteries: pairs of acts that differ only in one state. Specifically, let H_{L_i} $(i = 1, 2)$ be two constant horse lotteries that award, respectively, the v.N-M lottery L_i in all states. Let H_i $(i = 1, 2)$ be two s_j-called-off horse lotteries with $H_i(s_j) = L_i$ [and $H_1(s) = H_2(s)$ for $s \neq s_j$]. The Anscombe–Aumann (AA) axiom for state-independent utility reads:

AA Axiom 4. Provided $s_j \notin \mathbf{n}$, for each such quadruple of acts, $H_{L_1} \precsim H_{L_2}$ iff $H_1 \precsim H_2$.

(Recall, their Axiom 1 stipulates that preferences are weakly ordered, $\precsim$; hence, in their theory there is no difference between "potentially null" and "essentially null" states.)

This axiom requires that $\precsim$-preferences over "constant" acts (such as the H_{L_i}) are reproduced by *called-off* choices (the H_i) given each nonnull s_j. The unconditional preference for v.N-M lotteries is their conditional (that is, called-off) preference, given a nonnull state. (We discuss conditional partially ordered preferences in Section v.)

It is significant to understand that AA Axiom 4, though sufficient to create state-independent utilities when preference satisfies the usual ordering, independence and Archimedean conditions, does not preclude alternative expected utility representations by *state-dependent* utilities. Lemma 4.1 continues to apply, even in the presence of the extra axiom for state-independent utilities. Weak orderings that satisfy the independence, Archimedean and state-independent utility axioms admit a continuum of different probability/utility representations, each in accord with (4.1).

What the Anscombe–Aumann fourth axiom achieves, however, is to guarantee that precisely one probability/utility pair, among the set of all pairs $\{(p, U_j)\}$ indicated by Lemma 4.1, satisfies the more restrictive condition, (4.2). In Anscombe and Aumann's theory, as in Savage's theory, this probability/utility pair (p, U) is given priority over the others. That is, these theories select the one (and only one) expected state-independent utility representation of preference, in accordance with (4.2), and thereby fix a personal probability uniquely from $\precsim$-preferences.

We are not satisfied with a conventional resolution of the representation problem indicated by Lemma 4.1. If state-dependent utilities are plausible candidates for an agent's values, and we think sometimes they are, then the measurement question remains open despite the fourth axiom. What justification is there for a convention which gives priority to state-independent values? In two essays [Schervish, Seidenfeld and Kadane (1990, 1991)], we examine the case of weakly ordered preferences without the extra axiom for "state-independent" utility. Here, instead, we adopt the strategy of imposing a modified Axiom 4 and asking which probability/state-independent utility pairs agree with the partial order $\prec$. Unlike the Anscombe–Aumann or Savage theories, ours does *not* assert that these pairs of probability/(state-independent) utility functions identify the agent's degrees of beliefs and values.

We adapt Anscombe and Aumann's final axiom to our construction by restricting it to states which are not potentially null. This produces the following axiom:

HL Axiom 4. If s_k is not $\prec$-potentially null, then for each quadruple of acts H_{L_i}, $H_i (i = 1, 2)$ as described above, $H_{L_1} \prec H_{L_2}$ iff $H_1 \prec H_2$.

Suppose $\prec$ is a preference on horse lotteries subject to HL Axioms 1–4. Surprisingly, there may not exist a probability and *state-independent* utility agreeing with $\prec$ [according to (4.2)], even for simple acts. Moreover, the problem has nothing to do with existence of potentially null states. That is, even if no state is potentially null, the fourth axiom (HL Axiom 4) is *insufficient* for the existence of a probability/state-independent utility pair agreeing with $\prec$.

Example 4.1. Let $\mathbf{R} = \{r_*, r, r^*\}$ be three rewards and consider the set $\mathbf{H_R}$ of horse lotteries defined on the binary partition $\{s_1, s_2\}$. Next consider two probability/utility pairs (p^i, U^i) $(i = 1, 2)$, where $U^i(r_*) = 0$, $U^i(r^*) = 1$, $U^1(r) = 0.1$ and $U^2(r) = 0.4$; also, $p^1(s_1) = 0.1$ and $p^2(s_1) = 0.3$. Define $H_1 \prec H_2$ iff $p^i(s_1)U^i(L_{1,1}) + p^i(s_2)U^i(L_{1,2}) < p^i(s_1)U^i(L_{2,1}) + p^i(s_2)U^i(L_{2,2})$ $(i = 1, 2)$. Then, by Theorem 4, $\prec$ satisfies HL Axioms 1–3, and we claim it satisfies HL Axiom 4 as well. Moreover neither state is potentially null under $\prec$.

The proof that $\prec$ satisfies HL Axiom 4 is straightforward. We observe the following (expected utility) bounds on $\prec$-preferences for the constant horse lottery $\mathbf{r}$. $(\forall\, 0.1 > \varepsilon > 0)$, $(0.9 + \varepsilon)\mathbf{r}_* + (0.1 - \varepsilon)\mathbf{r}^* \prec \mathbf{r} \prec (0.6 - \varepsilon)\mathbf{r}_* + (0.4 + \varepsilon)\mathbf{r}^*$. However, the utilities U^i are state-independent and neither state is null for either p^i $(i = 1, 2)$. That is, using conditional preference (see Definition 34, $(\forall\, 0.1 > \varepsilon > 0)$ $(0.9 + \varepsilon)\mathbf{r}_* + (0.1 - \varepsilon)\mathbf{r}^* \prec_{\mathbf{s_j}} \mathbf{r} \prec_{\mathbf{s_j}} (0.6 - \varepsilon)\mathbf{r}_* + (0.4 + \varepsilon)\mathbf{r}^*$ $(j = 1, 2)$. The utility bounds for $\mathbf{r}$ reproduce in both families of s_j-called-off acts. Hence, $\prec$ satisfies HL Axiom 4.

According to Theorem 1, the two pairs (p^i, U^i) are the sole state-independent expected utilities agreeing with $\prec$ [according to (4.2)]. Next, we assert that $\prec$ may be extended to a strict partial order $\prec''$, also satisfying HL Axioms 2–4, but where $\prec''$ narrows the expected utility bounds for r, as follows: $0.9r_* + 0.1r^* \prec'' \mathbf{r} \prec'' 0.6r_* + 0.4r^*$.

We outline a general result for extending $\prec$ by forcing a new strict preference $H_1 \prec' H_2$, when $H_1 \sim H_2$. This contrasts with the extension created through Definition 20, which, instead, forces a new indifference relation.

Suppose H_1 and H_2 are elements of $\mathbf{H_R}$ that satisfy (1) $H_1 \sim H_2$ and (2) there do not exist two sequences $\{H_{i,n}\} \Rightarrow H_i$ $(i = 1, 2)$, where $\forall$ $(n = 1, \ldots)$, $H_{2,n} \prec H_{1,n}$. Create an extension $\prec'$ of $\prec$ as follows:

Definition ($\prec'$). $\forall$ $(H_a, H_b \in \mathbf{H_R})$, $H_a \prec' H_b$ if and only if either:

i. $H_a \prec H_b$ (so $\prec'$ extends $\prec$)

or

ii. $\exists$ $\{H_{a,n}\} \Rightarrow H_a$ and $\exists$ $\{H_{b,n}\} \Rightarrow H_b$ and $\exists$ $\{x_n\}$ with $\lim_{n\to\infty} \{x_n\} \neq 1$, $x_n H_{a,n} + (1 - x_n) H_2 \prec x_n H_{b,n} + (1 - x_n) H_1$.

Claim. $\prec'$ *satisfies* HL *Axioms* 1–3, *provided* $\prec$ *does. Also,* $H_1 \prec' H_2$.

We omit the proof which follows along similar lines for the demonstration of Lemma 3.3. Regarding HL Axiom 4, it suffices that $\prec'$ is formed by extending $\prec$ using a target set endpoint, for example, let $H_1 = v_* B + (1 - v_*)W$, where $\mathcal{T}(H_2) = [v_*, v^*]$ and this interval has interior, that is, $v_* < v^*$. Then $\prec'$ satisfies HL Axiom 4 too.

Last, for Example 4.1, apply the claim, twice over, first to force $0.9\mathbf{r}_* + 0.1\mathbf{r}^* \prec' \mathbf{r}$, then to force $\mathbf{r} \prec'' 0.6\mathbf{r}_* + 0.4\mathbf{r}^*$.

Consider the convex sets $\mathcal{V}$ and $\mathcal{V}''$ of *agreeing* utilities for $\prec$ and $\prec''$ provided by Corollary 3.2. (These utilities agree since $\mathbf{R}$ is finite.) Because $\prec''$ extends $\prec$, then $\mathcal{V}'' \subset \mathcal{V}$. A fortiori, each agreeing expected state-independent utility model for $\prec''$ also is one for $\prec$. However, by Theorem 1, there does not exist an agreeing expected state-independent utility model for $V'' \in \mathcal{V}''$, since $\mathcal{V}$ excludes all (that is, both) expected state-independent utility models for $\prec$. Nonetheless, $\prec''$ satisfies HL Axioms 1–4. This ends our discussion of Example 4.1.

IV.3. *Representation of* $\prec$ *in Terms of (Nearly) State-independent Utilities*

The four axioms HL Axioms 1–4 are insufficient for the existence of an agreeing state-independent utility. However, with the addition of a fifth axiom to regulate state-dependence for potentially null states, the resulting theory is sufficient for an agreeing "almost" state-independent utility. First, we make precise the notion of an "almost" state-independent utility.

Consider a set of probability/state-dependent utility pairs $\{(p, U_j)\}$, each pair agreeing with the partial order $\prec$ for simple acts, according to (4.1).

Definition 31. Say that $\prec$ *admits almost state-independent utilities for a set of n-rewards* $\{r_1, \ldots, r_n\}$ if, for each $\varepsilon > 0$, there exists a pair (p, U_j) that agrees with $\prec$ on simple acts (and almost agrees, otherwise), where for a set of states $S^{\#} = \{s_{j1}, \ldots, s_{jk}\}$, $p(S^{\#}) \geq 1 - \varepsilon$,

$$\max_{\substack{s_j, s_{j'} \in S^{\#} \\ 1 \leq i \leq n}} |U_j(r_i) - U_{j'}(r_i)| \leq \varepsilon.$$

Say $\prec$ *admits almost state-independent utilities* if it does so for each set of n-rewards, $n = 1, \ldots$.

Obviously, if (p, U) agree with $\prec$ and U is state-independent, then $\prec$ admits almost state-independent utilities.

There are two problems created by state-dependent utilities. First, given the partial order $\prec$, we would like to indicate probability bounds for an event E by $\prec$-preferences between a constant act of the form $H_x(s) = xB + (1 - x)W$ and the act $H_E(s) = B$ if $s \in E$, and $H_E(s) = W$ if $s \notin E$. That is, in general, we want the upper probability bound $p^*(E)$ to be the l.u.b.$\{x: H_E \prec H_x\}$ (or 1, if $H_E \sim \mathbf{B}$), and we want the lower bound, $p_*(E)$, to equal the g.l.b.$\{x: H_x \prec H_E\}$ (or 0, if $H_E \sim \mathbf{W}$). However, if such preferences are to indicate probability bounds, then we require that the rewards B and W carry state-independent utilities 1 and 0, respectively. Thus the first problem.

Second, if a state s_j is potentially null under $\prec$, then there are no $\prec$-preferences among pairs of acts called-off in case s_j does not obtain. Let $\mathbf{Hs_j}$ be the family of s_j-called-off acts that yield outcome W for all $s \notin s_j$. When s_j is a potentially null state, $\forall$ $(H_1, H_2 \in \mathbf{Hs_j})$, $H_1 \sim H_2$. Suppose $\precsim_V$ $(V \in \mathcal{V})$ extends $\prec$ (on simple acts) and let $\{(p, U_j)\}$ be the set of probability/(possibly) state-dependent utilities which represent $\precsim_V$ according to (4.1). Then, if state s_j is potentially null under $\prec$, unfortunately, HL Axioms 1–4 impose too few restrictions on U_j (the state-dependent utility, given state s_j) even when $p(s_j) > 0$. In particular, it may be that $V(\mathbf{r}_1) > V(\mathbf{r}_2)$, yet for all the U_j, $U_1(r_1) \leq U_1(r_2)$.

To resolve both these problems, we impose a fifth axiom – a requirement of "stochastic dominance" among lotteries. For each state s_j and each v.N-M lottery L_α, define the set of acts $\{H^\alpha_{j,m}: H^\alpha_{j,m}(s) = (1 - 2^{-m})W + (2^{-m})L_\alpha$, if $s \notin s_j$; $H^\alpha_{j,m}(s_j) = L_\alpha$ for state $s_j\}$ $(m = 1, \ldots)$. Observe that $(\forall j) \lim_{m\to\infty} \{H^\alpha_{j,m}\} = H_{j,\alpha} \in \mathbf{Hs_j}$. Moreover, $H_{j,\alpha}(s_j) = L_\alpha$. Then, we require the following axiom:

HL Axiom 5. For each two "constant" acts $H_{L\alpha}(s) = L_\alpha$ and $H_{L\beta}(s) = L_\beta$,

$$\forall (j, m)\left[H_{L\alpha} \prec H_{L\beta} \text{ iff } H^\alpha_{j,m} \prec H^\beta_{j,m}\right] \quad (j = 1, \ldots, n;\, m = 1, \ldots).$$

Thus, exactly when $\mathbf{L}_\beta$ is $\prec$-preferred to $\mathbf{L}_\alpha$ (as constant acts), HL Axiom 5 imposes a $\prec$-preference on sequences of pairs of lotteries, $(H^\alpha_{j,m}, H^b_{j,m})$ which converge to the s_j-called-off pair $(H_{j,\alpha}; H_{j,\beta})$. Thus, we obtain the constraint (Definition 21 of the Appendix) "$\neg(H_{j,\beta} \prec H_{j,\alpha})$."

Lemma 4.2. *Suppose* $\prec$ *satisfies* HL *Axioms* 1–5. *Then, for each* $V \in \mathcal{V}$ *(of Theorem* 3.1*) we may select (exactly) one pair* (p^V, U^V_j) *from the set of pairs* $\{(p, U_j)\}$ *provided by Corollary* 4.1*– where each pair represents* $\precsim_V$ *in accord with* (4.1) *– so that acts* **W** *and* **B** *have constant value and bound the state-dependent utilities of other rewards. In symbols,*

$$\forall (s_j) \forall (L_i, L_k \in \mathbf{L}_{\mathbf{R}-\{W,B\}}),\quad H_{L_i} \prec H_{L_k}$$
$$\textit{iff } 0 = U^V_j(W) \le U^V_j(L_i) \le U^V_j(L_k) \le U^V_j(B) = 1,$$

with at least one outside inequality strict for each s_j *such that* $p(s_j) > 0$, *and all inequalities strict for each* s_j *that is not* $\prec$*-potentially null.*

Definition 32. We call (p^V, U^V_j) the *standard representation* of V.

Thus, HL Axiom 5 (via HL Axiom 3) constrains state-dependent utilities of the rewards **W** and **B** in potentially null states, as desired. In the course of the proof of Theorem 6 (below), we explain how HL Axiom 5 also regulates the $\prec$-potentially null, state-dependent utilities of all v.N-M lotteries.

Of course, HL Axioms 1–5 are insufficient for guaranteeing existence of a state-independent utility agreeing with $\prec$. Counterexample 4.1 applies, that is, $\prec'$ satisfies all five axioms (since no states are $\prec'$-potentially null). However, as we show next, these axioms suffice for an almost state independent utility.

Theorem 6. *Assume that* $\prec$ *satisfies* HL *Axioms* 1–5.

i. *Then* $\prec$ *admits almost state-independent utilities.*

ii. *If* $\prec$ *has a countable basis* $\mathcal{B}$, *each* (p, U_j) *pair in Definition 31 agrees with* $\prec$.

There is a sufficient condition for the existence of a state-independent utility over the finite set $\{W, B, r_1, \ldots, r_i, \ldots, r_n\}$, using closure (at one endpoint, at least) of the target sets $J_i(r_i)$ defined in Definition 19.

Lemma 4.3. *If the target sets* $J_i(r_i)$ $(i = 1, \ldots, n)$ *are not open intervals, there exists a subset* $\mathcal{V}' \subset \mathcal{V}$ *of expected utilities for* $\prec$ *(agreeing on simple acts), where each* $V' \in \mathcal{V}'$ *is standardly represented by the set of pairs* $\{(p', U'_j)\}$ *according to* (4.1) *and where* $U'_j(r_i)$ *is state-independent* $(i = 1, \ldots, n)$.

Note. $\mathcal{V}'$ may fail to be convex. Also, results similar to Lemma 4.3, using different assumptions, appear in Rios Insua (1992). Related ideas appear in Nau (1992).

V. CONDITIONAL PREFERENCE AND CONDITIONAL PROBABILITIES

Let e be an event. (Recall, we equate the set state s_j with its elements.) Let H_1 and H_2 be a pair of e-called-off acts. Suppose $\prec$ satisfies HL Axioms 1 and 2.

Lemma 5.1. *Let* H'_1 *and* H'_2 *be another pair of e-called-off acts which agree with* H_1 *and* H_2 *(respectively) on e, that is,* $\forall$ $(s \notin e)$, $[H'_1(s) = H'_2(s)]$ *and* $\forall$ $(s \in e)$, $[H_1(s) = H'_1(s)$ and $H_2(s) = H'_2(s)]$: (i) $H_1 \prec H_2$ *iff* $H'_1 \prec H'_2$ *and* (ii) $H_1 \approx H_2$ *iff* $H'_1 \approx H'_2$.

Therefore, a $\prec$-preference (or $\approx$-indifference) among two e-called-off acts does not depend upon how they are called-off, that is, the preference (or indifference) does not depend upon how the acts agree with each other when e fails. This replicates the core of Savage's [(1954), page 23] "sure thing" postulate, P2, as that applies to our concept of a partially ordered preference.

Consider a (maximal) subset of $\mathbf{H_R}$, denoted by $\mathbf{H_e}$, where every two elements of $\mathbf{H_e}$ form an e-called-off pair. Obviously, each such family $\mathbf{H_e}$ of e-called-off acts is closed under convex combinations.

Definition 33. Define $\prec_e = \prec/\mathbf{H_e}$, the restriction of $\prec$ to the family of e-called-off acts in $\mathbf{H_e}$. We call $\prec_e$ the *conditional $\prec$-preference relation, given e*. (The preceding lemma insures this relation is well defined, that is, it depends on e but not on how acts are called-off.)

Note. The event e^c is essentially null with respect to the conditional preference $\prec_e$.

Definition 34. Also, for each pair of horse lotteries H_1 and H_2, say that H_2 is $\prec$-preferred to H_1 given e, provided that, for some pair H'_1 and H'_2 (and by Lemma 5.1, provided for all pairs) of e-called-off acts agreeing (respectively) with H_1 and H_2 on e, $H'_1 \prec_e H'_2$.

In light of Lemma 5.1(i) and because $\mathbf{H_e}$ is a subset of $\mathbf{H_R}$, the following result is immediate.

Theorem 7. *If* $\prec$ *(over* $\mathbf{H_R}$*) satisfies (a subset of)* HL *Axioms* 1–5, *then* $\prec_e$ *(over* $\mathbf{H_e}$*) also satisfies the same horse lottery axioms, at least.*

Theorem 7 prompts an interesting question: What is the relation between (i) the set of *conditional* probability/utility pairs $\{p(|e), U_{j\in e}\}$, given e, that arise from the representation of $\prec$ over the family of acts $\mathbf{H_R}$ and (ii) the set of probability/utility pairs $\{p_e, U_{e,j\in e}\}$ that represent the conditional preference $\prec_e$ over the restricted family of acts $\mathbf{H_e}$?

The following discussion of conditional indifference tells some of the answer.

Definition 35. Let $\approx_e$ be the *conditional $\approx$-indifference relation, given e*, defined by restricting $\approx$ to acts in the family $\mathbf{H_e}$. Then, say that horse lotteries H_1 and H_2 are $\approx$-indifferent, given e, provided that for some pair H'_1 and H'_2 (and by Lemma 5.1, provided for all such pairs) of e-called-off acts agreeing (respectively) with H_1 and H_2 on e, $H'_1 \approx_e H'_2$.

It is important, however, to see that $\approx_e$ is not always the same as the $\approx$-indifference relation (defined by Definition 8) induced by $\prec_e$ over acts solely in $\mathbf{H_e}$.

Definition 36. Denote by $\approx_{\mathbf{H_e}}$ the $\approx$-indifference relation over elements of $\mathbf{H_e}$, induced by $\prec_e$.

Of course $\approx_e$-indifference entails $\approx_{\mathbf{H}_e}$-indifference, but not conversely. H_1 and H_2 may be two e-called-off acts from a family $\mathbf{H}_e$ which satisfies $H_1 \approx_{\mathbf{H}_e} H_2$, but where, nevertheless, $H_1 \not\approx H_2$, that is, $H_1 \not\approx_e H_2$. We illustrate this phenomenon using a potentially null state which is not essentially null.

Example 5.1. Consider a binary partition $S = \{s_1, s_2\}$ and horse lotteries defined over a binary reward set $\mathbf{R} = \{W, B\}$. Suppose $\prec$ is created by the Pareto principle applied to expected utility inequalities from the following set of probability/(state-independent) utility pairs: $S = \{(p, U)\colon 1 \geq p(s_1) \geq 0.5;\ U(B) > U(W)\}$. Then, s_2 is potentially null: acts are $\sim$-incomparable whenever they belong to a common $\mathbf{H}_{s_2}$ family. [With $p(s_1) = 1$, all elements of $\mathbf{H}_{s_2}$ have equal expected utility.] Hence, $\prec_{s_2}$ is vacuous. So, based on $\prec_{s_2}$ restricted to a family $\mathbf{H}_{s_2}$, all pairs of (e-called-off) acts are $\approx_{\mathbf{H}_{s2}}$-indifferent. However, the pair of s_2-called-off acts (H_1, H_2), defined by $H_1(s_1) = H_2(s_1) = W$, $H_1(s_2) = W$ and $H_2(s_2) = B$, though $\sim$-incomparable are not $\approx$-indifferent: $H_1 \sim H_2$ and $H_1 \not\approx H_2$. This is shown as follows. Consider the act H_3 defined by $H_3(s_1) = B$ and $H_3(s_2) = W$. Observe that $0.5H_1 + 0.5H_3 \sim 0.7W + 0.3B$, whereas $0.7W + 0.3B \prec 0.5W + 0.5B = 0.5H_2 + 0.5H_3$. This shows that $H_1 \not\approx H_2$.

Returning to the question, above, we state our central result about conditional probabilities and conditional preferences.

Theorem 8.

i. *If (p, U_j) belongs to the set of probability/utility pairs representing $\prec$, then the pair $(p(\cdot|e), U_{j\in e})$ belongs to the set that represent the conditional preference $\prec_e$.*

ii. *Suppose that the pair $(p_e, U_{e,j\in e})$ belongs to the set representing the conditional preference $\prec_e$ with respect to the family $\mathbf{H}_e$. Then for some pair (p, U_j) in the set that represents $\prec$, $p(\cdot|e) = p_e$ and $U_{j\in e} = U_{e,j\in e}$, provided two conditions obtain:* (1) *The event e is not potentially null (with respect to $\prec$) and* (2) *the expected utility $V_e(\ )$ (with arguments from $\mathbf{H}_e$), corresponding to the pair $(p_e, U_{e,j\in e})$, does not use $\prec$-precluded target endpoints, as regulated by Definition* 21 (*of the Appendix*).

Next, we offer an example of Theorem 8, relating Bayes' updating to conditional preferences.

Example 5.2. Consider a partition into three states $\{s_1, s_2, s_3\}$ and acts involving the three rewards, W, r and B. Let $\mathbf{r}$ denote the constant act, with outcome r in each state. For $j = 1, 2$ and 3, define the three acts $H_j(s_j) = B$, $H_j(s_k) = W$ $(j \neq k)$ and also the three acts $H_{j,r}(s_j) = r$ and $H_{j,r}(s_k) = W$ $(j \neq k)$. Apart from the strict preferences that follow because $\mathbf{W}$ and $\mathbf{B}$ are, respectively, the "worst" and "best" acts, suppose also the agent reports these preferences:

$$0.5W + 0.5H_{3,r} \prec H_1 \prec H_{3,r} \prec H_2 \prec H_3 \prec \mathbf{r} \prec 0.5H_3 + 0.5\mathbf{r}.$$

We investigate the standardized, state-independent utility representations for these preferences. That is, with $U(W) = 0$, $U(B) = 1$, let $u = U(r)$, independent of the state s_j. If we denote by p_j the probability of state s_j, then the preferences above are modeled by each probability/utility pair $(p_1, p_2, p_3; u)$ satisfying $0 < 0.5p_3u < p_1 < p_3u < p_2 < u < 0.5p_3 + 0.5$.

For each $0 < u < 1$, it is possible to determine the set $\mathcal{P}(u)$ of all (p_1, p_2, p_3) that satisfy these inequalities. For example the set $\mathcal{P}(0.5)$ is shown in Figure 2. The union of all sets $\mathcal{P}(u) \times \{u\}$ such that $\mathcal{P}(u) \neq \varnothing$ is the set of all probability/utility pairs that agree with the strict preferences above. From this set, one can determine other preferences not listed above which must also hold if the axioms do. For example, it is required, though not obvious from the reported preferences, that $0.4B + 0.6W \prec \mathbf{r}$. [By contrast, it is obvious from the preferences above that $(1/3)\mathbf{B} + (2/3)\mathbf{W} \prec \mathbf{r}$.]

If we were to learn that, say, the event $E = \{s_1, s_2\}$ occurred, we can determine which preferences are implied in the conditional problem. The set of all pairs (q_1, u), where q_1 is a conditional probability of s_1 given E, is shown in Figure 3. Observe that, as provided by Theorem 8, the set of conditional probabilities from Figure 2 is exactly the set represented by the vertical line (at $u = 0.5$) in Figure 3. However, the set shown in Figure 3 is not convex since it contains the points (0.415, 0.293) and (0.455, 0.379), but does not contain the point (0.435, 0.336) = 0.5(0.415, 0.293) + 0.5(0.455, 0.379).

VI. CONCLUDING REMARKS

There is a burgeoning literature dealing with applications of sets of probabilities. Separate from work on robust Bayesian statistical analysis, they occur also in the following settings: as a rival account to strict Bayesian theory for representing uncertainty, such as in Levi's (1974,

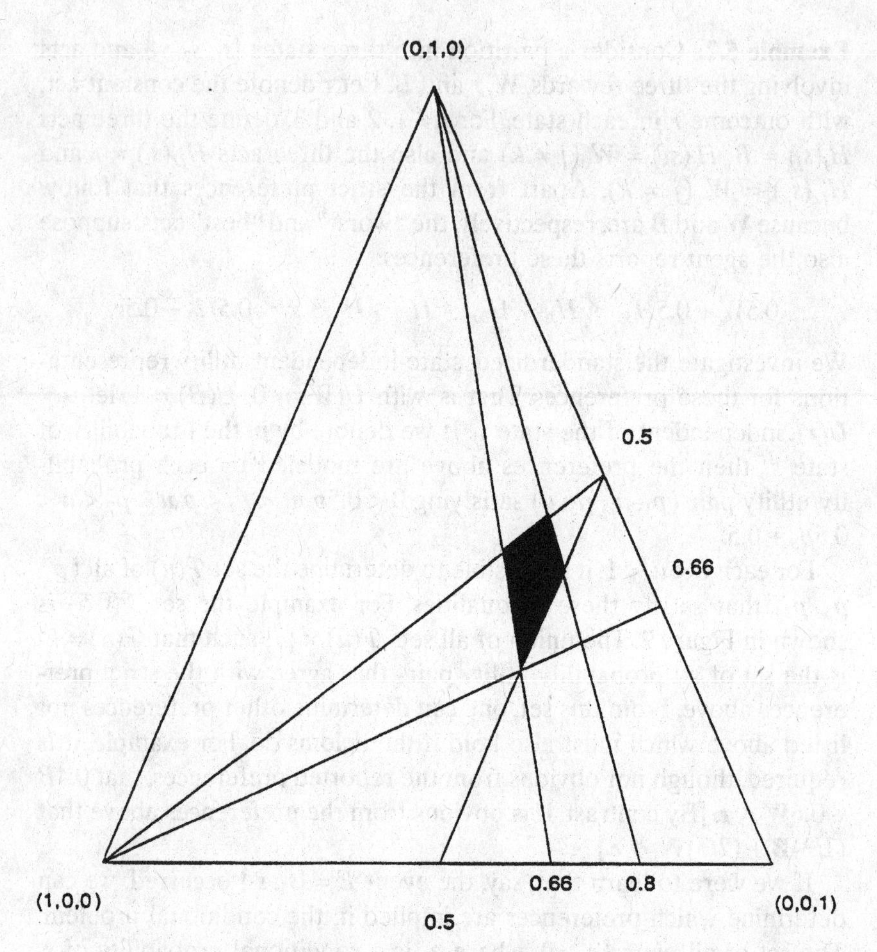

Figure 2. The set $p(0.5)$ in Example 5.2

1980) theory for Ellsberg's (1961) "paradox"; relating to indeterminate degrees of belief, as in Smith's (1961) theory of "medial odds" developed by Williams (1976), Giron and Rios (1980), Walley (1991) and Nau (1993); and as a method for capturing multiple "expert" opinions [Kadane and Sedransk (1980); Kadane (1986)]. In addition, sets of probabilities arise from incomplete elicitations, where some but not all of an agent's opinions are formalized by inequalities in probabilities and the question is what decisions are fixed by these partially reported degrees of belief; see Moskowitz, Wong and Chu (1988) and White (1986). Dual to sets of probabilities, the articles by Aumann (1962) and

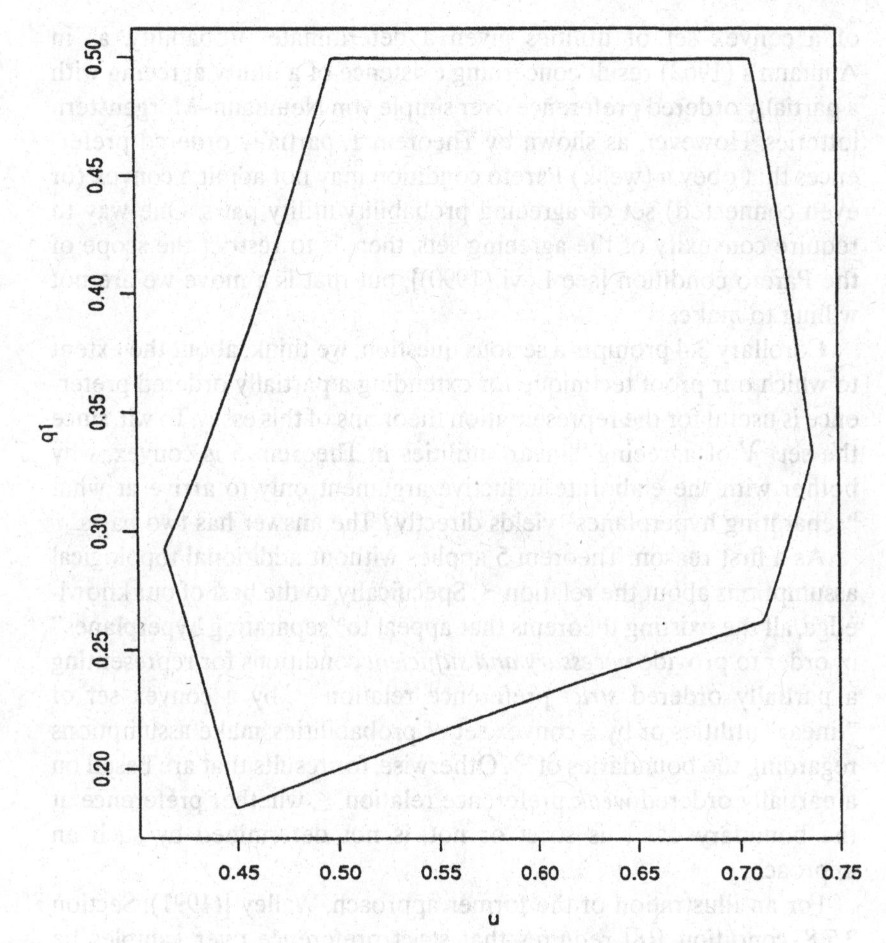

Figure 3. The set of conditional probabilities and utilities

Kannai (1963) explore the existence of ("linear") utilities for von Neumann–Morgenstern lotteries when probabilities are completely specified.

However, these efforts rely on convexity of the spaces of probabilities and utilities to arrive at their conclusions. From our point of view, this mathematical convenience is justified under an assumption, for example, that either the agent's probability or the agent's utility is fully determinate. For instance, in light of Corollary 3.4, convexity is appropriate for Bayesian robustness when a loss function is specified but probability is left indeterminate. Likewise, our theory endorses the use of a convex set of utilities given a determinate probability, as in Aumann's (1962) result concerning existence of a utility agreeing with a partially ordered preference over simple von Neumann–Morgenstern lotteries. However, as shown by Theorem 1, partially ordered preferences that obey a (weak) Pareto condition may not admit a convex (or even connected) set of agreeing probability/utility pairs. One way to require convexity of the agreeing sets, then, is to restrict the scope of the Pareto condition [see Levi (1990)], but that is a move we are not willing to make.

Corollary 3.4 prompts a serious question, we think, about the extent to which our proof technique for extending a partially ordered preference is useful for the representation theorems of this essay. To wit, since the set $\mathcal{V}$ of agreeing "linear" utilities in Theorem 5 is convex, why bother with the elaborate inductive argument only to arrive at what "separating hyperplanes" yields directly? The answer has two parts.

As a first reason, Theorem 5 applies without additional topological assumptions about the relation $\prec$. Specifically, to the best of our knowledge, all the existing theorems that appeal to "separating hyperplanes" in order to provide *necessary and sufficient* conditions for representing a partially ordered *strict* preference relation $\prec$ by a convex set of "linear" utilities or by a convex set of probabilities, make assumptions regarding the boundaries of $\prec$. Otherwise, for results that are based on a partially ordered *weak* preference relation $\precsim$, whether preference at the boundary of $\mathcal{V}$ is strict or not, is not determined by such an approach.

For an illustration of the former approach, Walley [(1991), Section 3.7.8, condition R8] requires that strict preference over gambles be "open" so that so-called strong separation leads to a representation in terms of sets of probabilities "closed" with respect to infimums. By avoiding "separating hyperplanes," we are able to sidestep this artifice. Surfaces of the set $\mathcal{V}$ need not have a simple topological character.

For an illustration of the latter approach, Giron and Rios (1980) use a reflexive, partial (quasi-Bayesian) preference relation, denoted in their paper by $\precsim$, which they represent with a closed, convex set of probabilities. They note [Giron and Rios (1980), footnote 3, page 20] that their method generates the same "quasi-Bayesian preorder" whether the so-called uncertainty set of probabilities (which they denote by K^*) or its closure ($\overline{K}^*$) is used. To explain our assertion about the loss of information at the boundary of $\mathcal{V}$, consider the following example involving preferences over acts using only two prizes, **W** and **B**.

Example 6.1. Define act H_E as $H_E(s) = B$ for $s \in E$, and $H_E(s) = W$ otherwise.

Case 1. The agent reports the strict preferences $x\mathbf{B} + (1 - x)\mathbf{W} \prec H_E$ for $0 < x \leq 0.6$ and noncomparability $x\mathbf{B} + (1 - x)\mathbf{W} \sim H_E$ for $0.6 < x \leq 1$.

Case 2. The agent reports the strict preferences $x\mathbf{B} + (1 - x)\mathbf{W} \prec H_E$ for $0 < x \leq 0.6$ and noncomparability $x\mathbf{B} + (1 - x)\mathbf{W} \sim H_E$ for $0.6 \leq x \leq 1$.

The (closed) target set of utilities for H_E is the same in both cases: $\mathcal{T}(H_E) = [0.6, 1]$. However, in the first case the lower bound is not a "candidate utility" (Definition 24), whereas in the second case it is. Therefore, by our construction, the representation for the agent's strict preferences in Case 1 is the set $\mathcal{P} = \{P{:}\, 0.6 < P(E) \leq 1\}$ and in the second case it is the closed set $\overline{\mathcal{P}} = \{P{:}\, 0.6 \leq P(E) \leq 1\}$.

By contrast, the Giron and Rios (1980) theory uses only a weak preference relation, $\precsim$. In both Cases 1 and 2 their theory entails

$$x\mathbf{B} + (1-x)\mathbf{W} \precsim H_E \quad \text{for } 0 < x \leq 0.6 \quad \text{and}$$

$$x\mathbf{B} + (1-x)\mathbf{W} \sim H_E \quad \text{for } 0.6 < x \leq 1.$$

[The weak preferences of Case 2 result from applying Giron and Rios' Axiom A5 (continuity). In particular, that axiom yields the conclusion $0.6\mathbf{B} + 0.4\mathbf{W} \precsim H_E$ from the premise $x\mathbf{B} + (1 - x)\mathbf{W} \precsim H_E$ $(0 < x < 0.6)$.] In their notation, the weak-preference relation does not distinguish between these two cases: where $K^* = \{p{:}\, 0.6 < p(E) \leq 1\}$) and $\overline{K}^* = \{p{:}\, 0.6 \leq p(E) \leq 1\}$, though our strict-preference does.

As a second reason for bypassing proof techniques using "separating hyperplanes," though the set $\mathcal{V}$ is convex, not so for the set of "linear" utilities that admit a decomposition as subjective (almost) state-independent utilities. We do not see how to show the existence of the set of agreeing probability/(almost) state-independent utility pairs, corresponding to Theorem 6, without exploring details about the surface of $\mathcal{V}$. In light of Theorem 5, we have no right to assume those surfaces are closed. By contrast, when $\mathcal{V}$ has sufficiently many closed faces, Lemma 4.3 gives a representation of $\prec$ in terms of sets of probability/state-independent utility pairs. Thus, we feel justified in our choice of an "inductive" proof technique by the increased content to the theorems reached.

APPENDIX

Proofs of Selected Results

A. RESULTS FROM SECTION II. Corollaries 2.1, 2.2 and 2.3 have elementary proofs.

Proof of Corollary 2.4. From left to right, argue indirectly and apply Corollary 2.3 for a contradiction. In the other direction, assume that $xH_1 + (1 - x)H \approx xH_2 + (1 - x)H$ for some $1 \geq x > 0$ and some lottery H. Also, assume $yH_1 + (1 - y)H_3 \prec (\succ)H_4$, with $1 \geq y > 0$. Then by HL Axiom 2, $\forall\ (1 \geq z > 0)$, $z(yH_1 + (1 - y)H_3) + (1 - z)H \prec (\succ)zH_4 + (1 - z)H$. Let $z\ x/(x + y - xy) > 0$ and then $1 > z$ (unless $x = y = 1$, in which case we are done). Last, define the term $w = y/(x + y - xy)$ and we know that $0 < w < 1$. Thus, we have $w(xH_1 + (1 - x)H) + (1 - w)H_3 \prec (\succ)zH_4 + (1 - z)H$. Since $(xH_1 + (1 - x)H) \approx (xH_2 + (1 - x)H)$, by Corollary 2.3 also we have $w(xH_2 + (1 - x)H) + (1 - w)H_3 \prec (\succ)zH_4 + (1 - z)H$. Again by HL Axiom 2, we may cancel the common factor $(1 - z)H$ from both sides, to yield $yH_2 + (1 - y)H_3 \prec (\succ)H_4$. By Corollary 2.3, $H_1 \approx H_2$. □

Proof of Corollary 2.5. Assume the premises and, by Corollary 2.3, show using HL Axiom 3 that $\forall\ (0 < x \leq 1, H_a, H_b)$ whenever $xM + (1 - x)H_a \prec (\succ)H_b$, then $xH + (1 - x)H_a \prec (\succ)H_b$. □

Lemma 2.1 has a straightforward proof.

Proof of Lemma 2.2. Without loss of generality, as L is discrete, write L as $\{P(r_n): P(r_i) \geq P(r_j) \text{ for } i \leq j\}$. Let

$$x_n = \sum_{i=1}^{n} P(r_i)$$

and define the simple lotteries $L_n = \{(1/x_n)P(r_i): i = 1, \ldots, n\}$. Then $\{H_{L_n}\} \Rightarrow H_L$. If for each $r_n \in \text{supp}(L)$, $\mathbf{r_n} \prec \mathbf{r}$, then by HL Axioms 1 and 2, $H_{L_n} \prec \mathbf{r}$ (or, if $\mathbf{r} \prec \mathbf{r}_n$, then $\mathbf{r} \prec H_{L_n}$). Thus, we have the desired conclusion: not $(\mathbf{r} \prec H_L)$ [or, alternatively, not $(H_L \prec \mathbf{r})$]. For if not, by HL Axiom 3 and transitivity of $\prec$, $(H_L \prec H_L)$. □

Proof of Corollary 2.6. On the contrary, if a utility V for acts is unbounded, then there are acts with infinite utility.

Just consider the discrete horse lottery H_∞, where $H_\infty(s_j) = \Sigma_i (2^{-i})P_{i,j}$ for a sequence of acts H_i such that $V(H_i) \geq 2^i$ $(i = 1, \ldots)$. Then, by the expected utility property, $V(H_\infty) = \infty$. The existence of such acts leads to a contradiction with the first two axioms, just as in the St. Petersburg paradox. Assume for convenience that $H_1 \prec H_2$. By axiom HL Axiom 2, $0.5H_1 + 0.5H_\infty \prec 0.5H_2 + .05H_\infty$. However, $V(0.5H_1 + 0.5H_\infty) = V(0.5H_2 + 0.5H_\infty) = V(H_\infty) = \infty$, which, if V agrees with $\prec$, entails the contrary result that $0.5H_1 + 0.5H_\infty \sim 0.5H_2 + 0.5H_\infty$. □

Proof of Lemma 2.3. The proof is indirect. Most of the work is done by HL Axiom 3. We present the argument for the case in which $\prec$ fails to be bounded above, using Axiom 3(b). By similar reasoning using Axiom 3(a) instead, the result obtains when $\prec$ fails to be bounded below.

Let $\{H_n: n = 1, \ldots\}$ be an *increasing chain* and suppose it is not bounded above, that is, $\lim_{n\to\infty} \sup\{x: (H_2 \prec xH_1 + (1 - x)H_n)\} = 1$. Choose a subsequence, also denoted by $\{H_n\}$, so that $x_n \geq 1 - 1/n$ and so that $H_2 \prec x_nH_1 + (1 - x_n)H_n$. However, $\{x_nH_1 + (1 - x_n)H_n\} \Rightarrow H_1$. Trivially, the constant sequence $\{H_2\} \Rightarrow H_2$. Also, $H_1 \prec H_2$ by assumption. Then by HL Axiom 3(b), $H_1 \prec H_2$, contradicting HL Axiom 1. □

Proof of Corollary 2.7. If not, then there is an unbounded increasing (or decreasing) $\prec_W$-chain of preferences amongst the set of rewards **R**. By Lemma 2.3, $\prec_W$ does not satisfy both Axioms 1 and 3. □

Proof of Corollary 2.8. Let H_1 and H_2 be a pair of acts which are "called-off" in case **n** does not obtain, that is, $\forall$ $(s \notin \mathbf{n})$, $H_1(s) = H_2(s)$. (Properties of "called-off" acts are examined in Section v.) Define k pairs of acts "called-off" in case s_{j_i} obtains, H_{1_i} and H_{2_i} $(i = 1, \ldots, k)$ as follows: Let l be a lottery. $\forall$ $(s \in s_{j_i})$, $[H_{1_i}(s) = H_1(s)$ and $H_{2_i}(s) = H_2(s)]$; $\forall$ $(s \in \mathbf{n}$ & $s \notin s_{j_i})$, $[H_{1_i}(s) = H_{2_i}(s) = L]$; $\forall$ $(s \notin \mathbf{n})$, $[H_{1_i}(s) = H_{2_i}(s) = H_1(s) = H_2(s)]$. By assumption, each $s_{j_i} \in \mathbf{n}$ is essentially null. Therefore, by iteration of Corollary 2.4 (and transitivity of $\approx$) $H'_1 \approx H'_2$, where $H'_1 = \Sigma_{i=1}^{k}(1/k)H_{1_i}$ and $H'_2 = \Sigma_{i=1}^{k}(1/k)H_{2_i}$. However, $H'_1 = (1/k)H_1 + (k - 1/k)H$ and, likewise, $H'_2 = (1/k)H_2 + (k - 1/k)\ H$ for act H defined by $\forall$ $(s \in \mathbf{n})$, $[H(s) = L]$ and $\forall$ $(s \notin \mathbf{n})$ $[H(s) = H_1(s) = H_2(s)]$. Then, by Corollary 2.4, the desired result obtains, $H_1 \approx H_2$. □

B. PROOF OF THEOREM 2. The extension from $\prec$ to $\prec'$ is given in steps, by adding the two new rewards one at a time. First, extend $\prec$ to a partial order $\prec^*$ on $\mathbf{H}_{\mathbf{R}\cup\{W\}}$, where W is left $\sim^*$-incomparable with elements of $\mathbf{H}_{\mathbf{R}}$. The definition of $\prec^*$ is introduced by a lemma that shows the extension is minimal.

Lemma 2.4. *Suppose H_1, H_2, H'_1 and $H'_2 \in \mathbf{H}_{\mathbf{R}}$ and are related as follows: $H_1(s_j) = x_jL_j + (1 - x_j)L_{1,j}$ and $H_2(s_j) = x_jL_j + (1 - x_j)L_{2,j}$, while $H'_1(s_j) = x_jL'_j + (1 - x_j)L_{1,j}$ and $H'_2(s_j) = x_jL'_j + (1 - x_j)L_{2,j}$. Then $H_1 \prec H_2$ iff $H'_1 \prec H'_2$.*

Proof. The lemma is immediate by HL Axiom 2 and the identity $0.5H_1 + 0.5H'_2 = 0.5H_2 + 0.5H'_1$. □

Now, let $H_i \in \mathbf{H}_{\mathbf{R}\cup\{W\}}$ be written $H_i(s_j) = x_{i,j}W + (1 - x_{i,j})L_{i,j}$, where $L_{i,j} \in \mathbf{H}_{\mathbf{R}}$ is well defined if and only if $x_{i,j} < 1$. Choose a reward $r \in \mathbf{R}$ and let $H^\#_i \in \mathbf{H}_{\mathbf{R}}$ be the act that results by substituting r for W in H_i. Thus, $H^\#_i(s_j) = x_{i,j}r + (1 - x_{i,j})L_{i,j}$. Lemma 2.4 shows this choice is arbitrary and, if $\prec^*$ is to extend $\prec$, it must satisfy the following:

Definition 15 ($\prec^*$). Given $H_1, H_2 \in \mathbf{H}_{\mathbf{R}\cup\{W\}}$, as expressed above, define the preference $\prec^*$ from $\prec$ by $H_1 \prec^* H_2$ iff $x_{1,j} = x_{2,j}$ (for all $s_j \notin \mathbf{n}$) and $H^\#_1 \prec H^\#_2$.

Lemma 2.5. *The order $\prec^*$ is identical with $\prec$ on $\mathbf{H}_{\mathbf{R}}$ and satisfies HL Axioms* 1–3.

Proof. If $H_1, H_2 \in \mathbf{H}_{\mathbf{R}}$, then $x_{1,j} = x_{2,j} = 0$, $H_1 = H^\#_1$, $H_2 = H^\#_2$ and thus $H_1 \prec^* H_2$ iff $H_1 \prec H_2$. Next, we show that $\prec^*$ satisfies the axioms. Consider all $H_1 \in \mathbf{H}_{\mathbf{R}\cup\{W\}}$.

HL Axiom 1 (Irreflexivity). If, on the contrary, for some H_1, $H_1 \prec^* H_1$, then $H^\#_1 \prec H^\#_1$, contradicting the irreflexivity of $\prec$.

Transitivity holds because if $H_1 \prec^* H_2$ and $H_2 \prec^* H_3$, then the corresponding three $H^\#_i$ acts $(i = 1, 2, 3)$ can be written $x_j(r) + (1 - x_j)L_{i,j}$. Since $\prec$ is transitive, $H^\#_i \prec H^\#_3$; thus, $H_1 \prec^* H_3$.

HL Axiom 2 (Independence). $\forall\, (0 < y \leq 1)$, $\forall\, H \in \mathbf{H}_{\mathbf{R}\cup\{W\}}$ $H_1 \prec^* H_2$ iff $x_{1,j} = x_{2,j}$ and $H^\#_1 \prec H^\#_2$ iff $yx_{1,j} + (1 - y)x_{3,j} = yx_{2,j} + (1 - y)x_{3,j}$ and $yH^\#_1 + (1 - y)\, H^\#_3 \prec yH^\#_2 + (1 - y)H^\#_3$ iff $yH_1 + (1 - y)H_3 \prec^* yH_2 + (1 - y)H_3$.

HL Axiom 3. Let $\{H_{1n} \prec^* H_{2n}\}$ be an infinite sequence of $\prec^*$-

preferences where $\{H_{1n}\} \Rightarrow H_1$, $\{H_{2n}\} \Rightarrow H_2$ and assume $H_2 \prec^* H_3$. Thus $H^{\#}_{1n} \prec H^{\#}_{2n}$, where $\{H^{\#}_{1n}\} \Rightarrow H^{\#}_1$ and $\{H^{\#}_{2n}\} \Rightarrow H^{\#}_2$. Since $H_2 \prec^* H_3$, then $H^{\#}_2 \prec H^{\#}_3$. By applying HL Axiom 3 to these $\prec$-preferences, we obtain $H^{\#}_1 \prec H^{\#}_3$. We derive $x_{1,j} = x_{3,j}$ from the equalities $x^n_{1,j} = x^n_{2,j}$ and $x_{2,j} = x_{3,j}$. Therefore, $H_1 \prec^* H_3$. The argument for HL Axiom 3(b) is similar. □

Lemma 2.5 shows, also, that if $\prec'$ is defined on $\mathbf{R} \cup \{W\}$, extends $\prec$ and satisfies the axioms, then it extends $\prec^*$. That is, $\prec^*$ is a minimal extension of $\prec$ to the domain $\mathbf{R} \cup \{W\}$. Next, we extend $\prec^*$ to a preference $\prec_W$ in which **W** serves as a least preferred (worst) act. (The reader is alerted to the fact that, though the partial order $\prec_W$ makes **W** a least preferred act with respect to elements of $\mathbf{H_R}$, it does not guarantee that W is, state by state, a least favorable reward. This feature is addressed in Section IV.1, where we consider state-dependent utilities for partial orders.)

Definition 16 ($\prec_W$). $\forall$ $(H_1, H_2 \in \mathbf{H}_{\mathbf{R}\cup\{W\}})$, $H_1 \prec_W H_2$ iff either (a) $H_1 \prec^* H_2$ or (b) $\exists$ $\{H_n \in \mathbf{H_R}\}$, $\exists$ $\{H_{1n}\} \Rightarrow H_1$, $\exists$ $\{H_{2n}\} \Rightarrow H_2$ and $\exists$ $(q_n\text{: } 0.5 \le q_n < 1)$ with $\lim_{n\to\infty}\{q_n\} = q$, $q < 1$, such that $\forall$ n, $q_nH_{1n} + (1 - q_n)H_n \prec^*$ (or $\approx^*$) $q_nH_{2n} + (1 - q_n)\mathbf{W}$.

Lemma 2.6.

1. *The partial order $\prec_W$ agrees with $\prec$ on $\mathbf{H_R}$.*
2. ***W** bounds $\prec^*$ from below; that is, $\forall$ $(H \in \mathbf{H_R})$, $\mathbf{W} \prec_W H$.*
3. *$\prec_W$ satisfies HL Axioms 1–3.*

Proof.

1. Let H_1 and H_2 belong to $\mathbf{H_R}$. If $H_1 \prec H_2$, then by Lemma 2.4, $H_1 \prec^* H_2$, and by clause (a) in Definition 16, $H_1 \prec_W H_2$. For the converse, if $H_1 \prec_W H_2$, then it is not by clause (b), since $\forall$ $(H_n \in \mathbf{H_R})$ and for all sufficiently large n, as $1 > q \ge 0.5$, $q_nH_{1n} + (1 - q_n)H_n \sim^*$ (and $\not\approx^*$) $q_nH_{2n} + (1 - q_n)\mathbf{W}$. Hence, it must be that clause (a) obtains. So, $H_1 \prec^* H_2$ and $H_1 \prec H_2$.
2. For each $H \in \mathbf{H_R}$, recall that $0.5H + 0.5\mathbf{W} \approx^* 0.5\mathbf{W} + 0.5H$. Then, in Definition 16(b), set $H_1 = \{H_{1n}\} = \mathbf{W}$, $H_2 = \{H_{2n}\} = H$, $H_n = H$ and $q_n = 0.5$. Thus, $\mathbf{W} \prec_W H$.
3. We verify the axioms individually:

HL Axiom 1 (Irreflexivity). On the contrary, suppose that $H_1 \prec_W H_1$. There are two cases to consider. If this $\prec_W$-relation results by Definition 16(a), then $H_1 \prec^* H_1$, contradicting Lemma 2.4. If we hypothesize that $H_1 \prec_W H_1$ results by Definitions 16(b), then we derive a contradiction as follows. Let $H'_{1n} = q_n H_{1n} + (1 - q_n)H_n$ and $H'_{2n} = q_n H_{2n} + (1 - q_n)\mathbf{W}$. A necessary condition for the relation $H'_{1n} \prec^*$ (or $\approx^*$)H'_{2n} to obtain is that $x_{H'_{1n}} = x_{H'_{2n}}$ (except on essentially null states). However, as both $\{H_{1n}\} \Rightarrow H_1$ and $\{H_{2n}\} \Rightarrow H_1$, while $\lim_{n\to\infty} q_n = q < 1$, this is impossible. That is, $\lim_{n\to\infty} x_{H'_{1n}} = \lim_{n\to\infty} q_n x_{H_{1n}} = 0$, while $x_{H'_{2n}} \geq (1 - q_n)$; hence, for all sufficiently large n, $x_{H'_{1n}} < x_{H'_{2n}}$.

HL Axiom 1 (Transitivity). Suppose both $H_1 \prec_W H_2$ and $H_2 \prec_W H_3$ obtain. There are four cases to consider depending upon which clause in Definition 16 is used for each $\prec_W$-preference. The argument is most complicated when Definition 16(b) is used twice; hence, we give the details for this case only. Thus, assume there are two sequences of *-relations:

$$H'_{1n} \prec^* (\text{or} \approx^*) H'_{2n} \quad \text{and} \quad H''_{2n} \prec^* (\text{or} \approx^*) H'_{3n}, \tag{B1}$$

where

$$H'_{1n} = q_n H_{1n} + (1 - q_n)H_n, \quad H'_{2n} = q_n H_{2n} + (1 - q_n)W,$$

$$H''_{2n} = q'_n \hat{H}_{2n} + (1 - q'_n)H'_n, \quad H'_{3n} = q'_n H_{3n} + (1 - q'_n)W$$

and where $\{H_{1n}\} \Rightarrow H_1$, $\{H_{2n}\} \Rightarrow H_2$, $\{\hat{H}_{2n}\} \Rightarrow H_2$, $\{H_{3n}\} \Rightarrow H_3$ and $\lim_{n\to\infty} \{q_n\} = q$, $\lim_{n\to\infty} \{q'_n\} = q'$, with $0.5 \leq q, q' < 1$. Then $\forall\ (0 < r_n < 1)$, $r_n H'_{1n} + (1 - r_n)H''_{2n} \prec^*$ (or $\approx^*$) $r_n H'_{2n} + (1 - r_n)H'_{3n}$. This is an $\prec^*$-preference, unless both equations of (B1) are $\approx^*$-indifferences. Choose r_n so that $r_n q_n = (1 - r_n)q'_n$. Since $\{H_{2n}\}$ and $\{\hat{H}_{2n}\}$ both converge to H_2, by HL Axiom 2, cancel the common acts in H'_{2n} and H''_{2n} (also common with acts in H_2). Then apply clause Definition 16(b) to obtain $H_1 \prec_W H_3$.

HL Axiom 2 (Independence). $\forall\ H \in \mathbf{H}_{\mathbf{R}\cup\{\mathbf{W}\}}$, $\forall\ 0 < x \leq 1$:

Case (a): $H_1 \prec^* H_2$ iff $xH_1 + (1 - x)H \prec^* xH_2 + (1 - x)H$ iff $xH_1 + (1 - x)H \prec_W xH_2 + (1 - x)H$.

Case (b)-(i): If $q_n H_{1n} + (1 - q_n)H_n \prec^*$ (or $\approx^*$)$q_n H_{2n} + (1 - q_n)\mathbf{W}$, then $r_n[q_n H_{1n} + (1 - q_n)H_n] + (1 - r_n)H \prec^*$ (or $\approx^*$) $r_n[q_n H_{2n} + (1 - q_n)\mathbf{W}] + (1 - r_n)H$. Write $r_n = x/(q_n + (1 - q_n)x)$. Then $xH_1 + (1 - x)H \prec_W xH_2 + (1 - x)H$ by Definition 16(b).

Case (b)-(ii): Suppose $xH_1 + (1 - x)H \prec_W xH_2 + (1 - x)H$. Let $\{H_{3n}\} \Rightarrow xH_1 + (1 - x)H$ and $\{H_{4n}\} \Rightarrow xH_2 + (1 - x)H$. Assume $q_n H_{3n} + (1 - q_n)H_n \prec^*$ (or $\approx^*$) $q_n H_{4n} + (1 - q_n)\mathbf{W}$. Apply HL Axiom 2 to cancel acts in H_{3n} and H_{4n} common with H. Regroup the remainders to yield a $\prec^*$-relation of the desired form for Definition 16(b): $q'_nH_{1n} + (1 - q'_n)H_n \prec^*$ (or $\approx^*$) $q'_nH_{2n} + (1 - q'_n)\mathbf{W}$, where $\{H_{1n}\} \Rightarrow H_1$ and $\{H_{2n}\} \Rightarrow H_2$. (A simple calculation shows that $\lim_{n\to\infty}\{q'_n\} = q' \geq 0.5$.) Thus $H_1 \prec^* H_2$.

Next, we give the details for HL Axiom 3(a). [Axiom 3(b) follows similarly.]

HL Axiom 3(a) (Archimedes). Assume $H_n \prec_W M_n$ and $M \prec_W N$, where $\{H_n\} \Rightarrow H$ and $\{M_n\} \Rightarrow M$. We are to show that $H \prec_W N$. Again, there are four cases to consider, depending upon how (infinitely many of) the first and the second $\prec_W$-preferences arise through Definition 16. The argument is most complicated in case clause 16(b) is used throughout.

That is, assume $\exists \{R_{n_m}, S_n \in \mathbf{H_R}\}, \exists \{H'_{n_m}\}, \exists \{M'_{n_m}\}, \exists \{M''_n\}$ and $\exists \{N'_n\}$ such that, $\forall n$, as $m \to \infty$, $\{H'_{n_m}\} \Rightarrow H_n$ and $\{M'_{n_m}\} \Rightarrow M_n$, while as $n \to \infty$, $\{M''_n\} \Rightarrow M$ and $\{N'_n\} \Rightarrow N$. Also assume $\exists \{q_{n_m}, s_n \geq 0.5\}$, so $\forall n$, $\lim_{m\to\infty} \{q_{n_m}\} = q_n < 1$ and $\lim_{n\to\infty} \{s_n\} = s < 1$. By Definition 16,

$$q_{n_m}H'_{n_m} + (1 - q_{n_m})R_{n_m} \prec^* (\text{or} \approx^*)\, q_{n_m}M'_{n_m} + (1 - q_{n_m})\mathbf{W},$$
$$S_nM''_n + (1 - s_n)S_n \prec^* (\text{or} \approx^*)\, s_nN'_n + (1 - s_n)\mathbf{W}. \quad \text{(B2)}$$

Since $\{H_n\} \Rightarrow H$ and $\{M_n\} \Rightarrow M$, $\forall n, \exists (m^* = m(n))$ so that, as $n \to \infty$, both $\{H'_{n_{m^*}}\} \Rightarrow H$ and $\{M'_{n_{m^*}}\} \Rightarrow M$. Moreover, we may choose (a subsequence of) these m^* so that $\lim_{n\to\infty} \{q_{n_{m^*}}\} = q$, $0.5 \leq q < 1$. Thus we have

$$q_{n_{m^*}}H'_{n_{m^*}} + (1 - q_{n_{m^*}})R_{n_{m^*}} \prec^* (\text{or} \approx^*) q_{n_{m^*}}M'_{n_{m^*}} + (1 - q_{n_{m^*}})\mathbf{W}. \quad \text{(B3)}$$

An application of the first two axioms to (B2) and (B3) yields

$$x_n[\text{left side(B2)}] + (1 - x_n)[\text{left side(B3)}]$$
$$\prec^* (\text{or} \approx^*)\, x_n[\text{right side(B2)}] + (1 - x_n)[\text{right side(B3)}]. \quad \text{(B4)}$$

Let $x_n = s_n/(s_n + q_{n_{m^*}})$. Then as both $\{M'_{n_{m^*}}\} \Rightarrow M$ and $\{M''_n\} \Rightarrow M$, we may cancel acts common to M on both sides of (B4) to yield $z_nH''_n + (1 - z_n)T_n \prec^*$ (or $\approx^*$) $z_nN'_n + (1 - z_n)\mathbf{W}$, where $\{H''_n\} \Rightarrow H$, $(N''_n) \Rightarrow N$, $T_n \in H_R$ and $\lim_{n\to\infty} \{z_n\} = z = sq/(s + q - sq)$. Last, $0.5 \leq z < 1$ because $0.5 \leq s, q < 1$. Therefore, by Definition 16(b), $H \prec_W N$ as required. □

Finally, Theorem 2 is concluded by repeating this construction in a dualized form: extend the preference $\prec_W$ to $\prec'$ by introducing the act $\mathbf{B}$ and making it most preferred in $\mathbf{H}_{\mathbf{R}\cup\{\mathbf{W}\}}$. □

C. PROOF OF THEOREM 3. We show (by induction) how to extend $\prec$ ($= \prec_0$) while preserving Axioms 2 and 3 over simple lotteries until the desired weak order is achieved. At stage i of the induction, the strategy is to identify a utility v_i for act $\tilde{H}_i \in \mathcal{H}$, where v_i is chosen (arbitrarily) from a (convex) set of target utilities for $\tilde{H}_i$, $\mathcal{T}_i(\tilde{H}_i)$. We create the partially ordered preference $\prec_i$ so that $\tilde{H}_i \approx_i v_iB + (1 - v_i)W$.

Begin with a function $\mathcal{T}$ which provides a set of target "utilities" for all elements of $\mathbf{H_R}$. We use $\mathbf{W}$ and $\mathbf{B}$ as the 0 and 1 of our utility. Assume $\{H_n\} \Rightarrow H$ and $H_n \in \mathbf{H_R}$. For $i = 1$, by Definitions 17 and 18, $v_1^*(H)$ is the lim inf of the quantities x_n for which $H_n \prec x_n\mathbf{B} + (1 - x_n)\mathbf{W}$ and $v_{1*}(H)$ is the lim sup of the quantities x_n for which $x_n\mathbf{B} + (1 - x_n)\mathbf{W} \prec H_n$. The two "utility" bounds, $v_*(H)$ and $v^*(H)$, do not depend upon which sequence $\{H_n\} \Rightarrow H$ is used. We show this for $v_*(H)$. The argument for $v^*(H)$ is the obvious dual.

Claim 1. *Let $\{H_n\} \Rightarrow H$ and $\{H'_n\} \Rightarrow H$. Then $v_*(H)$ is the same for both sequences.*

Proof. Suppose $v_*(H) = \limsup\{x_n: x_nB + (1 - x_n)W \prec H_n\}$. Then we show that $v_*(H) \le \limsup\{x_n: x_nB + (1 - x_n)W \prec H'_n\}$. This suffices, since by symmetry with $\{H'_n\}$, when $v'_*(H) = \limsup\{x_n: x_nB + (1 - x_n)W \prec H'_n\}$, then $v'_*(H) \le \limsup\{x_n: x_nB + (1 - x_n)W \prec H_n\}$; hence, $v_*(H) = v'_*(H)$. Since both sequences $\{H_n\}$ and $\{H'_n\}$ converge to act H, we can write each pair (H_n, H'_n) as the pair $(y_nK_n + (1 - y_n)M_n, y_nK_n + (1 - y_n)M'_n)$, where $\lim_{n\to\infty} y_n = 1$. Assume all but finitely many $y_n < 1$; else we are finished. Acts M_n and M'_n belong to $\mathbf{H_R}$ because H_n and H'_n do. Of course, neither of the two sequences of acts $\{M_n\}$ and $\{M'_n\}$ need be convergent, but $\{K_n\} \Rightarrow H$. For each n, define the act $N_n = y_nK_n + (1 - y_n)\mathbf{W}$. Clearly, $\{N_n\} \Rightarrow H$. It follows from the preference $\mathbf{W} \prec M_n$ that $N_n \prec H_n$ and from the preference $\mathbf{W} \prec M'_n$ that $N_n \prec H'_n$. By hypothesis, there exists a sequence $\{x_n\}$ such that $x_n\mathbf{B} + (1 - x_n)\mathbf{W} \prec H_n$ and $\lim_{n\to\infty}\{x_n\} = v_*(H)$. Let α_n be the maximum of 0 and $(x_n + y_n - 1)$. Since $x_n\mathbf{B} + (1 - x_n)\mathbf{W} \prec y_nK_n + (1 - y_n)\mathbf{B}$, then $\alpha_n\mathbf{B} + (1 - \alpha_n)\mathbf{W} \prec y_nK_n + (1 - y_n)\mathbf{W} = N_n$. Transitivity of $\prec$ yields $\alpha_n\mathbf{B} + (1 - \alpha_n)\mathbf{W} \prec H'_n$. However, as $\lim_{n\to\infty}\{y_n\} = 1$ and $\lim_{n\to\infty}\{x_n\} = v_*(H)$, then $\lim_{n\to\infty}\{\alpha_n\} = v_*(H)$. Thus, $v_*(H) \le \limsup\{x_n: x_n\mathbf{B} + (1 - x_n)\mathbf{W} \prec H'_n\}$. □

Observe that if $v^*(H_1) < v_*(H_2)$, then $H_1 \prec H_2$, by Axiom 3 and the fact that $\mathbf{W} \prec \mathbf{B}$. However, these "utility" bounds are merely sufficient, not necessary, for the $\prec$-preference $H_1 \prec H_2$.

Proof of Lemma 3.1. Note that $xB + (1 - x)W \prec yB + (1 - y)W$ whenever $x < y$.

i. Suppose, on the contrary, that $v^*(H) < v_*(H)$. Then by Corollary 2.5 applied twice over, $v^*(H)B + (1 - v^*(H))W \approx H \approx v_*(H)B + (1 - v_*(H))W$. Since $v^*(H) < v_*(H)$, also $v^*(H)B + (1 - v^*(H))W \prec v_*(H)B + (1 - v_*(H))W$, contradicting the $\approx$-relation between them, as just derived.
ii. This is immediate, by similar reasoning. □

Next, we show that $\prec$ may be extended to $\prec_H$, a strict partial order satisfying HL Axioms 2 and 3, in which $H \approx_H v_H B + (1 - v_H)W$ and where the utility v_H may be any value in the *interior* of the closed target set $\mathcal{T}(H)$. We resolve when an endpoint of the (closed) target set may be a utility afterward.

Definition 20. For $H \in \mathbf{H_R}$, let $v \in \operatorname{int}\mathcal{T}(H)$. [When $\mathcal{T}(H)$ has no interior, when $v_*(H) = v^*(H) = v$, then by Lemma 3.1(ii), $H \approx v\mathbf{B} + (1 - v)\mathbf{W}$. Thus it is appropriate that Definition 20 creates no extension of $\prec$. Then act H already has its "utility" fixed by $\prec$.] Define $\prec_H$ by

$$(H_1 \prec_H H_2) \text{ iff } \exists(0 < x < 1)\,\exists(G, G'),$$
$$xH_1 + (1 - x)G \prec xH_2 + (1 - x)G',$$

where G and G' are *symmetric mixtures* of H and $vB + (1 - v)W$.

Specifically, $\exists\, y$ with $G = yH + (1 - y)[vB + (1 - v)W]$ and $G' = y[vB + (1 - v)W] + (1 - y)H$.

Lemma 3.2. *$\prec_H$ extends $\prec$.*

Proof. Assume $H_1 \prec H_2$. Choose $y = 0.5$ in Definition 20, so $G = G'$. By Axiom 2, $xH_1 + (1 - x)G \prec xH_2 + (1 - x)G'$, so that $(H_1 \prec_H H_2)$. □

Lemma 3.3. *$\succ_H$ satisfies HL Axioms* 1–3.

Proof. We establish Axioms 1–3 separately.

HL Axiom 1 (Irreflexivity). Assume not $(H_1 \prec_H H_1)$. Then $xH_1 + (1 - x)G \prec xH_1 + (1 - x)G'$, which by Axiom 2 yields $G \prec G'$. By Definition 20 and another application of Axiom 2, either $H \prec vB + (1 - v)W$ or else $vB + (1 - v)W \prec H$. Either contradicts the relation $H \sim vB + (1 - v)W$. That follows from the assumption $v_* < v < v^*$.

HL Axiom 1 (Transitivity). Assume $(H_1 \prec_H H_2)$ and $(H_2 \prec_H H_3)$. Then we have

$$xH_1 + (1-x)G \prec xH_2 + (1-x)G' \quad \text{and}$$
$$wH_2 + (1-w)J \prec wH_3 + (1-w)J',$$

where both pairs (G, G') and (J, J') satisfy Definition 20. These equations may be combined to create $\forall\ z,\ z(xH_1 + (1-x)G) + (1-z)(wH_2 + (1-w)J) \prec z(xH_2 + (1-x)G') + (1-z)(wH_3 + (1-w)J')$. Choose $z/(1-z) = w/x$. By HL Axiom 2, we may cancel the common term zxH_2 [$= (1-z)wH_2$] from both sides and recombine the pairs (G, J) and (G', J') to yield $uH_1 + (1-u)K \prec uH_3 + (1-u)K'$. Thus, $H_1 \prec_H H_3$.

HL Axiom 2. Argue that $H_1 \prec_H H_2$ iff $yH_1 + (1-y)G \prec yH_2 + (1-y)G'$ iff $\forall\ (0 < z < 1)$, $z(yH_1 + (1-y)G) + (1-z)H_3 \prec z(yH_2 + (1-y)G') + (1-z)H_3$ iff $(\exists\ w)\ w(xH_1 + (1-x)H_3) + (1-w)G \prec w(xH_2 + (1-x)H_3) + (1-w)G'$ (choose $w = zy/x$) iff $xH_1 + (1-x)H_3 \prec_H xH_2 + (1-x)H_3$.

HL Axiom 3(a). Assume $\forall\ n\ (M_n \prec_H N_n)$, and $(N \prec_H O)$. Then show $(M \prec_H O)$.

1. Thus (a) $(x_nM_n + (1-x_n)G_n) \prec (x_nN_n + (1-x_n)G'_n)$ and also (b) $(yN + (1-y)J) \prec (yO + (1-y)J')$.

As Definition 20 applies to the pairs (G_n, G'_n), (J, J'), we may cancel (by Axiom 2) common terms to create:

2. Either (a) $u_nM_n + (1-u_n)H \prec u_nN_n + (1-u_n)(vB + (1-v)W)$ or (b) $u_nM_n + (1-u_n)(vB + (1-v)W) \prec u_nN_n + (1-u_n)H$.
3. Also, in addition, either (a) $wN + (1-w)H < wO + (1-w)(vB + (1-v)W)$ or (b) $wN + (1-w)(vB + (1-v)W) < wO + (1-w)H$. where $u_n \geq x_n$ and $w \geq y$.

At least one of 2(a) or 2(b) occurs infinitely often. Without loss of generality, assume 2(a) does. Since $v_*(H) < v < v^*(H)$, then $\lim\inf\{u_n\} = u > 0$, in this infinite subsequence. [Only here do we use the fact that v is an interior point of $\mathcal{T}(H)$. See Lemma 3.5 for additional remarks.]

Thus, we are justified in considering a convergent sequence of the form 2(a), also indexed by n, with coefficients converging to $u > 0$. We argue by cases: Assume 3(a) obtains. Using Axiom 2, we mix in the act H to both sides of 2(a) and the act $(vB + (1-v)W)$ to both sides of 3(a), yielding:

$$x[u_n M_n + (1-u_n)H] + (1-x)H \prec$$
$$x[u_n N_n + (1-u_n)(vB + (1-v)W)] + (1-x)H. \quad 4(a)$$

$$z[wN + (1-w)H] + (1-z)(vB + (1-v)W)$$
$$\prec z[wO + (1-w) \times (vB + (1-v)W)]$$
$$+ (1-z)(vB + (1-v)W). \quad 4(b)$$

Choose $xu = zw = q \neq 0$, $(1 - x) = z(1 - w)$. Note all of the following occur: the l.h.s. of 4(a) converges to the act $(qM + (1 - q)H)$; the r.h.s. of 4(b) is the act $(qO + (1 - q)(vB + (1 - v)W))$; the r.h.s. of 4(a) converges to the l.h.s. of 4(b). Then by HL Axiom 3(a), $(qM + (1 - q)H) \prec (qO + (1 - q)(vB + (1 - v)W))$, so by Definition 20, $M \prec_H O$.

In case 3(b) obtains, instead, we modify this argument by mixing the term $(vB + (1 - v)W)$ into 2(a) in case $u < w$ or into 3(b) in case $w < u$, so that (as above) the r.h.s. of 4(a) converges to the l.h.s. of 4(b) and so forth.

HL Axiom 3(b). This is verified just as HL Axiom 3(a) is.

Thus, $\prec_H$ satisfies the axioms. □

To complete our discussion of $\mathcal{T}(H)$, we explain when $\prec_H$ may be created using an endpoint of the target set. To motivate our analysis, consider when a partial order $\prec$ precludes an extension by a particular new preference or indifference.

Definition 21. Say that a *preference* for act H_a over act H_b *is precluded* by the partial order $\prec$, denoted as $\neg(H_b \prec H_a)$, if there exist two convergent sequences of acts $\{H_{a,n}\} \Rightarrow H_a$ and $\{H_{b,n}\} \Rightarrow H_b$, where $(\forall n)\ H_{a,n} \prec H_{b,n}$. [*Note:* $H_a \approx H_b$ or $H_a \prec H_b$ implies the condition $\neg(H_b \prec H_a)$.]

Definition 22. Say that *indifference* between acts H_a and H_b *is precluded* by the partial order $\prec$, denoted as $\neg(H_b \approx H_a)$, if assuming the relation $(H_a \approx H_b)$, the three axioms and the preferences $\prec$ all yield a preference precluded by $\prec$.

Example 3.2. We illustrate $\neg(H_b \approx H_a)$. Suppose $\prec$ satisfies the axioms and the following obtain. Let $H_a \sim H_b$. However, there exist two convergent sequences of acts $\{M_n\} \Rightarrow M$ and $\{N_n\} \Rightarrow N$ and coefficients $\{x_n$: $x_n > 0$, $\lim_{n\to\infty} x_n = 0\}$, where $x_n M_n + (1 - x_n)H_a \prec x_n N_n + (1 - x_n)H_b$. However, for some $y > 0$, $yN + (1 - y)H_a \prec yM + (1 - y)H_b$. Thus, $(H_a \approx H_b)$ entails $M_n \prec N_n$ and $N \prec M$. By Axiom 3, then $M \prec M$, which is a $\prec$-precluded preference since $M \approx M$ obtains whenever $\prec$ satisfies the axioms.

[We sketch a model for these $\prec$-preferences. Let $H_a = v_*B + (1 - v_*)W$ and $H_b = H$. Suppose $\mathcal{V}$ is a set of utilities $\{V_d: 1 > d > 0; V_d(H) = v_* + d$ and $V_d(M) - V_d(N) = d^{0.5}\}$. Consider $\prec_{\mathcal{V}}$, when $\mathcal{T}(H) = [v_*, v^*]$ yet $H_a \prec_{\mathcal{V}} H_b$. Let $\prec$ be as $\prec_{\mathcal{V}}$ except that $H_a \sim H_b$ is forced. $V_d(xM + (1 - x)v_*) \leq V_d(xN + (1 - x)H)$ entails that $x/(1 - x) \leq d^{0.5}$. Since d assumes each value in (0.1), $x = 0$ is a necessary condition for the preferences of Example 3.2.]

Claim 2. *If both $\neg(H_a \prec H_b)$ and $\neg(H_b \prec H_a)$, then $H_a \approx H_b$.*

Proof. When $\neg(H_a \prec H_b)$ and $\neg(H_b \prec H_a)$, then there exist pairs of convergent acts $\{H_{a,n}\}, \{H'_{a,n}\} \Rightarrow H_a$ and $\{H_{b,n}\}, \{H'_{b,n}\} \Rightarrow H_b$, where $(\forall\, n)$ $H_{a,n} \prec H_{b,n}$ and $H'_{b,n} \prec H'_{a,n}$. Then by Corollary 2.5, $H_a \approx H_b$. □

Example 3.2 illustrates that our axioms are not strong enough to ensure the preference $H_a \prec H_b$ when, for example $\neg(H_b \prec H_a)$ and $\neg(H_a \approx H_b)$. It so happens that when both the conditions $\neg(H_b \prec H_a)$ and $\neg(H_a \approx H_b)$ obtain and these two acts do *not* involve the distinguished rewards W and B, then each extension $\prec^*$ of $\prec$ which fixes "utilities" for H_a and H_b (and where $\prec^*$ arises by iteration of Definition 20) has the desired relation $H_a \prec^* H_b$. Our specific problem, however, is with the case when one of these two acts is a utility endpoint of the (closed) target set $\mathcal{T}(H)$, for example, let $H_a = v_*B + (1 - v_*)\mathbf{W}$ and $H_b = H$, as in the model for the $\prec$-preferences sketched in Example 3.2. We require an extra consideration, then, to determine whether, though $v_*B + (1 - v_*)W \sim H$, a combination of $\prec$-preferences arises which prohibits an extension $\prec_H$ of $\prec$ that assigns the "utility" v_* for H.

Our solution is to show how to extend the partial order $\prec$ to a partial order $\prec^+$ that includes all the so-called missing preferences $H_a \prec H_b$.

Definition 23. Define $\prec^+$ from $\prec$ by $H_a \prec^+ H_b$ iff $H_a \prec H_b$ or, both $\neg(H_b \prec H_a)$ and $\neg(H_a \approx H_b)$.

The next lemma establishes (very weak) conditions under which the $\prec^+$-closure of a partial order $\prec$ satisfies all three axioms. In particular, it is not necessary that $\prec$ satisfies HL Axiom 3. [The condition $\neg(H_b \prec$

H_a) is well defined according to Definition 23 even though $\prec$ is known only to satisfy HL Axioms 1 and 2. Specifically, the indifference relation $\approx$ is well defined and satisfies all those properties, e.g., Corollary 2.4, which depend on HL Axioms 1 and 2 alone.]

Lemma 3.4. *The partial order $\prec^+$ satisfies all three axioms provided $\prec$ satisfies the first two axioms,* HL *Axioms* 1 *and* 2, *and provided closure of $\prec$ under all three axioms does not produce a $\prec$-precluded preference.*

Proof. We verify the axioms separately.

HL Axiom 1 (Irreflexivity). Since $H \approx H$ obtains and $\prec$ yields no $\prec$-precluded preference (under the three axioms), no act H satisfies $H \prec^+ H$. That is, $H \approx H$ is not $\prec$-precluded.

HL Axiom 1 (Transitivity). Assume $H_a \prec^+ H_b$ and $H_b \prec^+ H_c$. Each of these $\prec^+$-preferences may arise two ways, according to Definition 23. We examine a general case: $\neg(H_b \prec H_a)$, $\neg(H_c \prec H_b)$, $\neg(H_a \approx H_b)$ and $\neg(H_b \approx H_c)$. We show that (i) $\neg(H_c \prec H_a)$ and (ii) $\neg(H_a \approx H_c)$.

i. From the two assumptions $\neg(H_b \prec H_a)$ and $\neg(H_c \prec H_b)$, we conclude that there exist convergent sequences $\{H_{a,n}\} \Rightarrow H_a$, $\{H_{b,n}\}$ and $\{H'_{b,n}\} \Rightarrow H_b$ and $\{H_{c,n}\} \Rightarrow H_c$, with $(\forall\, n)\ H_{a,n} \prec H_{b,n}$ and $H'_{b,n} \prec H_{c,n}$. Thus, by Axioms 1 and 2, $0.5H_{a,n} + 0.5H'_{b,n} \prec 0.5H_{b,n} + 0.5H_{c,n}$. Using HL Axiom 2 to cancel common terms in $H_{b,n}$ and $H'_{b,n}$, we obtain $\prec$-preferences of the form $H'_{a,n} \prec H'_{c,n}$, where $\{H'_{a,n}\} \Rightarrow H_a$ and $\{H'_{c,n}\} \Rightarrow H_c$. Thus, $\neg(H_c \prec H_a)$.
ii. Assume $H_a \approx H_c$. Because $H_{a,n} \prec H_{b,n}$ we may construct new convergent sequences $\{H'_{c,n}\} \Rightarrow H_c$ and $\{H''_{b,n}\} \Rightarrow H_b$, where $H'_{c,n} \prec H''_{b,n}$.

This exercise is done as follows. From the indifference $H_a \approx H_c$ conclude $(1/n)W + ([n-1]/n)H_c \prec (1/n)B + ([n-1]/n)H_a$. Then $0.5H_{a,n} + 0.5[(1/n)W + ([n-1]/n)H_c] \prec 0.5H_{b,n} + 0.5[(1/n)B + ([n-1]/n)H_a]$. Use Axiom 2 to cancel common terms involving act H_a.

We already have assumed $H'_{b,n} \prec H_{c,n}$. Then, since we are entitled to use HL Axiom 3 in determining the consequences of adopting $H_a \approx H_c$, by Corollary 2.5, from $H_a \approx H_c$ we derive $H_b \approx H_c$. $\neg(H_b \approx H_c)$ means that adding the $\approx$-indifference $H_b \approx H_c$ yields a $\prec$-precluded preference. Thus, adding $H_a \approx H_c$ to $\prec$ yields the same $\prec$-precluded preference. Hence, $\neg(H_a \approx H_c)$.

HL Axiom 2 (Independence). This axiom is easy to verify, since $\prec$ satisfies HL Axioms 1 and 2. We illustrate the argument from right to left. Suppose $xH_a + (1-x)H \prec^+ xH_b + (1-x)H$. We are to show that (i) $\neg(H_b \prec H_a)$ and (ii) $\neg(H_a \approx H_b)$.

i. We know that both $\neg(xH_b + (1-x)H \prec xH_a + (1-x)H)$ and $\neg(xH_a + (1-x)H \approx xH_b + (1-x)H)$. As in previous cases, we may assume existence of convergent sequences $\{H_{1,n}\} \Rightarrow xH_a + (1-x)H)$ and $\{H_{2,n}\} \Rightarrow xH_b + (1-x)H)$, where $H_{1,n} \prec H_{2,n}$. Use HL Axiom 2 to cancel common terms (involving act H) in each pair $H_{1,n}$ and $H_{2,n}$. The results are $\prec$-preferences of the form $H_{a,n} \prec H_{b,n}$, with $\{H_{a,n}\} \Rightarrow H_a$ and $\{H_{b,n}\} \Rightarrow H_b$. Thus, $\neg(H_b \prec H_a)$.

ii. By Corollary 2.4, from the assumption $H_a \approx H_b$ it follows that $xH_a + (1-x)H \approx xH_b + (1-x)H$, which yields a $\prec$-precluded preference as $\neg(xH_a + (1-x)H \approx xH_b + (1-x)H)$.

HL Axiom 3(a). Assume $M_n \prec^+ N_n$ and $N \prec^+ O$, where $\{M_n\} \Rightarrow M$ and $\{N_n\} \Rightarrow N$. We need to show that (a) $\neg(O \prec M)$ and (b) $\neg(M \approx O)$.

(a) Thus $\neg(N_n \prec M_n)$, $\neg(M_n \approx N_n)$, $\neg(O \prec N)$ and $\neg(N \approx O)$. As in previous cases, assume each of these $\prec$-precluded preferences arises from corresponding sequences of $\prec$-preferences. That is, for each n there is a pair of convergent sequences $\lim_{j\to\infty}\{M_{n,j}\} \Rightarrow M_n$ and $\{N_{n,j}\} \Rightarrow N_n$, where $M_{n,j} \prec N_{n,j}$. Also, there is a pair of convergent sequences $\{N'_n\} \Rightarrow N$ and $\{O_n\} \Rightarrow O$, where $N'_n \prec O_n$. Since $\{M_n\} \Rightarrow M$ and $\{N_n\} \Rightarrow N$, for each n we may choose a value j_n so that $\lim_{n\to\infty} \{M_{n,j_n}\} \Rightarrow M$ and $\{N_{n,j_n}\} \Rightarrow N$. Of course, $M_{n,j_n} \prec N_{n,j_n}$. Then, $0.5M_{n,j_n} + 0.5N'_n \prec 0.5N_{n,j_n} + 0.5O_n$. Use HL Axiom 2 to cancel terms common to act N, yielding $\prec$-preferences sufficient for $\neg(O \prec M)$.

(b) If we assume $M \approx O$, then (because $M_{n,j_n} \prec N_{n,j_n}$) there are sequences $\{O'_n \Rightarrow O$ and $\{N''_n\} \Rightarrow N$, with $O'_n \prec N''_n$. Since $N'_n \prec O_n$, using Axiom 3, by Corollary 2.5, then $N \approx O$. However, $\neg(N \approx O)$. Hence, assuming $M \approx O$ entails some $\prec$-precluded preference. Therefore, $\neg(M \approx O)$. HL Axiom 3(b) is demonstrated in the identical fashion. □

In the next definition, based on Lemma 3.4, we indicate whether either endpoint of $J(H)$ is eligible as a utility for H when extending $\prec$ to form $\prec_H$.

Definition 24. Say that $v_*(H)$ is a *candidate utility* for H if $v_*\mathbf{B} + (1 - v_*)\mathbf{W} \sim^+ H$. Likewise, $v^*(H)$ is a *candidate utility* for H if $H \sim^+ v^*\mathbf{B} + (1 - v^*)\mathbf{W}$.

We conclude our discussion of the extension $\prec_H$ for the special case when it is generated by a target set endpoint provided, of course, the endpoint is a candidate utility for H. The idea behind the extension is that as it stands, Definition 20 fails with $v = v_*$ or $v = v^*$ only because the resulting partial order is incomplete with respect to Axiom 3. (See Lemma 3.5, below.) Then, in light of Lemma 3.4, the +-closure (using Definition 23) corrects the omissions. (See Lemma 3.6.)

When extending $\prec$ with a candidate utility, $v = v_*$ or $v = v^*$, that is, using an endpoint of $J(H)$, we define the extension $\prec_H$ in two steps, as follows: Analogous with Definition 20, let G and G' be *symmetric mixtures* of H and $vB + (1 - v)W$.

Definition 25. Define $H_1 \prec_\# H_2$ iff $\exists$ $(0 < x < 1)$ $\exists$ (G, G'), $xH_1 + (1 - x)G \prec xH_2 + (1 - x)G'$, and let $\prec_H$ result by closing $\prec_\#$ using Definition 23, that is, $\prec_H = \prec_\#^+$.

Lemma 3.5. *The partial order $\prec_\#$ extends $\prec$ and satisfies axioms HL Axioms* 1 *and* 2.

Proof. Since v is a candidate utility, $v\mathbf{B} + (1 - v)\mathbf{W} \sim H$. Then, as Definition 25 duplicates Definition 20, the proofs from Lemmas 3.2 and 3.3 apply to show that $\prec_\#$ extends $\prec$ and satisfies the first two axioms. $\square$

Lemma 3.6. *The partial order $\prec_\#^+$ extends $\prec$ and satisfies all three axioms.*

Proof. In light of Lemma 3.5, the result follows by Lemma 3.4 once we show that $\prec_\#$ may be closed under the axioms without generating a $\prec_\#$-precluded preference. Note that $\prec_\#$ extends $\prec$ by some, but not necessarily all, preferences entailed (by the three axioms) from the $\approx$-indifference $H \approx vB + (1 - v)W$. Then, since v is a candidate utility, closing $\prec_\#$ under the three axioms does not lead to a $\prec$-precluded preference. We claim, next, that it does not lead to a $\prec_\#$-precluded preference either. Suppose, on the contrary, it does. Suppose, for example, closing $\prec_\#$ with the axioms results in a relation $H_b \prec_\# H_a$, where also $\neg(H_b \prec_\# H_a)$. The former means that adding $H \approx vB + (1 - v)W$ to the set of $\prec$-preferences entails (by the axioms) that $H_b \prec H_a$. The latter requires that, for two convergent sequences $\{H_{a,n}\}$ and $\{H_{b,n}\}$, $(H_{a,n} \prec_\# H_{b,n})$. Thus, adding $H \approx vB + (1 - v)W$ to the set of $\prec$-preferences entails (by the axioms) that $(H_{a,n} \prec H_{b,n})$. By HL Axiom 3, these lead to a $\prec$-precluded preference $(H_b \prec H_b)$. Then v is not a candidate utility for H with respect to $\prec$, a contradiction. □

Thus, with Definition 20, we have indicated how to extend $\prec$ to $\prec_H$, where act H is assigned a utility v from the interior of its target set $\mathcal{T}(H)$, and with Definition 25, how to extend to $\prec_H$ using a candidate utility endpoint.

We interject two simple but useful results about $\approx_H$-indifferences. The first confirms that the extension $\prec_H$ preserves $\approx$-indifferences. The second shows that the extension $\prec_H$ makes act H $\approx_H$-indifferent with its assigned utility v.

Lemma 3.7. *If $M \approx N$, then $M \approx_H N$.*

Proof. Suppose $M \approx N$ and that $xM + (1-x)H_3 \prec_H H_4$. We are to show that $xN + (1-x)H_3 \prec_H H_4$. By Definition 23, $[y(xM + (1-x)H_3) + (1-y)G] \prec [yH_4 + (1-y)G']$. After rearranging terms, by Corollary 23, $[y(xN + (1-x)H_3) + (1-y)G] \prec [yH_4 + (1-y)G']$, so that $xN + (1-x)H_3 \prec_H H_4$. □

Lemma 3.8. $H \approx_H v\mathbf{B} + (1-v)\mathbf{W}$.

Proof. Since $\mathbf{W} \prec \mathbf{B}$, we have the following:

$$\begin{aligned}&(1/n)W + [(n-1)/n][0.5H + 0.5(vB + (1-v)W)]\\ &\quad\prec (1/n)B + [(n-1)/n][0.5H + 0.5(vB + (1-v)W)].\end{aligned}$$

This equation may be written as $x_nH_n + (1-x_n)(vB + (1-v)W) \prec x_nM_n + (1-x_n)H$, where $\{x_n\} \to 0.5$, $\{H_n\} \Rightarrow H$ and $\{M_n\} \Rightarrow (vB + (1-v)W)$. By Definition 20, $H_n \prec_H M_n$. Similarly, it may be written $x_nM'_n + (1-x_n)(H) \prec x_nH'_n + (1-x_n)(vB + (1-v)W)$, where $\{x_n\} \to 0.5$ $\{H'_n\} \Rightarrow H$ and $\{M'_n\} \Rightarrow (vB + (1-v)W)$. By Definition 20, $M'_n \prec_H H'_n$. Then, by Corollary 2.5, $H \approx_H v\mathbf{B} + (1-v)\mathbf{W}$. □

We iterate Definition 20 (or Definition 25) in a denumerable sequence of extensions of $\prec$.

Definition 26. Define the set $\mathcal{H} = \{\hat{H}_i^k : \hat{H}_i^k(s_j) = r_1$ if $j \neq k$, and $\hat{H}_i^k(s_k) = r_i\}$. Let $\mathbf{r}_1$ denote the constant act that yields reward r_1 in each state, so that $\mathbf{r}_1 \in \mathcal{H}$.

Lemma 3.9. *$\mathcal{H}$ is countable and finite if* **R** *is finite.*

The proof is obvious.

[$\mathcal{H}$ remains countable even when π is a *denumerable* partition. Then it follows from HL Axiom 3 that personal probabilities over π are σ-additive. That is, HL Axiom 3 entails "continuity": $\lim_{n\to\infty} p\{\cap E_n\} = p\{\lim_{n\to\infty} \cap E_n\}$. In the light of Fishburn's (1979), page 139, result Theorem 10.5, we *conjecture* that our central theorems, e.g., Theorems 3 and 6, carry over to countably infinite partitions. However, this is not evident, e.g., our proof of Claim 1 (for Theorem 3) does not apply when π is infinite. Our use of finite partitions avoids mandating σ-additivity of personal probability.]

Hereafter, we enumerate $\mathcal{H}$ with a single subscript *i*. At stage *i* of the induction, $\prec_i$ is obtained by choosing a target utility v_i for act $\tilde{H}_i \in \mathcal{H}$, denoted $V(\tilde{H}_i) = v_i$. Here $v_i \in \mathcal{T}_i(\tilde{H}_i)$ and $\mathcal{T}_i(\cdot)$ identifies sets of target utilities, based on $\prec_{i-1}$. By Lemma 3.7, extensions preserve utilities already assigned, so that all utilities fixed by stage *i* are well defined over stages $j \geq i$. Next, we show that each simple act has its "utility" V determined by a finite subset of $\mathcal{H}$.

Lemma 3.10. *If $H \in \mathbf{H_R}$ is a simple act, then there is a (finite) stage $\prec_m$ such that $\mathcal{T}_m(H)$ is a unit set, that is, by stage $\prec_m$, H is assigned a precise utility $V(H)$.*

Proof. First we verify that V has the expected utility property over elements of $\mathcal{H}$. Consider $\tilde{H}_a, \tilde{H}_b \in \mathcal{H}$. Without loss of generality, let $b = \max\{a, b\}$. Both $\tilde{H}_a$ and $\tilde{H}_b$ have their respective utilities by stage $\prec_b$. That is, $\tilde{H}_a \approx_b v_a\mathbf{B} + (1 - v_a)\mathbf{W}$ and $\tilde{H}_b \approx_b v_b\mathbf{B} + (1 - v_b)\mathbf{W}$. By Corollary 2.4, $x\tilde{H}_a + (1 - x)\tilde{H}_b \approx_b x(v_a\mathbf{B} + (1 - v_a)\mathbf{W}) + (1 - x)(v_b\mathbf{B} + (1 - v_b)\mathbf{W})$. Hence, $V(x\tilde{H}_a + (1 - x)\tilde{H}_b) = xV(\tilde{H}_a) + (1 - x)V(\tilde{H}_b)$.

Next, write $H(s_j) = \Sigma_{i=1}^{k_j} P_j(r_i)$. Define the act $H'_j(s) = H(s)$ if $s = s_j$; otherwise $H'(s) = r_1$. Since H is simple, each H'_j is a finite combination of $\tilde{H}_i \in \mathcal{H}$. Specifically, $H'_j = \Sigma_{i=1}^{k_j} P_j(\tilde{H}^j_i)$, where $\tilde{H}^j_i(s) = r_i$ if $s = s_j$ and $\tilde{H}^j_i(s) = r_1$ otherwise. Observe that $(1/n)H + (n - 1/n)\mathbf{r}_1 = \Sigma(1/n)H'_j$. Thus the utility $V(H)$ is determined once $V(\mathbf{r}_1)$ and the n values $V(H'_j)$ are fixed, all of which occurs after finitely many elements of $\mathcal{H}$ are assigned their utilities. □

We create a weak order $\precsim_\omega$ from the partial orders $\prec_i$ $(i = 1, \ldots)$ using the fact that each $H \in \mathbf{H_R}$ is a limit point of simple horse lotteries. For $H \in \mathbf{H_R}$ consider a sequence $\{H_n\} \Rightarrow H$, where H_n is a simple act. Let $V(H) = \lim_{n\to\infty} V(H_n)$. Then:

Lemma 3.11. *$V(H)$ is well defined.*

Proof. We show that if $\{H_n\} \Rightarrow H$, then $\lim_{n\to\infty} V(H_n)$ exists and is unique. Assume $\{H_n\} \Rightarrow H$ and $\{H'_n\} \Rightarrow H$, where all these acts belong to $\mathbf{H_R}$. Without loss of generality, since the simple acts form a dense subset of $\mathbf{H_R}$ under the topology of pointwise convergence, suppose that each of H_n, H'_n is simple. Then write H_n as $y_nK_n + (1 - y_n)M_n$ and H'_n as $y_nK_n + (1 - y_n)M'_n$, where $\lim_{n\to\infty} y_n = 1$ and each of K_n, M_n and M'_n is a simple act in $\mathbf{H_R}$. By Lemma 3.10, $V(H_n) - V(H'_n) = (1 - y_n)[V(M_n) - V(M'_n)]$. Since $\lim_{n\to\infty} y_n = 1$ and V is in the unit interval $[0, 1]$, $\lim_{n\to\infty} V(H_n) - V(H'_n) = 0$. □

The next lemma establishes that V has the expected utility property for all $H \in \mathbf{H_R}$.

Lemma 3.12. *If H_a, $H_b \in \mathbf{H_R}$, then $V(xH_a + (1 - x)H_b) = xV(H_a) + (1 - x)V(H_b)$.*

Proof. Consider two sequences $\{H_{a,n}\} \Rightarrow H_a$ and $\{H_{b,n}\} \Rightarrow H_a$, where each of $H_{a,n}$ and $H_{b,n}$ is simple and belongs to $\mathbf{H_R}$. Then, for each n, the act $x(H_{a,n}) + (1 - x)H_{b,n}$ is simple and belongs to $\mathbf{H_R}$. It is evident that $\{x(H_{a,n}) + (1 - x)H_{b,n}\} \Rightarrow xH_a + (1 - x)H_b$. By Lemma 3.10, $V(xH_{a,n} + (1 - x)H_{b,n}) = xV(H_{a,n}) + (1 - x)V(H_{b,n})$. Then by Lemma 3.11, $V(xH_a + (1 - x)H_b) = xV(H_a) + (1 - x)V(H_b)$. □

Last, define the weak order $\precsim_\omega$ for $H \in \mathbf{H_R}$ using the utilities fixed by V:

Definition 27. $(H_1 \precsim_\omega H_2)$ iff $\mathbf{V}(H_1) \leq \mathbf{V}(H_2)$.

We complete the proof of Theorem 3:

i. That $\precsim_\omega$ is a weak order over elements of $\mathbf{H_R}$ follows simply by noting that V is real-valued. By Lemma 3.12, it satisfies the independence axiom. The Archimedean axiom also is a simple consequence of Lemmas 3.10 and 3.11, that is, if $\{M_n\} \Rightarrow M$, $\{N_n\} \Rightarrow N$ and

$M_n \prec_\omega N_n$, then $V(M) \leq V(N)$. Next, let H_a and H_b be simple, that is, each with finite support. Suppose $(H_a \prec H_b)$. According to Lemma 3.10, the utilities $V(H_a)$ and $V(H_b)$ are determined by some stage k of the induction, where k is the maximum index of the (finitely many) elements of $\mathcal{H}$ in the combined supports of H_a and H_b. Lemma 3.2 establishes that $\prec_k$ extends $\prec$. Then $(L_1 \prec_k L_2)$ and thus $V(H_a) < V(H_b)$. Therefore, $\precsim_\omega$ extends $\prec$ for simple lotteries.

ii. We argue that V almost agrees with $\prec$, that is, if $(H_1 \prec H_2)$, then $(H_1 \precsim_V H_2)$. Here is a simple lemma about the changing endpoints of target sets which completes the theorem.

Lemma 3.13. *For every act $H \in \mathbf{H_R}$ and stage $j = 2, \ldots,$* (i) $v_{j-1*}(H) \leq v_{j*}(H) \leq v_j^*(H) \leq v_{j-1}^*(H)$ *and* (ii) $\lim_{j\to\infty} v_{j*}(H) = v_j^*(H) = V(H)$.

Proof.

i. Since $\prec_j$ extends $\prec_{j-1}$, any sequence of $j-1$ stage preferences $H_n \prec_{j-1} x_nB + (1 - x_n)W$ also obtain at stage j. Thus, by Definitions 20 and 25 and Lemma 3.2(i), $v_{j-1*}(H) \leq v_{j*}(H) \leq v_j^*(H) \leq v_{j-1}^*(H)$.

ii. For each act $\tilde{H}_i \in \mathcal{H}$, $\forall\ (j > i)$, $v_{j*}(\tilde{H}_i) = v_j^*(\tilde{H}_i) = V(\tilde{H}_i) = v_i$. Hence, (ii) is obvious for all simple lotteries. Assume H is not simple. It is easy to find a convergent sequence of simple acts in $\mathbf{H_R}$, $\{K_n\} \Rightarrow H$, where $v_{n*}(K_n) = v_n^*(K_n) = V(K_n)$ and $H = y_nK_n + (1 - y_n)M_n$. The sequence of acts M_n, though elements of $\mathbf{H_R}$, need not converge. Since H is not simple, $y_n < 1$. Then, $y_nK_n + (1 - y_n)\mathbf{W} \prec y_nK_n + (1 - y_n)M_n \prec y_nK_n + (1 - y_n)\mathbf{B}$. As each $\prec_n$ extends $\prec$, we have $y_nV(K_n) < v_{n*}(H) \leq v_n^*(H) < y_nV(K_n) + (1 - y_n)$. However, $\lim_{n\to\infty} y_n = 1$ and, by Lemma 3.8, $\lim_{n\to\infty} V(K_n) = V(H)$. Thus, $\lim_{j\to\infty} v_{j*}(H) = \lim_{j\to\infty} v_j^*(H) = V(H)$. □

Finally, if $(H_1 \prec H_2)$, since for each n, $\prec_n$ extends $\prec$, we have that $v_{n*}(H_1) \leq v_n^*(H_2)$. Then by Lemma 3.13, $V(H_1) \leq V(H_2)$. □

D. OTHER RESULTS FROM SECTION III

Proof of Corollary 3.1. The extensions $\prec_i$ created in Theorem 3 rely on the existence at stage $i-1$ of a nonempty target set $\mathcal{T}_i(\tilde{H}_i)$, only for the acts $\tilde{H}_i \in \mathcal{H}$. However, $\mathcal{T}_i(\cdot)$ is defined on all of $\mathbf{H_R}$, including the nonsimple acts. Hence, we can amend the sequence of extensions of $\prec$

to fix utilities for any countable set of acts, in addition to fixing utilities for each element of $\mathcal{H}$. Just modify the argument of Theorem 3 to assign utilities to the countable set $\mathcal{H} \cup \mathcal{B}$. □

In connection with Example 3.1, for instance, we can introduce acts H_a and H_b into a well ordering of $\mathcal{H}$, for example, $\{\tilde{H}_1, H_a, \tilde{H}_2, H_b, \tilde{H}_3, \ldots\}$, so that by stage 4 of the sequence of extensions, $k_1 = V(H_a) < V(H_b) = k_2$, which precludes the undesired limit stage in which $V(r_i) = 0.25$ $(i = 1, \ldots)$.

Theorem 4 is easily demonstrated.

Proof of Theorem 5. That $\phi \neq \mathcal{V} \subseteq \mathcal{Z}^S$ is part (i) of Theorem 4. For the converse, argue indirectly. If $Z \in \mathcal{Z}^S / \mathcal{V}$ then let $\tilde{H}_k$ be the first element of $\mathcal{H}$ (that is, let k be the least integer) for which $Z(\tilde{H}_k) \notin \mathcal{T}_k(\tilde{H}_k)$, even though $v_1 = Z(\tilde{H}_1), \ldots, v_{k-1} = Z(\tilde{H}_{k-1})$ for acts $\tilde{H}_1, \ldots, \tilde{H}_{k-1}$. Then Z agrees with $\prec_{k-1}$, since $\prec_{k-1}$ is the result of extending $\prec$ by the conditions $\tilde{H}_i \approx v_i B + (1 - v_i)W$ $(i = 1, \ldots, k-1)$. That is, expand each $\prec_{k-1}$-preference into a $\prec$-preference. The former follows from the latter by adding a set of $k - 1$ assumptions $\{\tilde{H}_i \approx v_i B + (1 - v_i)W: (i = 1, \ldots, k-1)\}$ to $\prec$. But these $k - 1$ conditions are satisfied under Z, and Z agrees with $\prec$ on simple acts. Hence, it must be that either $Z(\tilde{H}_k) = v_k = v_{k*}(\tilde{H}_k)$ and $\mathcal{T}_k(\tilde{H}_k)$ is open at the lower end or else $Z(\tilde{H}_k) = v_k = v_k^*(\tilde{H}_k)$ and $\mathcal{T}_k(\tilde{H}_k)$ is open at the upper end. However, if the target set is open and if an endpoint v_k of $\mathcal{T}_k(\tilde{H}_k)$ is not a candidate utility for $\tilde{H}_k$, then adding $\tilde{H}_k \approx v_k B + (1 - v_k)W$ to $\prec_{k-1}$ produces a $\prec_{k-1}$-precluded preference. Since Z agrees with $\prec_{k-1}$, Z does not agree with any $\prec_{k-1}$-precluded preference. Thus, Z cannot assign act $\tilde{H}_k$ the utility v_k, which contradicts the assumption $Z(\tilde{H}_k) = v_k$. □

E. RESULTS FROM SECTION IV. The proof of Lemma 4.1 is immediate after Theorem 13.1 of Fishburn (1979).

Proof of Lemma 4.2. Recall the strict preferences $\mathbf{W} \prec H \prec \mathbf{B}$, whenever $W, B \notin \text{supp}(H)$. Hence, for each V, we may standardize the (expected) utility of act $\mathbf{W}$ as 0 and the (expected) utility of act $\mathbf{B}$ as 1, where all other acts (not involving W and B) have (expected) cardinal utilities in the open interval (0, 1). Next, define a set of simple, called-off acts $\{H_{i,j} \in \mathbf{Hs}_j\}$, which yield the lottery outcome $L_i \in \mathbf{L}_{\mathbf{R}\text{-}\{W,B\}}$ in state s_j and outcome W in all other states. In keeping with this notation, let $H_{w,j} = W$ and let $H_{B,j}$ be the $\mathbf{Hs}_j$ act with outcome B in state s_j.

Recall, for each j, $\lim_{m\to\infty} \{H^i_{m,j}\} \to H_{i,j}$ and $H^w_{m,j} = H_{w,j} = \mathbf{W}$. Then, whenever $H_{L_i} < H_{L_k}$ (by HL Axiom 5), $\mathbf{W} \prec H^i_{mj} \prec H^k_{mj} \prec H^B_{mj}$. Hence, by the Archimedean HL Axiom 3 (as in Lemma 2.3), we have the restriction $\neg(H_{B,j} \prec H_{k,j} \prec H_{i,j} \prec \mathbf{W})$. Moreover, this constraint obtains also for each extension of $\prec$, including all the limit extensions $\precsim_V$ since these $\prec$-preferences involve simple acts. Then, for each V, $\mathbf{W} \precsim_V H_{k,j} \precsim_V H_{i,j} \precsim_V H_{B,j}$. Trivially, either $\mathbf{W} \approx_V H_{B,j}$ or else $\mathbf{W} \prec_V H_{B,j}$. The upshot is that, for each V, one of two circumstances obtains:

Case 1. If $\mathbf{W} \approx_V H_{B,j}$, $H_{a,j} \approx_V H_{\beta,j}$ and s_j is null under $\precsim_V$, so $p(s_j) = 0$.

Case 2. If $\mathbf{W} \prec_V H_{B,j}$, then s_j is V-nonnull and for each representation of V as an expected, state-dependent utility [in accord with condition (4.1)], $U_j(W) \leq U_j(L_i) \leq U_j(L_k) \leq U_j(B)$, with at least one of the outside inequalities strict. However, since the U_j are defined only up to a similarity transformation, without loss of generality choose $U_j(W) = 0$ and $U_j(B) = 1$ and rescale p accordingly. □

Proof of Lemma 4.3. Without loss of generality (Corollary 3.3), let the denumerable sequence $\mathcal{H} = \{H_i\}$ of simple horse lotteries, used to create the set $\mathcal{V}$ of extensions for $\prec$, take $\{H_{r_1}, \ldots, H_{r_n}\}$ as its initial segment: the constant acts that award r_i in each state. Suppose the interval $\mathcal{T}_1(r_1)$ is *not* open, for example, $\mathcal{T}_1(r_1) = [v_{1*}, v_1^*)$. Then $0 < v_{1*}$. Extend $\prec$ according to the condition $H_{r_1} \approx_1 v_{1*}\,\mathbf{B} + (1 - v_{1*})\mathbf{W}$. That is (by Definition 2.3), $H_1 \prec_1 H_2$ iff $xH_1 + (1 - x)G_1 \prec xH_2 + (1 - x)G_1'$, where G_1 and G_1' are constant acts, symmetric mixtures of outcomes r_1 and $v_{1*}\,B + (1 - v_{1*})W$.

We show that each $V \in \mathcal{V}$ which extends $\prec_1$ (where V is standardly represented by the set of pairs $\{(p, U_j)\}$ according to condition (4.1)) carries only state-independent utilities for r_1. That is, for each such U_j, $U_j(r_1) = v_{1*}$ if s_j is p-nonnull. To verify this claim, define act H_j as follows:

$$H_{-\varepsilon,j}(s_j) = (v_{1*} - \varepsilon)B + (1 - [v_{1*} - \varepsilon])W \quad \text{and}$$

$$H_{-\varepsilon,j}(s) = W \quad \text{for } s \notin s_j.$$

If state s_j is not $\prec$-potentially null then, since v_{1*} is the lower bound of $\mathcal{T}_1(r_1)$, by HL Axiom 4, we have $\forall\,(v_{1*} > \varepsilon > 0)$, $H_{-\varepsilon,j} \prec H_{1,j}$. [Recall, $H_{1,j}(s_j) = r_1$ and $H_{1,j}(s) = W$ when $s \notin s_j$.] By the Archimedean condition HL Axiom 3, letting $\varepsilon \to 0$, we find that these $\prec$-preferences create the constraint $\neg(H_{1,j} \prec H_{\varepsilon=0,j})$, which applies also to each extension of $\prec$. Thus, if s_j is not $\prec$-potentially null, each V which extends $\prec_1$ has v_{1*} as a lower bound on the state-dependent utility $U_j(r_1)$. Likewise, by appeal to HL Axiom 5 in case s_j is not $\prec$-potentially null, it follows that $\neg(H_{1,j} \prec H_{\varepsilon=0,j})$ and this applies also to all extensions of $\prec$. So, again, v_{1*} is a lower bound on the state-dependent utility $U_j(r_1)$ for cases where s_j is $\prec$-potentially null but V-nonnull and $\precsim_V$ extends $\prec$ (on simple acts). (*Note*: Here we use axiom HL Axiom 5 to regulate the state-dependent utility of lotteries in $\prec$-potentially null states.) Because $V(r_1) = v_{1*}$ for each $\precsim_V$ that extends $\prec_1$ on simple acts, v_{1*} also is an upper bound on all such V-nonnull state-dependent utilities $U_j(r_1)$. This is so because $v_{1*} = V(r_1)$ is the p-expectation of $U_j(r_1)$. Hence, each $\precsim_V$ that so extends $\prec_1$ assigns to reward r_1 the state-independent utility v_{1*}.

Next, assume that $J_2(r_2)$ is not an open interval, for example, let $J_2(r_2) = (v_{2*}, v_2^*]$, and we know $v_2^* < 1$. Thus, $H_{r_2} \prec_1 (v_2^* + \varepsilon)\mathbf{B} + (1 - [v_2^* + \varepsilon])\mathbf{W}$. Extend $\prec_1$ to $\prec_2$ by introducing the $\approx_2$-condition $H_{r_2} \approx_2 v_2^*\mathbf{B} + (1 - v_2^*)\mathbf{W}$. That is, define $\prec_2$ by $H_1 \prec_2 H_2$ iff $xH_1 + (1-x)G_2 \prec_1 xH_2 + (1-x)G_2'$, where G_2 and G_2' are constant horse lotteries, which are symmetric mixtures of acts H_{r_2} and $v_2^*\mathbf{B} + (1 - v_2^*)\mathbf{W}$.

To see that all $\precsim_V$-extensions of $\prec_2$ impose a state-*independent* utility on r_2, that is, to show $U_j(r_2) = v_2^*$, it suffices to demonstrate that $v_2^*B + (1 - v_2^*)W$ serves as an upper utility bound for r_2 over all $\prec_1$, s_j-called-off preferences, called-off if s_j fails. In other words, we are to establish that, for each state s_j, the constraint $\neg(H_{v_2^*+\varepsilon,j} \prec_1 H_{2,j})$ applies to $\prec_1$ and its extensions. Then, by the reasoning we used above, since $V(r_2) = v_2^*$ for all $\precsim_V$ which extend $\prec_2$ (on simple acts), v_2^* also is a lower utility bound for each state-dependent utility $U_j(r_2)$, and thus $U_j(r_2)$ is state-independent. That is, since $V(r_2) = v_2^*$ is the p-expectation of quantities, none of which is greater than v_2^*, then $U_j(r_2) = v_2^*$ if $P(s_j) > 0$.

To establish that v_2^* is such a state-independent upper bound, expand each of the relevant $\prec_1$-preferences, to wit, $\forall\ (1 - v_2^* > \varepsilon > 0)$ expand $H_{r_2} \prec_1 (v_2^* + \varepsilon)\mathbf{B} + (1 - [v_2^* + \varepsilon])\mathbf{W}$, into its respective $\prec$-preference: $\exists\, x_\varepsilon > 0, \exists\, (G_{1\varepsilon}, G_{1\varepsilon}'), x_\varepsilon H_{r_2} + (1 - x_\varepsilon)G_{1\varepsilon} \prec x_\varepsilon[(v_2^* + \varepsilon)\mathbf{B} + (1 - [v_2^* + \varepsilon])\mathbf{W}] + (1 - x_\varepsilon)G_{1\varepsilon}'$.

Each pair $(G_{1\varepsilon}, G_{1\varepsilon}')$ is a symmetric mixture of acts H_{r_1} and $v_{1*}\,\mathbf{B} + (1 - v_{1*})\mathbf{W}$. These $\prec$-preferences are between constant horse lottery acts. By appeal to HL Axiom 4 in case s_j is not $\prec$-potentially null, or by appeal to HL Axiom 5 in case s_j is $\prec$-potentially null, we arrive at a constraint for called-off acts involving the two lottery outcomes $x_\varepsilon r_2 + (1 - x_\varepsilon)G_{1\varepsilon}$ and $x_\varepsilon[(v_2^* + \varepsilon)B + (1 - [v_2^* + \varepsilon])W] + (1 - x_\varepsilon)G_{1\varepsilon}'$. Specifically, we obtain the restriction $\neg(H_{x+\varepsilon,j} \prec H_{x_2,j})$ – a constraint on all extensions of $\prec$ – where

$$H_{x+\varepsilon,j}(s_j) = x_\varepsilon[(v_2^* + \varepsilon)B + (1 - [v_2^* + \varepsilon])W] + (1 - x_\varepsilon)G'_{1\varepsilon} \quad \text{and}$$
$$H_{x+\varepsilon,j}(s) = W \quad \text{if } s \notin s_j,$$

$$H_{x2,j}(s_j) = x_\varepsilon r_2 + (1 - x_\varepsilon)G_{1\varepsilon} \quad \text{and} \quad H_{x2,j}(s) = W \quad \text{if } s \notin s_j.$$

However, each $\precsim_V$ extension of $\prec_1$ (on simple acts) assigns to r_1 the state-independent utility v_{1*}. Thus, each extension assigns $G_{1\varepsilon}$ and $G'_{1\varepsilon}$ this same state-independent utility v_{1*}. Then, as the constraint $\neg(H_{x+\varepsilon,j} \prec_1 H_{x_2,j})$ obtains, so too does the constraint which results at the limit, when $\varepsilon = 0$, and terms $G_{1\varepsilon}$ and $G'_{1\varepsilon}$ are canceled according to HL Axiom 2. Hence, each $\precsim_V$ extension of $\prec_1$ has the quantity v_2^* as an upper bound on the state-dependent utility $U_j(r_2)$ of r_2, provided s_j is not null under $\precsim_V$. Therefore, since V is a weighted average of U_j values, $U_j(r_2) = v_2^*$ for each $\precsim_V$ that extends $\prec_2$ on simple acts.

Proceed this way through the first n stages in the extension of $\prec$ (using Theorem 3), by choosing for the ith stage either the condition $H_{r_i} \approx_i v_{i*}\mathbf{B} + (1 - v_{i*})\mathbf{W}$ or the condition $H_{r_i} \approx_i v_i^*\mathbf{B} + (1 - v_i^*)\mathbf{W}$, as $\mathcal{T}_i(r_i)$ is closed below or above (respectively). Then the set $\mathcal{V}'$ of extensions for $\prec_n$ provides the requisite subset of $\mathcal{V}$. [Note: $\mathcal{V}'$ may fail to be convex when $\mathcal{T}_i(r_i)$ is a closed interval, as in the example for Theorem 1. Then either endpoint may be chosen, but not values in between.] □

F. PROOF OF THEOREM 6. The proof of Theorem 6(i) is based on the idea of the proof of Lemma 4.3. The argument is by induction on the number of rewards, that is, on the length of the initial segment of $\{r_1, r_2, \ldots\}$. The method is a straightforward epsilon–delta technique of fixing the degree of state-dependence to be tolerated and then choosing target set values sufficiently close to a boundary of the target sets to force agreement with the allowed tolerance for state-dependent utilities.

The proof of Theorem 6(ii) follows the argument of Corollary 3.1; that is, use the countable set $\{\mathcal{R} \cup \mathcal{B}\}$ in forming the extensions of $\prec$, subject to the following modification in the ordering of $\{\mathcal{R} \cup \mathcal{B}\}$: Fix k, which determines the initial segment of $\mathcal{R}$, $\{r_1, \ldots, r_k\}$, over which the almost state-independent utilities are to be provided. Given a nonsimple act $H \in \mathcal{B}$, insert it into the sequence of extensions based on $\mathcal{H}$ only after these k-many rewards have been assigned their utilities. This method ensures that assigning utilities to the nonsimple acts in $\mathcal{B}$ does not interfere with using the boundary regions of the target sets of the

k-many rewards, $\{r_1, \ldots, r_k\}$, to locate their almost state-independent utilities. For interesting discussion of this point, see Section 5 of Nau (1993). □

Two remarks help to explain the content of Theorem 6. First, in light of Example 4.1, it may be that for each $\varepsilon > 0$, $\prec$ admits an almost state-independent utility, but (corresponding to $\varepsilon = 0$) there is no agreeing probability/state-independent utility pair in the limit. That is, the limit (as $\varepsilon \to 0$) of the (nested) sets of agreeing, almost state-independent utilities is empty. Second, Definition 31 requires only that $\prec$ admit almost state-independent utilities for each finite set of n-many rewards. Obviously, by increasing n, we can form sequences of (nested) sets of probability/utility pairs. However, Definition 31 does not provide for an almost state-independent utility covering infinitely many rewards simultaneously. We do not yet know whether, given our five axioms, there exists a nonempty limit (as $n \to \infty$) to these nested sets.

G. RESULTS FROM SECTION V

Proof of Lemma 5.1.

i. By HL Axiom 2, $H_1 \prec H_2$ iff $0.5H_1 + 0.5H'_1 \prec 0.5H_2 + 0.5H'_1$. Regrouping terms on the r.h.s. of the second $\prec$ relation, we obtain $H_1 \prec H_2$ iff $0.5H_1 + 0.5H'_1 \prec 0.5H_1 + 0.5H'_2$. Another application of HL Axiom 2 yields the desired result: $H_1 \prec H_2$ iff $H'_1 \prec H'_2$.
ii. Suppose $H_1 \approx H_2$. By Corollary 2.3, it suffices to show that $xH'_1 + (1-x)H_3 \prec H_4$ iff $xH'_2 + (1-x)H_3 \prec H_4$. By HL Axiom 2, $xH'_1 + (1-x)H_3 \prec H_4$ iff $z[xH'_1 + (1-x)H_3] + (1-z)H_1 \prec zH_4 + (1-z)H_1$ $(0 < z \leq 1)$. Since $H_1 \approx H_2$, by Corollary 2.3, substituting H_2 for H_1 on the l.h.s., the biconditional reads: iff $z[xH'_1 + (1-x)H_3] + (1-z)H_2 \prec zH_4 + (1-z)H_1$. Let $zx = 1 - z$, that is, $z = (1+x)^{-1}$. Then regrouping terms in H'_1 and H_2, the biconditional reads: iff $z[xH'_2 + (1-x)H_3] + (1-z)H_1 \prec zH_4 + (1-z)H_1$. Another application of HL Axiom 2 yields the desired result. □

Proof of Theorem 8. Part (i) is immediate as $\mathbf{H_e}$ is a subset of $\mathbf{H_R}$. Specifically, if a weak order $\precsim_V$ (of Theorem 3) agrees with $\prec$, it agrees with $\prec_e$. That is, consider the e-called-off family $\mathbf{H_e}$, where $H(s) = W$ if $s \notin e$ and $\prec_e$ is the restriction of $\prec$ to $\mathbf{H_e}$. Let H_1 and H_2 be simple acts that belong to $\mathbf{H_e}$. If $H_1 \prec_e H_2$, then $H_1 \prec H_2$ and therefore $V(H_1) \prec V(H_2)$. Let the expected utility V be given by the probability/(state-dependent) utility pair (p, U_j). As $U_j(W) = 0$ and $H_i(s) = W$ for $s \notin e$ $(i = 1, 2)$, then $\Sigma_{s_j \in e}\, p(s_j)U_j(L_{1j}) < \Sigma_{s_j \in e}\, p(s_j)U_j(L_{2j})$. Hence, $(p_e, U_{j \in e})$ agrees with $\prec_e$.

For part (ii), without loss of generality (Lemma 5.1), continue with the e-called-off family $\mathbf{H_e}$ determined by fixing $H(s) = W$ if $s \notin e$. Define the act $\mathbf{B_e} \in \mathbf{H_e}$ by $\mathbf{B_e}(s) = B$ if $s \in e$. With respect to $\prec_e$, $\mathbf{B_e}$ serves as the "best" act and $\mathbf{W}$ serves as the "worst." Thus, for $H \in \mathbf{H_e}$, $V(H|e)V(\mathbf{B_e}) = V(H)$. Let $V_e(\cdot)$ agree with $\prec_e$ over the set $\mathbf{H_e}$. Assume $V_e(\cdot)$ differs from each conditional expected utility $V(\cdot|e)$ $(V \in \mathcal{V})$. In particular, with $\mathcal{H}_e$ ordered for applying Theorem 3 to $\prec_e$, let $H_z \in \mathcal{H}_e$ satisfy the following condition: For each $V \in \mathcal{V}$ such that $V_e(H_i) = V(H_i|e)$ $(i = 1, \ldots, z-1)$, $V_e(H_z) \neq V(H_z|e)$. That is, H_z is the first e-called-off act, where V_e differs from each $V(\cdot|e)$, $V \in \mathcal{V}$. Without loss of generality, according to Corollary 3.3, put the first z-elements of $\mathcal{H}_e$ as the initial segment of $\mathcal{H}$. Thus, H_z is the zth element in this reordering of $\mathcal{H}$.

By hypothesis, for some $V \in \mathcal{V}$, $V_e(H_i) = V(H_i|e)$ $(i = 1, \ldots, z-1)$. Then mimic the first $z-1$ extensions of $\prec_e$ in the first $z-1$ extensions of $\prec$. That is, provided e is not potentially null so that $\mathbf{W} \prec \mathbf{B_e}$, use Definition 20 to extend $\prec$ to $\prec_{z-1}$ with symmetric mixtures of the $z-1$ act pairs: H_i and $V_e(H_i)\mathbf{B_e} + (1 - V_e(H_i))\mathbf{W}$. Also by hypothesis, $\prec_{z-1}$ cannot be extended to $\prec_z$ using Definition 20 with symmetric mixtures of H_z and $V_e(H_z)\mathbf{B_e} + (1 - V_e(H_z))\mathbf{W}$.

Next, we show that $V_e(H_z)$ is an endpoint of the conditional target set $\mathcal{T}_z(H_z)$, defined using mixtures of $\mathbf{B_e}$ and $\mathbf{W}$. Argue indirectly: either $H_z \prec_{z-1} V_e(H_z)\mathbf{B_e} + (1 - V_e(H_z))\mathbf{W}$ or else $V_e(H_z)\mathbf{B_e} + (1 - V_e(H_z))\mathbf{W} \prec_{z-1} H_z$. We give the analysis for the former case. (The reasoning for the latter case is parallel.) Expand the $\prec_{z-1}$-preference into its equivalent $\prec$-preference. Thus, for $i = 1, \ldots, z-1$, there exist $x_i \geq 0$, $x_z > 0$, $\Sigma_i x_i + x_z = 1$, such that

$$x_1G_1 + \ldots + x_{z-1}G_{z-1} + x_zH_z$$
$$\prec x_1G_1' + \ldots + x_{z-1}G_{z-1}' + x_z[V_e(H_z)\mathbf{B_e} + (1 - V_e(H_z))\mathbf{W}],$$

where the pairs (G_i, G_i') are symmetric mixtures of H_i and $V_e(H_i)\mathbf{B_e} + (1 - V_e(H_i))\mathbf{W}$. However, as this $\prec$-preference involves elements of $\mathcal{H}_e$ only, then $x_1G_1 + \ldots + x_{z-1}G_{z-1} + x_zH_z \prec_e x_1G_1' + \ldots + x_{z-1}G_{z-1}' + x_z[V_e(H_z) \times \mathbf{B_e} + (1 - V_e(H_z))\mathbf{W}]$. Thus $V_e(H_z) \notin \mathcal{T}_{e,z}(H_z)$ – a contradiction with the assumption that $V_e(\cdot)$ agrees with $\prec_e$. Hence, $V_e(H_z)$ is a *precluded* endpoint of the conditional $\mathcal{T}_z(H_z)$ according to preferences $\prec_{z-1}$, but it is not precluded from $\mathcal{T}_{e,z}(H_z)$ according to the subset of preferences in $\prec_{e,z-1}$. □

REFERENCES

Anscombe, F. J. and Aumann, R. J. (1963). A definition of subjective probability. *Ann. Math. Statist.* **34** 199–205.

Aumann, R. J. (1962). Utility theory without the completeness axiom. *Econometrica* **30** 445–462.

Aumann, R. J. (1964). Utility theory without the completeness axiom: a correction. *Econometrica* **32** 210–212.

Berger, J. (1985). *Statistical Decision Theory and Bayesian Analysis*, 2nd ed. Springer, New York.

deFinetti, B. (1937). La prévision: ses lois logiques, ses sources subjectives. *Ann. Inst. H. Poincaré* **7** 1–68.

DeGroot, M. (1974). Reaching a consensus. *J. Amer. Statist. Assoc.* **69** 118–121.

Ellsberg, D. (1961). Risk, ambiguity, and the Savage axioms. *Quart. J. Econom.* **75** 643–669.

Fishburn, P. C. (1979). *Utility Theory for Decision Making*. Krieger, New York.

Fishburn, P. C. (1982). *The Foundations of Expected Utility*. Reidel, Dordrecht.

Giron, F. J. and Rios, S. (1980). Quasi-Bayesian behaviour: A more realistic approach to decision making? In *Bayesian Statistics* (J. M. Bernardo, M. H. DeGroot, D. V. Lindley and A. F. M. Smith, eds.) 17–38. Univ. Valencia Press.

Hartigan, J. A. (1983). *Bayes Theory*. Springer, New York.

Hausner, M. (1954). Multidimensional utilities. In *Decision Processes* (R. M. Thrall, C. H. Coombs and R. L. Davis, eds.) 167–180. Wiley, New York.

Kadane, J. B., Ed. (1984). *Robustness of Bayesian Analysis*. North-Holland, Amsterdam.

Kadane, J. B. (1986). Progress toward a more ethical method for clinical trials. *Journal of Medicine and Philosophy* **11** 385–404.

Kadane, J. B. and Sedransk, N. (1980). Toward a more ethical clinical trial. In *Bayesian Statistics* (J. M. Bernardo, M. H. DeGroot, D. V. Lindley and A. F. M. Smith, eds.) 329–338. Univ. Valencia Press.

Kannai, Y. (1963). Existence of a utility in infinite dimensional partially ordered spaces. *Israel J. Math.* **1** 229–234.

Levi, I. (1974). On indeterminate probabilities. *J. Philos.* **71** 391–418.

Levi, I. (1980). *The Enterprise of Knowledge*. MIT Press.

Levi, I. (1990). Pareto unanimity and consensus. *J. Philos.* **87** 481–492.

Moskowitz, H., Wong, R. T. and Chu, P.-Y. (1988). Robust interactive decision-analysis (RID): Concepts, methodology, and system principles. Paper 948, Krannert Graduate School of Management, Purdue Univ.

Nau, R. F. (1992). Indeterminate probabilities on finite sets. *Ann. Statist.* **20** 1737–1767.

Nau, R. F. (1993). The shape of incomplete preferences. Paper 9301, The Fuqua School of Business, Duke Univ.

Ramsey, F. P. (1931). Truth and probability. In *The Foundations of Mathematics and Other Logical Essays* (R. B. Braithwaite, ed.) 156–198. Kegan, Paul, Trench, Trubner and Co. Ltd., London.

Rios Insua, D. (1990). *Sensitivity Analysis in Multi-Objective Decision Making*. Springer, New York.

Rios Insua, D. (1992). On the foundations of decision making under partial information. *Theory and Decision* **33** 83–100.

Savage, L. J. (1954). *The Foundations of Statistics*. Wiley, New York.

Schervish, M. J., Seidenfeld, T. and Kadane, J. B. (1990). State-dependent utilities. *J. Amer. Statist. Assoc.* **85** 840–847. [Chapter 2.2, this volume]

Schervish, M. J., Seidenfeld, T. and Kadane, J. B. (1991). Shared preferences and state-dependent utilities. *Management Sci.* **37** 1575–1589. [Chapter 2.3, this volume]

Seidenfeld, T., Kadane, J. B. and Schervish, M. J. (1989). On the shared preferences of two Bayesian decision makers. *J. Philos.* **86** 225–244. [Chapter 1.1, this volume]

Seidenfeld, T. and Schervish, M. J. (1983). A conflict between finite additivity and avoiding Dutch Book. *Philos. Sci.* **50** 398–412. [Chapter 2.4, this volume]

Seidenfeld, T., Schervish, M. J. and Kadane, J. B. (1990). Decisions without ordering. In *Acting and Reflecting* (W. Sieg, ed.) 143–170. Kluwer, Dordrecht. [Chapter 1.2, this volume]

Smith, C. A. B. (1961). Consistency in statistical inference and decision. *J. Roy. Statist. Soc. Ser. B* **23** 1–25.

Szpilrajn, E. (1930). Sur l'extension de l'ordre partiel. *Fund. Math.* **16** 386–389.

von Neumann, J. and Morgenstern, O. (1947). *Theory of Games and Economic Behavior*, 2nd ed. Princeton Univ. Press.

Walley, P. (1991). *Statistical Reasoning with Imprecise Probabilities.* Chapman and Hall, London.

White, C. C. (1986). A posteriori representations based on linear inequality descriptions of a priori and conditional probabilities. *IEEE Trans. Systems Man Cybernet.* **16** 570–573.

Williams, P. (1976). Indeterminate probabilities. In *Formal Methods in the Methodology of Empirical Sciences* (M. Przelecki, K. Szaniawski and R. Wojcicki, eds.) 229–246. Reidel, Dordrecht.

PART 2

The Truth About Consequences

2.1

Separating Probability Elicitation from Utilities

JOSEPH B. KADANE AND
ROBERT L. WINKLER

ABSTRACT

This chapter deals with the separation of probability elicitation from utilities. We show that elicited probabilities can be related to utilities not just through the explicit or implicit payoffs related to the elicitation process, but also through other stakes the expert may have in the events of interest. We study three elicitation procedures – lotteries, scoring rules, and promissory notes – and show how the expert's utility function and stakes in the events can influence the resulting probabilities. Particularly extreme results are obtained in an example involving a market at equilibrium. The applicability of a no-stakes condition and some implications for probability elicitation are discussed. Let π represent an expert's probability for an event A, and let p denote the elicited probability from some elicitation procedure. We determine the value of p that maximizes the expert's expected utility. When utility is linear in money, $p = \pi$ for all of the procedures studied here. Under nonlinear utility, the lottery procedure still yields $p = \pi$ as long as the expert has no other stakes involving the occurrence or nonoccurrence of A (the no-stakes condition). With the scoring-rule and promissory-note procedures, the no-stakes condition is no longer sufficient for $p = \pi$ in the presence of nonlinear utility. If the no-stakes condition holds and the elicitation-related payoffs approach 0, then $p = \pi$ in the limit. For all three procedures,

This research was sponsored by U.S. Office of Naval Research Contracts N00014-82-0622 and N00014-85-K-0539 (Kadane) and National Science Foundation Grants ATM-8507495 and IST-8600788 (Winkler). The authors are grateful for helpful comments from Gary Chamberlain, Morris DeGroot, Robert Nau, Teddy Seidenfeld, Ross Shachter, and participants at seminars given by Kadane at Carnegie-Mellon University and Winkler at Duke University.

Reprinted with permission from the *Journal of the American Statistical Association*, 83, no. 402 (June 1988): 357–363.

the combination of nonlinear utility and other stakes can lead to values of p other than π. Furthermore, an analysis of the promissory-note procedure in a market setting gives a very extreme result: In a complete market at equilibrium for such promissory notes, the elicited probability depends on the market price, not on π. Is the no-stakes condition reasonable? We suggest that it often is not, since experts are likely to have significant stakes already, particularly in important situations. Moreover, it may be difficult to determine exactly what those stakes are (and perhaps to obtain accurate information about the expert's utility function). This creates somewhat of a dilemma for probability elicitation, implying that, at least in theory, it is difficult to separate probability elicitation from utilities.

I. INTRODUCTION

The elicitation of experts' judgments in probabilistic form is often important in Bayesian inference and decision analysis, and a key question is whether or not elicited probabilities are influenced by experts' utility functions. Some elicitation methods involve real or hypothetical payoffs and thus may be subject to utility-related effects. Furthermore, even in the absence of specific elicitation-related payoffs, implicit rewards may cause utility-related effects. We find that the existence and extent of these effects depends on other stakes that the expert may have in the events of interest. The effects are eliminated or tend to be reduced under a no-stakes condition stating that, ignoring elicitation-related payoffs, the fortune of the expert is independent of the events for which probabilities are being elicited.

The purpose of this chapter, then, is to study conditions under which probability elicitation can be separated from utilities and to study the impact of utilities when such separation is not possible. In Section II we consider the elicitation of probabilities via lotteries and show that either linear utility or the no-stakes condition is sufficient to provide assessed probabilities consistent with an expert's judgments. Violations of the no-stakes condition, however, may cause systematic shifts in elicited probabilities if the expert's utility function is nonlinear. Alternative elicitation devices, using scoring rules and promissory notes, are studied in light of the no-stakes condition and the expert's utility function in Sections III and IV. Section V shows how elicitation in a market at equilibrium can lead to extreme violations of the no-stakes condition and thus to startling and unreasonable results in terms of elicited probabilities. In Section VI we ask when the no-stakes condition might be reasonable and find that it seems of doubtful validity in many circumstances. Some implications for probability elicitation and concluding comments are presented in Section VII.

II. LOTTERIES

When lotteries are used in probability elicitation, the general idea is for the expert to choose between two lotteries with identical payoffs. We consider the elicitation of a probability by an expert for a single event A. In the first lottery, the expert receives a reward r if A occurs and receives a "non-reward" n otherwise. In the second lottery, the expert receives r with probability p and n with probability $1 - p$, where p is an "accepted" probability from a standard device (e.g., a probability wheel). We assume without loss of generality that $n = 0$ and $r > 0$, although all that is required is $r \neq n$. The expert is asked: What value of p makes you indifferent between the two lotteries? This indifference value of p is then taken as the expert's assessed probability for A.

Let $g(f|A)$ and $g(f|\overline{A})$ represent the probability distributions of the expert's fortune given A and its complement $\overline{A}$, respectively, *not* including any payoffs from the lotteries used in the elicitation of the expert's probability for A. In addition, let U denote the expert's utility function for his or her fortune, including elicitation-related payoffs. Then the expert's expected utility if the first lottery is selected is

$$\mathrm{E}[U(f)|\,\text{Lottery 1}] = \pi \int U(f+r) g(f|A)\, df + (1-\pi) \int U(f) g(f|\overline{A})\, df, \tag{1}$$

where π represents the expert's probability that A will occur. If the second lottery is chosen, the expected utility is

$$\begin{aligned}\mathrm{E}[U(f)|\,\text{Lottery 2}] = p\Big[\pi \int U(f+r) g(f|A)\, df &+ (1-\pi) \int U(f+r) g(f|\overline{A})\, df\Big] \\ &+ (1-p)\Big[\pi \int U(f) g(f|A)\, df \\ &+ (1-\pi) \int U(f) g(f|\overline{A})\, df\Big].\end{aligned} \tag{2}$$

The expert's assessed probability is the value of p for which

$$\mathrm{E}[U(f)|\,\text{Lottery 1}] = \mathrm{E}[U(f)|\,\text{Lottery 2}]. \tag{3}$$

Equating (1) with (2)and solving for p yields the following expression for the assessed odds ratio:

$$p/(1-p) = c\pi/(1-\pi), \tag{4}$$

where

$$c = \frac{\int [U(f+r) - U(f)]g(f \mid A)df}{\int [U(f+r) - U(f)]g(f \mid \overline{A})df}. \tag{5}$$

From (4), we see that the assessed odds ratio $p/(1 - p)$ accurately reflects the expert's judgments, as represented by $\pi/(1 - \pi)$, iff $c = 1$.

If U is linear (i.e., the expert is risk neutral), then $c = 1$; however, if we allow the possibility of U being nonlinear, then c clearly depends on $g(f \mid A)$ and $g(f \mid \overline{A})$ as well as U. For example, suppose that $g(f \mid A)$ and $g(f \mid \overline{A})$ are degenerate, placing probability 1 at f_A and $f_{\overline{A}}$, respectively. From (5), we have

$$c = \frac{U(f_A + r) - U(f_A)}{U(f_{\overline{A}} + r) - U(f_{\overline{A}})}.$$

For U increasing and strictly concave, indicating a risk-averse expert, $c < 1$ if $f_A > f_{\overline{A}}$ and $c > 1$ if $f_A < f_{\overline{A}}$. Generalizing this result beyond the degenerate case, we can say that for risk-averse experts, $c < 1$ if $g(f \mid A)$ stochastically dominates $g(f \mid \overline{A})$ and $c > 1$ if $g(f \mid \overline{A})$ stochastically dominates $g(f \mid A)$. For risk-taking experts (i.e., experts with increasing, strictly convex U), the inequalities on c are reversed.

The foregoing discussion shows that an expert's elicited probabilities and utilities can be intertwined. Under what conditions can we separate the elicitation process from utilities and say something about c without making any assumptions about the expert's utility function? A sufficient condition for $c = 1$ is

$$g(f \mid A) = g(f \mid \overline{A}) \quad \text{for all } f. \tag{6}$$

This condition implies that, apart from payoffs involving the elicitation-related lotteries, no part of the expert's fortune is contingent on whether or not A occurs. Independence of f and the events or variables for which probabilities are being elicited can be thought of in terms of the expert having no stakes in these events or variables other than stakes that might be created as part of the elicitation process through lotteries or other devices. Thus, for convenience, we refer to (6) as the *no-stakes condition*. The no-stakes condition is close in spirit to Ramsey's (1931) notion of ethical neutrality.

Discussions of lottery-based probability-elicitation methods seem to focus solely on elicitation-related payoffs and thus to assume implicitly that the no-stakes condition holds. For example, LaValle (1978, p. 79) stated that the expert's probability "should *not* depend on the choice of reference consequences" (r and n). From (5), c may well depend on r (and the choice of $n = 0$) if the no-stakes condition is not satisfied.

To investigate the behavior of c for small r, we expand c from (5) in a Taylor series in r around 0, as follows:

$$c = c_0 + rc_1 + O(r^2), \tag{7}$$

where

$$c_0 = \int U'(f)g(f|A)df \Big/ \int U'(f)g(f|\overline{A})df \tag{8}$$

is the limiting value of c as r approaches 0 and

$$c_1 = \frac{c_0}{2}\left[\frac{\int U''(f)g(f|A)df}{\int U'(f)g(f|A)df} - \frac{\int U''(f)g(f|\overline{A})df}{\int U'(f)g(f|\overline{A})df}\right].$$

We can rewrite c_1 as

$$c_1 = (c_0/2)[E_{h_{\overline{A}}}(w) - E_{h_A}(w)], \tag{9}$$

with

$$w(f) = -U''(f)/U'(f) \tag{10}$$

representing the Pratt–Arrow risk-aversion function (Pratt 1964),

$$h_A(f) = U'(f)g(f|A)\Big/\int U'(f)g(f|A)df, \tag{11}$$

and

$$h_{\overline{A}}(f) = U'(f)g(f|\overline{A})\Big/\int U'(f)g(f|\overline{A})df. \tag{12}$$

From (7)–(9),

$$c = c_0\{1 + (r/2)[E_{h_{\overline{A}}}(w) - E_{h_A}(w)]\} + O(r^2). \tag{13}$$

Whether $c < 1$, $c = 1$, or $c > 1$ for r close to 0 depends on U, $g(f|A)$, and $g(f|\overline{A})$. For example, if the expert exhibits constant risk aversion through an exponential utility function, then the term in square brackets in (13) is 0 and c is approximately c_0 for small r. On the other hand, if $w(f)$ is not constant, then the sign of the term in square brackets in (13) depends on the relationship between $g(f|A)$ and $g(f|\overline{A})$. Finally, making r small does not mean that we are only concerned about the shape of U over a very limited domain; the entire set of values of f and $f + r$ implied by $g(f|A)$ and $g(f|\overline{A})$ is relevant. The size of this set reflects the various uncertainties related to the expert's fortune.

With lottery methods, the case for small stakes is not as strong as it is for the scoring rules and promissory notes discussed in Sections III and IV. In fact, LaValle (1978) claimed that "as a matter of convenience, [we] should select reference outcomes [r and n] which are easily conceptualizable and *sufficiently distinct* [emphasis ours] to encourage [the expert] to think *seriously* about the [lotteries]" (p. 80). Whatever the difference between r and n, the no-stakes condition is important to guarantee that $c = 1$ and thus $p = \pi$ in the presence of nonlinearities in the expert's utility function.

III. SCORING RULES

An alternative operational procedure for eliciting an expert's probability involves scoring rules. The expert receives a payoff equal to a score S from a scoring rule that is strictly proper in the sense that the expected score is maximized iff the reported value is exactly the expert's probability π. A frequently used strictly proper scoring rule is quadratic, with

$$S = -r(x - p)^2, \tag{14}$$

where p is the expert's stated probability of A, r is a positive constant, and x is an indicator variable corresponding to the occurrence ($x = 1$) or nonoccurrence ($x = 0$) of A.

The expert's expected utility as a function of p under the scoring scheme given by (14) is

$$EU(p) = \pi \int U\left[f - r(1-p)^2\right] g(f|A)df + (1-\pi)\int U(f - rp^2)g(f|\overline{A})df. \tag{15}$$

To maximize expected utility, the expert sets the derivative of $EU(p)$ with respect to p equal to 0, as follows:

$$\pi(1-p)\int U'[f-r(1-p)^2]g(f|A)df$$
$$-(1-\pi)p\int U'(f-rp^2)g(f|\overline{A})df=0,$$

which simplifies to

$$p/(1-p)=c\pi/(1-\pi) \tag{16}$$

with

$$c=\frac{\int U'[f-r(1-p)^2]g(f|A)df}{\int U'(f-rp^2)g(f|\overline{A})df}. \tag{17}$$

From (16), the expert's stated probability p equals π iff $c = 1$. As in Section II, $c = 1$ if U is linear. In the absence of linearity, however, the no-stakes condition is not sufficient for $c = 1$ with scoring rules; large values of r and/or probabilities near 0 or 1 can cause discrepancies between $U'[f-r(1-p)^2]$ and $U'(f-rp^2)$. For a strictly risk-averse individual, U' is strictly decreasing. Therefore, from (16) and (17), a risk-averse expert with (6) satisfied will have $c < 1$ when $\pi > \frac{1}{2}$ and $c > 1$ when $\pi < \frac{1}{2}$. As a result, risk aversion causes p to move away from π toward $\frac{1}{2}$. For a risk taker satisfying (6), the movement is in the opposite direction, toward 0 when $\pi < \frac{1}{2}$ and toward 1 when $\pi > \frac{1}{2}$. If the risk taking is extreme enough, the second-order condition is violated and the expert moves all the way to $p = 0$ or $p = 1$. As in Section II, potential nonlinearities in U must be considered not just over the range of elicitation-related payoffs [in this case, possible values of S from (14)], but over the entire range of possible fortunes $f + S$. Furthermore, violations of the no-stakes condition can cause c in (17) to deviate considerably from 1 just as they can cause c in (5) to differ from 1.

When r approaches 0, implying that the payoff from the scoring rule is of little consequence, then the no-stakes condition is sufficient for $c = 1$ to hold approximately. In the limit, we have

$$\lim_{r\to 0} c = c_0 = \frac{\int U'(f)g(f|A)df}{\int U'(f)g(f|\overline{A})df}, \tag{18}$$

which equals 1 if the no-stakes condition holds. Expanding c in a Taylor series in r around 0, we find that

$$c = c_0 + rc_1 + O(r^2), \tag{19}$$

where

$$c_1 = \left[c_0 - (1-p)^2 \frac{\int U''(f) g(f|A) df}{\int U'(f) g(f|A) df} + p^2 \frac{\int U''(f) g(f|\overline{A}) df}{\int U'(f) g(f|\overline{A}) df} \right]. \tag{20}$$

Thus from (19) and (20),

$$c = c_0 \left\{ 1 + r\left[(1-p)^2 E_{h_A}(w) - p^2 E_{h_{\overline{A}}}(w) \right] \right\} + O(r^2), \tag{21}$$

with w, h_A, and $h_{\overline{A}}$ given by (10)–(12). This expression for c is similar to that in (13) for the lottery method, with changes in the weights given to the two expectations in the term in square brackets. When U is exponential with $w(f) = w > 0$,

$$c = c_0[1 + r(1-2p)w] + O(r^2).$$

For small r, $c > c_0$ if $\pi < \frac{1}{2}$ and $c < c_0$ if $\pi > \frac{1}{2}$.

The limiting value of c, c_0, is the same for lotteries and scoring rules. Insofar as the separation of probability elicitation from utilities is concerned, a major difference between lotteries and scoring rules is that the no-stakes condition by itself is sufficient for $c = 1$ with lotteries but not with scoring rules. In that sense, stronger assumptions are needed to obtain $c = 1$ when scoring rules are needed. Of course, with either method, c can differ considerably from 1 if the no-stakes condition is not satisfied.

IV. PROMISSORY NOTES

DeFinetti (1974) used promissory notes in the elicitation of probabilities. Consider a promissory note that pays r if A occurs and nothing otherwise, where $r > 0$. This promissory note is identical to the lottery involving A in Section II, but we will not compare it with another lottery. Instead, we ask the expert: What is the largest price you will pay for the promissory note (i.e., the price that makes you indifferent between buying the note and not buying it)? Denote the indifference price by q. Then the expert's elicited probability for A is taken to be $p = q/r$.

The analysis of the use of promissory notes is summarized briefly here; for more details, see Kadane and Winkler (1987). The expert's odds ratio $\pi/(1-\pi)$ is related to $p/(1-p)$, the odds ratio implied by the indifference price q, by

$$p/(1-p) = c\pi/(1-\pi), \tag{22}$$

as in (4) and (16). In this case,

$$c = \frac{\int \left[\frac{U(f-q+r)-U(f)}{r-q} \right] g(f|A)df}{\int \left[\frac{U(f)-U(f-q)}{q} \right] g(f|\overline{A})df}, \tag{23}$$

and the limiting value of c is the same as that provided by lotteries and scoring rules:

$$\lim_{r \to 0} c = c_0 = \frac{\int U'(f)g(f|A)df}{\int U'(f)g(f|\overline{A})df}. \tag{24}$$

An expansion of c yields

$$c = c_0\{1-(r/2)[sE_{h_{\overline{A}}}(w)+(1-s)E_{h_A}(w)]\} + O(r^2), \tag{25}$$

where s is the limit, as $r \to 0$, of q/r.

If the no-stakes condition is satisfied, $c_0 = 1$ but c may differ from 1, as in the case of scoring rules. For instance, from (23), $c < 1$ for risk-averse experts satisfying (6). Thus the promissory-note approach will understate the odds in favor of A for a risk-averse expert with no other stakes involving A. Similarly, the odds in favor of A will be overstated by a risk-taking expert without other stakes involving A. In the presence of other stakes, c might move toward 1 or further away from 1, depending on the nature of $g(f|A)$ and $g(f|\overline{A})$. Clearly, violations of the no-stakes condition can have an important impact on p for any of the three methods we have considered in Sections II–IV.

V. ELICITATION IN A MARKET SETTING

The importance of the no-stakes condition in the elicitation of probabilities can be especially notable in a market setting. In this section we consider a particular type of market with risk-averse participants. If the market is at equilibrium and the promissory-note approach is used to elicit the probabilities of all of the participants, then they would all appear to have the same probability. This startling result provides another illustration of how crucial non-elicitation-related uncertainties concerning an expert's fortune can be.

Suppose that a market exists of the following type: An auctioneer announces a price for promissory notes of the type considered in Section IV. Each person indicates how much he or she wishes to buy or sell at that price. If desired purchases exceed desired sales, the auctioneer raises the price, and conversely. This continues until the auctioneer finds a price at which desired sales and desired purchases are equal. Then all transactions take place in those amounts at that price, q^*. Furthermore, suppose that the participants in the market are risk averse in the sense that they put infinite negative utility on the outcome of being completely broke. For example, the logarithmic utility $U(f) = \log f$ satisfies this condition. Then all individuals whose probabilities of A are less than q^*/r will sell some promissory notes, but not to the extent of their entire fortunes, and all individuals for whom $\pi > q^*/r$ will buy some notes, but again, will not spend their entire fortunes doing so. Those with $\pi = q^*/r$ will be indifferent, and we can suppose that they neither buy nor sell.

Now suppose that after such an auction we enter the room and attempt to use the promissory-note technique to discover the personal probability for A of a participant in the market. We ask what price would leave that person indifferent between buying a note and not buying it. Only the price $q = q^*$ can be given, since if any other price, say $q > q^*$, is given, the person would have bought more promissory notes at the price q^* in the auction. Similarly, if $q < q^*$, the person would have sold more. Thus q^* is given by everyone in the market. Since promissory notes on $\overline{A}$ are valued at $r - q^*$ by everyone (otherwise arbitrage is possible), everyone has the same price ratio of A to $\overline{A}$, $q^*/(r - q^*)$. Following the standard procedure with promissory-note elicitation (i.e., ignoring other stakes in A), we would, therefore, infer that everyone has the same probability q^*/r. Thus, in a market at equilibrium, contrary to the subjective view of probability, objective (i.e., interpersonal) probability appears to exist! This appearance, of course, is due to the stakes that the individuals in the market already have in the event of interest.

The aforementioned market is complete, which means that the events under consideration, A and $\overline{A}$, exhaust the sure event. Thus the sure event is one of the "securities" in the market. An incomplete market may not yield the result that everyone has the same price, and hence the same inferred probability, if the promissory-note scheme is used. As Leamer (1986) noted, however, in market settings "there is ... little information in the market price about differences in ...

individuals' pre-exchange valuations" (p. 218). Focusing on market-related and game-theoretic considerations, Leamer claimed that probabilities cannot be elicited economically and accurately.

VI. IS THE NO-STAKES CONDITION REASONABLE?

We would like to feel that elicited probabilities accurately reflect an expert's judgments. The analysis in this essay suggests that experts may have some incentive to give responses that imply probabilities differing from their judgments. Violations of the no-stakes condition can play an important role in these differences. Yet, in eliciting probabilities, we seldom ask whether, or to what extent, the expert is already making bets on the very stochastic events for which the expert's probabilities are to be elicited.

Are experts likely to already have significant stakes relating to the events of interest? Elicitations of probabilities of experts/stakeholders are done all the time in risk analysis and decision analysis. Experts are elicited about their probabilities concerning health effects associated with pollutants such as carbon monoxide (Keeney, Sarin, and Winkler 1984), even though these experts may have staked varying proportions of their scientific reputations on the outcome. Other experts are asked about their probabilities for nuclear war (Press 1985), in which we all have a large stake (What is the meaning of "You win $10 if a nuclear war occurs"?), as pointed out in an example given by DeGroot (1970, pp. 74–75). When probabilities are elicited in applications of decision analysis, experts may have vested interests in the decision that will be made as a result of the analysis. When scientists assess the chances of success of research proposals, their own investments in certain research directions and strategies may influence the assessments despite good intentions to remain neutral.

In financial circles, opinions about future prospects for individual investments, for the stock market as a whole, and about the economy are available from numerous "experts," and these opinions are sometimes expressed in probabilistic form. Surely most of these experts have significant stakes in how the stock market and the economy perform, both through their own personal investments and through the income they may derive from their predictions. Of course, finance is an area in which dependence among different investments has been studied in detail and allowed for in the determination of diversified portfolios through portfolio analysis (e.g., Markowitz 1959). Miller (1978) pointed out, however, that standard portfolio theory takes into consideration dependence among investments but generally ignores dependence between these investments and an investor's earned income. For example, those investors whose earned incomes may be closely related to the economy or the stock market (e.g., workers who might be laid off in a recession or stockbrokers who may find their commissions dwindling when the economy and the market are down) might be better off to diversify with non-stock-market investments even if they feel that the prospects for the market are reasonably good. If they agree with this line of reasoning, probabilities that might be inferred from their investment behavior could differ considerably from their actual judgments. But this is precisely the issue that we are concerned with when we ask if an expert's fortune is independent of events for which probabilities are being elicited!

When scientific and practical matters are at stake, we all have various previous "investments" in the outcomes. These may be difficult to untangle; nonetheless, they bear an important weight in the future bets we might make. Without studying these stakes, with no conflict-of-interest statement, elicitations could make serious errors. Moreover, the problem may be most serious precisely when the most important inferences and decisions are being made.

VII. ELICITATION WITHOUT ASSUMING THE NO-STAKES CONDITION

The purpose of this section is constructive, to consider ways of eliciting probabilities without the no-stakes condition. The spirit of the section is similar to that of the following quotation:

> A selling point for DMUU (decision making under uncertainty) modeling is that it permits the decomposition of the problem so that, for example, the modeling of uncertainty can be separated from the modeling of preferences. When a probability assessor has a vested interest in the eventual outcomes, this separation may be hard to maintain. How serious is this problem, and how can assessment procedures be designed to minimize the problem? (Winkler 1982, p. 523)

It is natural to ask questions such as these: What are the implications and proposed prescriptions for practice in terms of how probability elicitation should be designed so as to minimize the interaction with utility? What should be done under the worst circumstances with vested interests? We have no definitive answers, and we regard these as valid questions for further research. The discussion in this section is therefore brief; rather than go into great detail, we touch upon points of interest and offer some suggestions regarding elicitation.

If we know the expert's utility function U and we also know $g(f|A)$ and $g(f|\overline{A})$, then we can use the results in Sections II–IV to make the appropriate adjustments in the values provided by the expert. That is, we do not assume that the odds ratio equals the price ratio; instead, we use our information about U, $g(f|A)$, and $g(f|\overline{A})$ to calculate c from (5), (17), or (23). The observed $p/(1-p)$ can then be divided by c to arrive at the odds ratio.

Unfortunately, we generally do not know the expert's utility function and all of the many uncertainties related to the expert's fortune. We can assess the expert's utility function via standard procedures (e.g., Keeney and Raiffa 1976) that can be used when the relevant consequences are multiattribute in nature as well as in the basic case when only utility for money is being considered. Obtaining information about the expert's non-elicitation-related stakes in the event of interest, however, is a much more difficult task. The expert may be understandably reluctant to reveal these stakes, and some of the connections between the events and the expert's fortune may be so complicated that even the expert cannot completely unravel them. A full Bayesian approach would involve the consideration of a distribution representing our uncertainty about the stakes. This distribution could be used, together with the expert's responses, to generate a distribution for the expert's probability π. In this approach we are admitting that we are uncertain about the expert's non-elicitation-related stakes (and possibly about the expert's U as well) and that, as a result, we cannot be certain about π. But eliciting our own distribution for the expert's stakes and utilities is by no means an easy task.

The elicitation problem of interest here is related to the more general problem of designing appropriate incentives. The study of scoring rules is within the framework of incentives, and elicitation procedures such as lotteries and promissory notes can be thought of as attempts to provide the expert with inducements intended to encourage the expert to reveal probabilistic judgments. When the no-stakes condition is not reasonable, we must adjust the incentive plan to allow for the non-elicitation-related stakes. For example, consider the elicitation of a sales manager's probabilities for next year's sales. We want to get an accurate picture of the prospects for next year, but we also want the incentive plan to inspire the manager to work hard to maximize sales. For some work along these lines, see Sarin and Winkler (1980). Also relevant are the notion of state-dependent preferences (Drèze 1985) and the growing body of literature on "agency theory" in finance and economics (e.g., Pratt and Zeckhauser 1985). Finally, questions of incentives and "small worlds" versus "large worlds" are relevant to the design of organizations. What do various designs and incentives imply for how decisions made at various levels of the organization (small worlds) relate to the goals of the organization as a whole (large world)? Some thoughts of Hogarth (1982) are relevant in this direction.

Multiple elicitations may help to reduce the problem of systematic shifts in the assessed probabilities. Such shifts may cause incoherence, and adjustments in the elicited probabilities might be made in consultation with the expert (reconciling inconsistencies) or might be made by the analyst. In using multiple elicitations, we should keep in mind potential dependence among the payoffs related to the various elicitations.

The discussion in this section has assumed that we are using elicitation techniques such as lotteries, scoring rules, and promissory notes. Another approach, of course, is to abandon such techniques in favor of alternative approaches that attempt to avoid or at least reduce the utility-related difficulties we have encountered with these methods. A common approach (e.g., DeGroot 1970) is to use "is more likely than" as a primitive relation between events. The elicitation process can consist of comparisons of events to see which one is more likely, using reference events and not necessarily using payoffs.

The foundational work of Savage (1954) avoids the criticism here by use of the device of "consequences," which are not random variables and over which individuals have preferences. This device proves awkward, however, when one wants to use Savage's theory in "small worlds," restricted just to the events of immediate interest, avoiding references to the "grand world" that anticipates all future uncertainties and possible actions (see Savage 1954, pp. 82–91). As Seidenfeld and Schervish (1983) pointed out, consequences are a crucial device for permitting finitely additive probabilities to be implied by a theory of preference.

Our focus in this article is on the elicitation of probabilities;

however, similar issues could affect the elicitation of utilities. For example, Fukuba and Ito (1984) discussed some difficulties that may arise in utility elicitation when the decision maker's initial wealth is stochastic. Fundamental problems of elicitation remain to be solved.

REFERENCES

deFinetti, B. (1974), *Theory of Probability* (Vol. 1), New York: John Wiley.

DeGroot, M. H. (1970), *Optimal Statistical Decisions*, New York: McGraw-Hill.

Drèze, J. (1985), "Decision Theory With Moral Hazard and State-Dependent Preferences," Discussion Paper 8545, Université Catholique de Louvain, CORE.

Fukuba, Y., and Ito, K. (1984), "The So-Called Expected Utility Theory Is Inadequate," *Mathematical Social Sciences*, 7, 1–12.

Hogarth, R. M. (1982), "Decision Making in Organizations *and* the Organization of Decision Making," unpublished manuscript, University of Chicago, Center for Decision Research, Graduate School of Business.

Kadane, J. B., and Winkler, R. L. (1987), "DeFinetti's Methods of Elicitation," in *Probability and Bayesian Statistics*, ed. R. Viertl, New York: Plenum, pp. 279–284.

Keeney, R. L., and Raiffa, H. (1976), *Decisions With Multiple Objectives: Preferences and Value Tradeoffs*, New York: John Wiley.

Keeney, R. L., Sarin, R. K., and Winkler, R. L. (1984), "Analysis of Alternative National Ambient Carbon Monoxide Standards," *Management Science*, 30, 518–528.

LaValle, I. H. (1978), *Fundamentals of Decision Analysis*, New York: Holt, Rinehart & Winston.

Leamer, E. E. (1986), "Bid–Ask Spreads for Subjective Probabilities," in *Bayesian Inference and Decision Techniques: Essays in Honor of Bruno de Finetti*, eds. P. Goel and A. Zellner, Amsterdam: North-Holland, pp. 217–232.

Markowitz, H. M. (1959), *Portfolio Selection: Efficient Diversification of Investments*, New York: John Wiley.

Miller, E. M. (1978), "Portfolio Selection in a Fluctuating Economy," *Financial Analysts Journal*, 34, No. 3, 77–83.

Pratt, J. W. (1964), "Risk Aversion in the Small and in the Large," *Econometrica*, 32, 122–136.

Pratt, J. W., and Zeckhauser, R. J. (eds.) (1985), *Principals and Agents: The Structure of Business*, Boston: Harvard Business School.

Press, S. J. (1985), "Multivariate Group Assessment of Probabilities of Nuclear War," in *Bayesian Statistics 2*, eds. J. M. Bernardo, M. H. DeGroot, D. V. Lindley, and A. F. M. Smith, Amsterdam: North-Holland, pp. 425–455.

Ramsey, F. P. (1931), *The Foundations of Mathematics and Other Logical Essays*, London: Kegan Paul.

Sarin, R. K., and Winkler, R. L. (1980), "Performance-Based Incentive Plans," *Management Science*, 26, 1131–1144.

Savage, L. J. (1954), *The Foundations of Statistics*, New York: John Wiley.
Seidenfeld, T., and Schervish, M. J. (1983), "A Conflict Between Finite Additivity and Avoiding Dutch Book," *Philosophy of Science*, 50, 398–412. [Chapter 2.4, this volume]
Winkler, R. L. (1982), "Research Directions in Decision Making Under Uncertainty," *Decision Sciences*, 13, 517–533.

2.2

State-dependent Utilities

MARK J. SCHERVISH, TEDDY SEIDENFELD,
AND JOSEPH B. KADANE

ABSTRACT

Several axiom systems for preference among acts lead to a unique probability and a state-independent utility such that acts are ranked according to their expected utilities. These axioms have been used as a foundation for Bayesian decision theory and subjective probability calculus. In this chapter we note that the uniqueness of the probability is relative to the choice of what counts as a constant outcome. Although it is sometimes clear what should be considered constant, in many cases there are several possible choices. Each choice can lead to a different "unique" probability and utility. By focusing attention on state-dependent utilities, we determine conditions under which a truly unique probability and utility can be determined from an agent's expressed preferences among acts. Suppose that an agent's preference can be represented in terms of a probability P and a utility U. That is, the agent prefers one act to another iff the expected utility of that act is higher than that of the other. There are many other equivalent representations in terms of probabilities Q, which are mutually absolutely continuous with P, and state-dependent utilities V, which differ from U by possibly different positive affine transformations in each state of nature. We describe an example in which there are two different but equivalent state-independent utility representations for the same preference structure. They differ in which acts count as constants. The acts involve receiving different amounts of

This research was reported, in part, at the Indo-United States Workshop on Bayesian Analysis in Statistics and Econometrics, December 1988. The research was supported by National Science Foundation Grants DMS-8805676 and DMS-8705646 and Office of Naval Research Contract N00014-88-K0013. The authors thank Morris DeGroot, Bruce Hill, Irving LaValle, Isaac Levi, and Herman Rubin for helpful comments during the preparation of this article. They especially thank the associate editor of the *Journal of the American Statistical Association* for the patience and care that was given to this submission.

Reprinted with permission from the *Journal of the American Statistical Association*, 85, no. 411 (September 1990): 840–847.

one or the other of two currencies, and the states are different exchange rates between the currencies. It is easy to see how it would not be possible for constant amounts of both currencies to have simultaneously constant values across the different states. Savage (1954, sec. 5.5) discovered a situation in which two seemingly equivalent preference structures are represented by different pairs of probability and utility. He attributed the phenomenon to the construction of a "small world." We show that the small world problem is just another example of two different, but equivalent, representations treating different acts as constants. Finally, we prove a theorem (similar to one of Karni 1985) that shows how to elicit a unique state-dependent utility and does not assume that there are prizes with constant value. To do this, we define a new hypothetical kind of act in which both the prize to be awarded and the state of nature are determined by an auxiliary experiment.

I. INTRODUCTION

Expected utility theory is founded on at least one of several axiomatic derivations of probabilities and utilities from expressed preferences over acts (Anscombe and Aumann 1963; deFinetti 1974; Ramsey 1926; Savage 1954). These derivations allow for the simultaneous existence of a unique personal probability over the states of nature and a unique (up to positive affine transformations) utility function over the prizes such that the acts are ranked by expected utility. For example, suppose that there are n states of nature that form the set $S = \{s_1, \ldots, s_n\}$ and m prizes in the set $Z = \{z_1, \ldots, z_m\}$. An example of an act is a function f mapping S to Z. That is, if $f(s_i) = z_j$, then we receive prize z_j if state s_i occurs. (We will consider more complicated acts later.) Now suppose that there is a probability over the states such that $p_i = \Pr(s_i)$ and that there is a utility U over prizes. By saying that acts are ranked by expected utility, we mean that we strictly prefer act g to act f iff

$$\sum_{i=1}^{n} p_i U[f(s_i)] < \sum_{i=1}^{n} p_i U[g(s_i)]. \qquad (1)$$

If we allow the utilities of prizes to vary conditionally on which state of nature occurs, we can rewrite Equation (1) as

$$\sum_{i=1}^{n} p_i U_i[f(s_i)] < \sum_{i=1}^{n} p_i U_i[g(s_i)], \qquad (2)$$

where $U_i(z_j)$ is the utility of prize z_j given that state s_i occurs. Without restrictions, however, on the degree to which U_i can differ from $U_{i'}$ for $i \neq i'$, the uniqueness of the personal probability no longer holds. For example, let $q_1, \ldots, q_n$ be another probability over the states such that $p_i > 0$ iff $q_i > 0$. Then for an arbitrary act f,

$$\sum_{i=1}^{n} q_i V_i[f(s_i)] = \sum_{i=1}^{n} p_i U_i[f(s_i)],$$

where $V_i(\cdot) = p_i U_i(\cdot)/q_i$ when $q_i > 0$ (V_i can be arbitrary when $q_i = 0$). In this case, it is impossible to determine an agent's personal probability by studying his or her preferences for acts. Rubin (1987) noted this and developed an axiom system that does not lead to a separation of probability and utility. Arrow (1974) considered the problem for insurance (a footnote credits Rubin with raising this same issue in an unpublished 1964 lecture).

DeGroot (1970) began his derivation of expected utility theory by assuming that the concept of "at least as likely as" is an undefined primitive. This allows the construction of probability without reference to preferences. DeGroot also needs to introduce preferences among acts, however, to derive a utility function. In Section II, we will examine von Neumann and Morgenstern's (1947) axiomatization, along with Anscombe and Aumann's (1963) extension, to see how it attempts to avoid the non-uniqueness problem just described. In Section III, we look at Savage's system with the same goal in mind. Section IV provides a critical examination of the theory of deFinetti (1974). In Section V, we give an example illustrating the problem's persistence despite the best efforts of those who have derived the theories. While reviewing an example from Savage in Section VI, we see how close he was to discovering the non-uniqueness problem in connection with his own theory. In Section VII, we describe a method for obtaining a unique personal probability and state-dependent utility based on a proposal of Karni, Schmeidler, and Vind (1983).

II. STATE-INDEPENDENT UTILITY

Following von Neumann and Morgenstern (1947), we generalize the concept of act introduced in Section I by allowing randomization. That is, suppose that the agent is comfortable declaring probabilities for an auxiliary experiment the results of which he or she believes would in no way alter his or her preferences among acts. Furthermore, assume that this auxiliary experiment has events with arbitrary probabilities (e.g., it may produce a random variable with continuous distribution). Define a *lottery* as follows: If $A_1, \ldots, A_m$ is a partition of the possible outcomes of the auxiliary experiment with $\alpha_j = \Pr(A_j)$ for each j, then the lottery $(\alpha_1, \ldots, \alpha_m)$ awards prize z_j if A_j occurs. We assume that the choice of the partition events $A_1, \ldots, A_m$ does not affect the lottery. That is, if $B_1, \ldots, B_m$ is another partition such that $\Pr(B_j) = \alpha_j$ for each j also, then the lottery that awards prize z_j when B_j occurs is, to the agent, the *same lottery* as the one described in terms of the A_j. In this way, a lottery is just a simple probability distribution over prizes that is independent of the state of nature. Any two lotteries that award the prizes with the same probabilities are considered the same lottery. If L_1 and L_2 are the two lotteries $(\alpha_1, \ldots, \alpha_m)$ and $(\beta_1, \ldots, \beta_m)$, respectively, then for $0 \le \lambda \le 1$, we denote by $\lambda L_1 + (1 - \lambda)L_2$ the lottery $[\lambda\alpha_1 + (1 - \lambda)\beta_1, \ldots, \lambda\alpha_m + (1 - \lambda)\beta_m]$.

If we consider only lotteries, we can introduce some axioms for preferences among lotteries. For convenience, we will henceforth assume that there are two lotteries such that the agent has a strict preference for one over the other. Otherwise, the preference relation is trivial and no interesting results are obtained.

Axiom 1 (Weak Order). There is a weak order, $\succcurlyeq$, among lotteries such that $L_1 \succcurlyeq L_2$ iff L_1 is not strictly preferred to L_2.

This axiom requires that weak preference among lotteries be transitive, reflexive, and connected. If we define *equivalence* to mean "no strict preference in either direction," then equivalence is transitive also.

Definition 1. Assuming Axiom 1, we say that L_1 *is equivalent to* L_2 (denoted by $L_1 \sim L_2$) if $L_1 \succcurlyeq L_2$ and $L_2 \succcurlyeq L_1$. We say L_2 *is strictly preferred to* L_1 (denoted by $L_1 \prec L_2$) if $L_1 \succcurlyeq L_2$ but not $L_2 \succcurlyeq L_1$.

The axiom that does most of the work is one that entails stochastic dominance.

Axiom 2 (Independence). For each L, L_1, L_2, and $0 < \alpha < 1$, $L_1 \succcurlyeq L_2$ iff $\alpha L_1 + (1 - \alpha)L \succcurlyeq \alpha L_2 + (1 - \alpha)L$.

A third axiom is often introduced to guarantee that utilities are real valued.

Axiom 3 (Archimedean). If $L_1 \prec L_2 \prec L_3$, then there exists $0 < \alpha < 1$ such that $L_2 \sim (1 - \alpha)L_1 + \alpha L_3$.

Axiom 3 prevents L_3 from being infinitely better than L_2 and L_1 from being infinitely worse than L_2.

With axioms equivalent to these three, von Neumann and Morgenstern (1947) proved that there exists a utility over prizes U such that $(\alpha_1, \ldots, \alpha_m) \succcurlyeq (\beta_1, \ldots, \beta_m)$ iff $\Sigma_{i=1}^{m} \alpha_i U(z_i) \leq \Sigma_{i=1}^{m} \beta_i U(z_i)$. This utility is unique up to positive affine transformation. In fact, it is quite easy (and useful for the example in Sec. v) to construct the utility function from the stated preferences. Pick an arbitrary pair of lotteries L_0 and L_1 such that $L_0 \prec L_1$. Assign these lotteries the utilities $U(L_0) = 0$ and $U(L_1) = 1$. For all other lotteries L, the utilities are assigned as follows: If $L_0 \succcurlyeq L \succcurlyeq L_1$, $U(L)$ is that α such that $(1 - \alpha)L_0 + \alpha L_1 \sim L$. If $L \prec L_0$, then $U(L) = -\alpha/(1 - \alpha)$, where $(1 - \alpha)L + \alpha L_1 \sim L_0$ (hence $\alpha \neq 1$). If $L_1 \prec L$, then $U(L) = 1/\alpha$, where $(1 - \alpha)L_0 + \alpha L \sim L_1$. The existence of these α values is guaranteed by Axiom 3 and their uniqueness follows from Axiom 2.

To handle acts in which the prizes vary with the state of nature, Anscombe and Aumann (1963) introduced a fourth axiom designed to say that the preferences among prizes did not vary with the state. Before stating this axiom, we introduce a more general act, known as a *horse lottery*.

Definition 2. A function mapping states of nature to lotteries is called a horse lottery.

That is, if H is a horse lottery such that $H(s_i) = L_i$ for each i, then if state s_i occurs, the prize awarded is the prize that lottery L_i awards. The L_i can be all different or some (or all) the same. If $H(s_i) = L$ for all i, then we say $H = L$. If H_1 and H_2 are two horse lotteries such that $H_j(s_i) = L_i^{(j)}$ for each i and J and if $0 \leq \alpha \leq 1$, then we denote by $\alpha H_1 + (1 - \alpha)H_2$ the horse lottery H such that $H(s_i) = \alpha L_1^{(i)} + (1 - \alpha)L_2^{(i)}$ for each i. Axioms 1 and 2, when applied to preferences among horse lotteries, imply that the choice of an act has no effect on the probabilities of the state of nature.

Definition 3. A state of nature s_i is called *null* if for each pair of horse lotteries H_1 and H_2 satisfying $H_1(s_j) = H_2(s_j)$ for all $j \neq i$, $H_1 \sim H_2$. A state is called *non-null* if it is not null.

Axiom 4 (State-Independence). For each non-null state s_i, each pair of lotteries (L_1, L_2), and each pair of horse lotteries H_1 and H_2 satisfying

$H_1(s_j) = H_2(s_j)$ for $j \neq i$, $H_1(s_i) = L_1$, and $H_2(s_i) = L_2$, we have $L_1 \prec L_2$ iff $H_1 \prec H_2$.

Axiom 4 says that a strict preference between two lotteries is reproduced for every pair of horse lotteries that differ only in some non-null state, and their difference in that state is that each of them equals one of the two lotteries. With this setup, Anscombe and Aumann (1963) proved the following theorem.

Theorem 1 (Anscombe and Aumann). Under Axioms 1–4, there exist a unique probability P over the states and utility U over prizes (unique up to positive affine transformation) such that $H_1 \succcurlyeq H_2$ iff

$$\sum_{i=1}^{n} P(s_i)U[H_1(s_i)] \leq \sum_{i=1}^{n} P(s_i)U[H_2(s_i)],$$

where for each lottery $L = (\alpha_1, \ldots, \alpha_m)$, $U(L)$ stands for $\Sigma_{j=1}^{m}\alpha_j U(z_j)$.

Even when the four axioms hold, there is no requirement that the utility function U be the same, conditional on each state of nature. As we did when we constructed Equation (2), we could allow $U_i(z_j) = a_i U(z_j) + b_i$, where each $a_i > 0$. Then we could let $Q(s_i) = a_i P(s_i)/\Sigma_{k=1}^{n} a_k P(s_k)$. It would now be true that $H_1 \succcurlyeq H_2$ iff

$$\sum_{i=1}^{n} Q(s_i)U_i[H_1(s_i)] \leq \sum_{i=1}^{n} Q(s_i)U_i[H_2(s_i)].$$

The uniqueness of the probability in Theorem 1 depends on the use of a state-independent utility U. Hence one cannot determine an agent's probability from his or her stated preferences unless one assumes that the agent's utility is state-independent. This may not seem like a serious difficulty when Axiom 4 holds. We will see in Section v, however, that the problem is more complicated.

III. SAVAGE'S POSTULATES

Savage (1954) gave a set of postulates that do not rely on an auxiliary randomization to extract probabilities and utilities from preferences. Rather, they rely on the use of prizes that can be considered "constant" across states. Savage's most general acts are functions from states to prizes. Because he did not introduce an auxiliary randomization, he required that there be infinitely many states. The important features of Savage's theory, for this discussion, are the first three postulates and a few definitions. Some of the axioms and definitions are stated in terms of *events*, which are sets of states. Savage's postulates are consistent with the axioms of Section II in that they provide models for preference by maximizing expected utility.

The first postulate is the same as Axiom 1. The second postulate requires a definition of conditional preference.

Definition 4. Let B be an event. We say that $f \succcurlyeq g$ *given* B iff

- $f' \succcurlyeq g'$ for each pair f' and g' such that $f'(s) = f(s)$ for all $s \in B$, $g'(s) = g(s)$ for all $s \in B$, and $f'(s) = g'(s)$ for all $s \notin B$
- and $f' \succcurlyeq g'$ for every such pair or for none.

The second postulate is an analog of Axiom 2 (see Fishburn 1970, p. 193).

Postulate 2. For each pair of acts f and g and each event B, either $f \succcurlyeq g$ given B or $g \preceq f$ given B.

Savage has a concept of *null event* that is similar to the concept of null state from Definition 3.

Definition 5. An event B is *null* if for every pair of acts f and g, $f \succcurlyeq g$ given B. An event B is *non-null* if it is not null.

Savage's third postulate concerns acts that are constant, such as $f(s) = z$ for all s, where z is a single prize. For convenience, we will call such an act f by the name z also.

Postulate 3. For each non-null event B and each pair of prizes z_1 and z_2 (considered as constant acts), $z_1 \succcurlyeq z_2$ iff $z_1 \succcurlyeq z_2$ given B.

Savage's definition of probability relies on Postulate 3.

Definition 6. Suppose that A and B are events. We say that A *is at least as likely as* B if for each pair of prizes z and w, with $z \prec w$, we have $f_B \succcurlyeq f_A$, where $f_A(s) = w$ if $s \in A$, $f_A(s) = z$ if $s \notin A$, $f_B(s) = w$ if $s \in B$, and $f_B(s) = z$ if $s \notin B$.

Postulate 2 guarantees that with f_A and f_B as defined in Definition 6, either $f_B \succcurlyeq f_A$ no matter which pair of prizes z and w one chooses (as long as $z \prec w$) or $f_A \succcurlyeq f_B$ no matter which pair of prizes one chooses.

Postulate 3 says that the *relative* values of prizes cannot change between states. Savage (1954, p. 25) suggested that problems in locating prizes that satisfy this postulate may be solved by a clever redescription. For example, rather than describing prizes as "receiving a bathing suit" and "receiving a tennis racket" (whose relative values change, depending on which of the two states "picnic at the beach" or "picnic in the park" occurs), Savage suggested that the prizes might be "a refreshing swim with friends," "sitting alone on the beach with a tennis racket," and so on. We do not see how to carry out such redescriptions, however, while satisfying Savage's structural assumption that each prize is available as an outcome under each state. (What does it mean to receive the prize "sitting alone on the beach with a tennis racket" when the state "picnic in the park" occurs?)

Our problem, however, is deeper than this. Definition 6 assumes that the *absolute* values of prizes do not change from state to state. For example, suppose that A and B are disjoint and the value of z is 1 for the states in A and 2 for the states in B. Similarly, suppose that the value of w is 2 for the states in A and 4 for the states in B. Then even if A is more likely than B, but is not twice as likely, we would get $f_A \prec f_B$ and we would conclude, by Definition 6, that B is more likely than A. The example in Section v (using just one of the currencies) and our interpretation of Savage's "small worlds" problem (in Sec. vi) suggest that it might be very difficult to find prizes with the property that their "absolute" values do not change from state to state even though their "relative" values remain the same from state to state.

IV. DEFINETTI'S GAMBLING APPROACH

deFinetti (1974) assumed that there is a set of prizes with numerical values such that utility is linear in the numerical value. That is, a prize numbered 4 is worth twice as much as a prize numbered 2. More specifically, to say that utility is linear in the numerical values of prizes, we mean the following: For each pair of prizes, (z_1, z_2) with $z_1 \prec z_2$, and each $0 \leq \alpha \leq 1$, the lottery that pays z_1 with probability $1 - \alpha$ and pays z_2 with probability α (using the auxiliary randomization of Sec. II) is equivalent to the lottery that pays $(1 - \alpha)z_1 + \alpha z_2$ for sure. Using such a set of prizes, deFinetti supposed that an agent will accept certain gambles that pay these prizes. If f is an act, to gamble on f means to accept a contract that pays the agent $c[f(s) - x]$ when state s occurs, where c and x are some values. A negative outcome means that the agent has to pay out, whereas a positive outcome means that the agent gains some amount.

Definition 7. The *prevision* of an act f is the number x that one would choose so that all gambles of the form $c[f - x]$ would be accepted for all small values of c, both positive and negative.

If an agent is willing to gamble on each of several acts, then it is assumed that he or she will also gamble on them simultaneously. (For a critical discussion of this point, see Kadane and Winkler 1988; Schick 1986.)

Definition 8. A collection of previsions for acts is *coherent* if for each finite set of the acts, say $f_1, \ldots, f_n$ with previsions $x_1, \ldots, x_n$, respectively, and each set of numbers $c_1, \ldots, c_n$, we have

$$\sup_{\text{all } s} \sum_{i=1}^{n} c_i[f_i(s) - x_i] \geq 0.$$

Otherwise, the previsions are *incoherent*.

deFinetti (1974) proved that a collection of previsions of bounded acts is coherent iff there exists a finitely additive probability such that the prevision of each act is its expected value. This provides a method of eliciting probabilities by asking an agent to specify previsions for acts, such as $f(s) = 1$ if $s \in A$ and $f(s) = 0$ if $s \notin A$. The prevision of such an act f would be its probability if the previsions are coherent. As plausible as this sounds, the following example casts doubt on the ability of deFinetti's program to elicit probabilities accurately.

V. AN EXAMPLE

Let the set of available prizes be various amounts of dollars. We suppose that there are three states of nature (which we will describe in more detail later) and that the agent expresses preferences that satisfy the axioms of Section II and Savage's postulates. Furthermore, suppose that the agent's utility for money is linear. That is, for each state i, $U_i(\$cx) = cU_i(\$x)$. In particular, $U_i(\$0) = 0$. We now offer the agent three horse lotteries, H_1, H_2, and H_3, whose outcomes are

State of Nature

	s_1	s_2	s_3
H_1	\$1	\$0	\$0
H_2	\$0	\$1	\$0
H_3	\$0	\$0	\$1

Suppose that the agent claims that these three horse lotteries are equivalent. If we assume that the agent has a state-independent utility, the expected utility of H_i is $U(\$1)P(s_i)$. It follows from the three horse lotteries' being equivalent that $P(s_i) = 1/3$ for each i.

Next we alter the set of prizes to be various Japanese yen amounts. Suppose that we offer the agent three yen horse lotteries, H_4, H_5, and H_6, whose outcomes are

State of Nature

	s_1	s_2	s_3
H_4	¥100	¥0	¥0
H_5	¥0	¥125	¥0
H_6	¥0	¥0	¥150

If the agent claims that these three horse lotteries are equivalent, and if we assume that he or she uses a state-independent utility for yen prizes, then $P(s_1)U(¥100) = P(s_2)U(¥125) = P(s_3)U(¥150)$. Supposing that the agent's utility is linear in yen, as it was in dollars, we conclude that $P(s_1) = 1.25P(s_2) = 1.5P(s_3)$. It follows that $P(s_1) = 0.4054$, $P(s_2) = 0.3243$, and $P(s_3) = 0.2703$. It would be incoherent for the agent to express both sets of equivalences, since he or she is apparently now committed to two different probability distributions over the three states. This is not correct, however, as we now see.

Suppose that the three states of nature represent three different exchange rates between dollars and yen. s_1 = {\$1 is worth ¥100}, s_2 = {\$1 is worth ¥125}, and s_3 = {\$1 is worth ¥150}. Suppose further that the agent can change monetary units at the prevailing rate of exchange without any penalty. As far as this agent is concerned, H_i and H_{3+i} are worth exactly the same for $i = 1, 2, 3$ because in each state the prizes awarded are worth the same amount. A problem arises in this example: The two probability distributions were constructed under incompatible assumptions. The discrete uniform probability was constructed under the assumption that $U(\$1)$ is the same in all three states, whereas the other probability was constructed under the assumption that $U(¥100)$ was the same in all three states. Clearly these cannot both be true given the nature of the states. Both Theorem 1 and Savage's theory are saved because preference can be represented by expected utility *no matter which of the two assumptions one makes*. Unfortunately, this same fact forces the uniqueness of the probability to be relative to the choice of which prizes count as constants in terms of utility. There are two different representations of the agent's preferences by probability and state-independent utility. What is state-independent in one representation, though, is state-dependent in the other.

If we allow both types of prizes at once, we can calculate the marginal exchange rate for the agent. That is, we can ask, "For what value x will the agent claim that \$1 and ¥x are equivalent?" This question can be answered by using either of the two probability–utility representations, and the answers will be the same. First, with dollars having constant value, the expected utility of a horse lottery paying \$1 in all three states is $U(\$1)$. The expected value of the horse lottery paying ¥x in all three states is

$$\begin{aligned}\frac{U_1(¥x)+U_2(¥x)+U_3(¥x)}{3} &= \frac{1}{3}\left[\frac{x}{100}U(\$1)+\frac{x}{125}U(\$1)+\frac{x}{150}U(\$1)\right]\\ &= 0.008222xU(\$1),\end{aligned}$$

using the linearity of utility and the state-specific exchange rates. By setting this expression equal to $U(\$1)$, we obtain $x = 121.62$. Equivalently, we can calculate the exchange rate assuming that yen have constant value over states. The act paying ¥x in all states has expected utility $U(¥x) = 0.01xU(¥100)$. The act paying \$1 in all states has expected utility

$$\begin{aligned}&0.4054U_1(\$1)+0.3243U_2(\$1)+0.2703U_3(\$1)\\ &= 0.4054U(¥100)+0.3243U(¥125)+0.2703U(¥150)\\ &= U(¥100)[0.4054+0.3243\times 1.25+0.2703\times 1.5]\\ &= 1.2162U(¥100).\end{aligned}$$

Setting this equal to $0.01xU(¥100)$ yields $x = 121.62$, which is the same exchange rate as calculated earlier.

The implications of this example for elicitation are staggering. Suppose that we attempt to elicit the agent's probabilities over the three states by offering acts in dollar amounts and using deFinetti's gambling approach from Section IV. The agent has utility that is linear in both dollars and yen without reference to the states, hence deFinetti's program will apply. To see this, select two prizes, such as \$0 and \$1, to have utilities 0 and 1, respectively. Then for $0 < x < 1$, $U(\$x)$ must be the value c that makes the following two lotteries equivalent: $L_1 = \$x$ for certain, and $L_2 = \$1$ with probability c and \$0 with probability $1 - c$. Assuming that dollars have constant utility, it is obvious that $c = x$. Assuming that yen have constant utility, the expected utility of L_1 is $1.2162xU(¥100)$ and the expected utility of L_2 is $cU(¥121.62)$. These two are the same iff $x = c$. Similar arguments work when x is not between 0 and 1 and when the two prizes with utilities 0 and 1 are yen prizes. Now suppose that the agent actually uses the state-independent utility for dollars and the discrete uniform distribution to rank acts, but the eliciter does not know this. The eliciter will try to elicit the agent's probabilities for the states by offering gambles in yen (linear in utility). For example, the agent claims that the gamble $c(f - 40.54)$ would be accepted for all small values of c, where $f(s) = ¥150$ if $s = s_3$ and ¥0 otherwise. The reason for this is that since ¥150 equals \$1 when s_3 occurs, the winnings are \$1 when s_3 occurs, which has a probability of 1/3. The marginal exchange rate is ¥121.62 for \$1, so the appropriate amount to pay (no matter which state occurs), to win \$1 when s_3 occurs, is \$1/3, which equals ¥121.62/3 = ¥40.45. Realizing that utility is linear in yen, the eliciter now decides that $\Pr(s_3)$ must equal $40.54/150 = 0.2703$. Hence the eliciter elicits the wrong probability, even though the agent is coherent!

The expressed preferences satisfy the four axioms of Section II, all of Savage's postulates, and deFinetti's linearity condition, but we are still unable to determine the probabilities of the states based only on preferences. The problem becomes clearer if we allow both dollar and yen prizes at the same time. It is impossible, however, for a single utility to be state-independent for all prizes. That is, Axiom 4 and Postulate 3 would no longer hold. Things are more confusing in deFinetti's framework, because there is no room for state-dependent utilities. The agent appears to have two different probabilities for the same event, even though there would be no incoherency.

VI. SAVAGE'S "SMALL WORLDS" EXAMPLE

In section 5.5 of Savage (1954), the topic of *small worlds* is discussed. An anomaly occurs in this discussion, and Savage implies that it is an effect of the construction of the small world. In this section, we briefly introduce small worlds and then explain why we believe that the anomaly discovered by Savage is actually another example of the non-uniqueness illustrated in Section v. It is a mere coincidence that it arose in the discussion of small worlds. We show how precisely the same effect arises without any mention of small worlds.

A small world can be thought of as a description of the states of nature in which each state can actually be partitioned into several smaller states, but we do not actually do the partitioning when making comparisons between acts. For a mathematical example, Savage mentioned the following case. Consider the unit square $S = \{(x, y)\}:0 \leq x, y \leq 1\}$ as the finest possible partition of the states of nature. Suppose, however, that we consider as states the subsets $\bar{x} = \{(x, y)\}:0 \leq y \leq 1\}$ for each $x \in [0, 1]$. Savage discovered the following problem in this example: It is possible to define small world prizes in a natural way and for preferences among small world acts to satisfy all of his axioms and, at the same time, consistently define prizes in the "grand world" consisting of the whole square S. It is possible, however, for the preferences among small world acts to be consistent with the preferences among grand world acts in such a way that the probability measure determined from the small world preferences is not the marginal probability measure over the sets $\bar{x}$ induced from the grand world probability. As we will see, the problem that Savage discovered results from using different prizes as constants in the two problems. It is not due to the small world but will actually appear in the grand world as well.

Any grand world act can be considered a small world prize. In fact, the very reason for introducing small worlds is to deal with the case in which what we count as a prize is actually worth different amounts depending on which of the subdivisions of the small world state of nature occurs. Therefore, we let the grand world prizes be non-negative numbers and the grand world acts all bounded measurable functions on S. The grand world probability is uniform over the square and the grand world utility is the numerical value of the prize. To guarantee that Savage's axioms hold in the small world, choose the small world prizes to be 0 and positive multiples of a single function h. Assuming that $U(h) = 1$, the small world probability of a set $\overline{B} = \{\overline{x}: x \in B\}$ is (from Savage 1954, p. 89) $Q(\overline{B}) = \int_{\overline{B}} q(x)\, dx$, where

$$q(x) = \frac{\int_0^1 h(x, y)dy}{\int_0^1\int_0^1 h(x, y)dy\, dx}. \tag{3}$$

Unless $\int_0^1 h(x, y)dy$ is constant as a function of x, Q will not be the marginal distribution induced from the uniform distribution over S. Even if $\int_0^1 h(x, y)dy$ is not constant, however, the ranking of small world acts is consistent with the ranking of grand world acts. Let $ch(\cdot, \cdot)$, considered as a small world prize, be denoted by $\overline{c}$. Let $U(\overline{c}) = c$ denote the small world utility of small world prize $\overline{c}$. If $\overline{f}$ is a small world act, then for each $\overline{x}$, $\overline{f}(\overline{x}) = \overline{c}$ for some c. The expected small world utility of f is $\int_0^1 U[\overline{f}(\overline{x})]q(x)\, dx$. Let the grand world act f corresponding to $\overline{f}$ be defined by $f(x, y) = \overline{f}(\overline{x})h(x, y)$. It follows from Equation (3) that

$$U[\overline{f}(\overline{x})q(x) = \frac{\int_0^1 f(x, y)\, dy}{\int_0^1\int_0^1 h(x, y)\, dy\, dx}.$$

Hence the expected small world utility of $\overline{f}$ is

$$\int_0^1 \frac{\int_0^1 f(x, y)dy}{\int_0^1\int_0^1 h(x, y)dy\, dx}\, dx,$$

which is just a constant times the grand world expected utility of f. Hence small world acts are ranked in precisely the same order as their grand world counterparts, even though the small world probability is not consistent with the grand world probability.

We claimed that the inconsistency of the two probabilities is due to the choice of "constants" and not to the small worlds. To see this, let the grand world constants be 0 and the positive multiples of h. Then an act f in the original problem becomes an act f^* with $f^*(x, y) = f(x, y)/h(x, y)$. That is, the prize that f^* assigns to (x, y) is the number of multiples of $h(x, y)$ that $f(x, y)$ is. We define the new probability for B, a two-dimensional Borel set,

$$R(B) = \frac{\int_B h(x,y)dy\,dx}{\int_S h(x,y)dy\,dx}.$$

The expected utility of f^* is now

$$\frac{\int_S f^*(x,y)h(x,y)dy\,dx}{\int_S h(x,y)dy\,dx} = \frac{\int_S f(x,y)dy\,dx}{\int_S h(x,y)dy\,dx}.$$

This is just a constant times the original expected utility. Hence acts are ranked in the same order by both probability–utility representations. Both representations are state-independent but each one is relative to a different choice of constants. The constants in one representation have different utilities in different states in the other representation. Both representations satisfy Savage's axioms, however. (Note that the small world probability constructed earlier is the marginal probability associated with the grand world probability R, so Savage's small world problem evaporates when the definition of constant is allowed to change.) Remember that the uniqueness of the probability–utility representation for a collection of preferences is relative to what counts as a constant. To use Savage's notation in the example of Section v, suppose that we use yen gambles to elicit probabilities. Instead, however, of treating multiples of ¥1 as constants, we treat multiples of gamble $f(s_1) = ¥100$, $f(s_2) = ¥125$, $f(s_3) = ¥150$ as constants. Then we will elicit the discrete uniform probability rather than the nonuniform probability.

VII. HOW TO ELICIT UNIQUE PROBABILITIES AND UTILITIES SIMULTANEOUSLY

There is one obvious way to avoid the confusion of the previous examples – elicit a unique probability without reference to preferences. This is DeGroot's (1970) approach. It requires that the agent have an understanding of the primitive concept "at least as likely as" in addition to the more widely understood primitive "is preferred to." Some decision theorists prefer to develop the theory solely from preference without reference to the more statistical primitive "at least as likely as"; they need an alternative to the existing theories in order to separate probability from utility.

Karni et al. (1983; see also Karni 1985) proposed a scheme for simultaneously eliciting probability and state-dependent utility. Essentially, in addition to stating preferences among horse lotteries, an agent is asked to state preferences among horse lotteries under the assumption that he or she holds a particular probability distribution over the states (explicitly, they say on p. 1024, "contingent upon a strictly positive probability distribution p' on S"). They also require the agent to compare acts with different "contingent" probabilities. Karni (1985) described these (in a slightly more general setting) as *prize–state lotteries* that are functions $\hat{f}$ from $Z \times S$ to $\mathcal{R}^+$ such that $\Sigma_{\text{all}(z,s)}\hat{f}(z, s) = 1$ and the probability $\hat{f}(z, s)$ for each z and s is understood in the same sense as the probabilities involved in the lotteries of Section II. That is, the results of a prize–state lottery are determined by an auxiliary randomization. The agent is asked to imagine that the state of nature could be chosen by the randomization scheme rather than by the forces of nature. This is intended to remove the uncertainty associated with how the state of nature is determined so that a pure utility can be extracted by using Axioms 1–3 applied to a preference relation among prize–state lotteries.

For example, suppose that the agent in Section V expresses a strict preference for the prize–state lottery that awards \$1 in state 2 with probability 1 [$\hat{f}(\$1, s_2) = 1$] over $\hat{g}(\$1, s_1) = 1$. This preference would not be consistent with a state-independent utility for dollar prizes; however, it would be consistent with a state-independent utility in yen prizes.

The pure utility elicited in this fashion is a function of both prizes and states, so it is actually a state-dependent utility. As long as the preferences among prize–state lotteries are consistent with the preferences among horse lotteries, the elicited state-dependent utility can then be assumed to be the agent's utility. There will then be a unique probability such that $H_1 \preccurlyeq H_2$ iff the expected utility of H_1 is at most as large as the expected utility of H_2. The type of consistency that Karni et al. (1983) require between the two sets of preferences is rather more complicated than necessary. The following simple consistency axiom will suffice.

Axiom 5 (Consistency). For each non-null state s and each pair $(\hat{f}_1, \hat{f}_2)$ of prize–state lotteries satisfying $\Sigma_{\text{all}\, z}\hat{f}_i(z, s) = 1$ and some pair of horse lotteries H_1 and H_2 satisfying $H_1(s_i) = H_2(s_i)$ for all $s_i \neq s$ and $H_1(s) = f_1$ and $H_2(s) = f_2$, we have $H_1 \preccurlyeq H_2$ iff $\hat{f}_1 \preccurlyeq \hat{f}_2$, where f_1 and f_2 are lotteries that correspond to $\hat{f}_1$ and $\hat{f}_2$ as follows: $f_i = [\hat{f}_i(z_1, s), \ldots, \hat{f}_i(z_m, s)]$, $i = 1, 2$, in the notation of Section II.

This just says that preferences among prize–state lotteries with all of their probabilities on the same state must be reproduced as preferences between horse lotteries that differ only in that common state.

Theorem 2. Suppose that there are n states of nature and m prizes. Assume that preferences among horse lotteries satisfy Axioms 1–3. Also assume that preferences among prize–state lotteries satisfy Axioms 1–3. Finally, assume that Axiom 5 holds. Then there exists a unique probability P over the states and a utility $U: Z \times S \to \mathcal{R}$, unique up to positive affine transformation, satisfying the following:

1. $H_1 \succcurlyeq H_2$ iff

$$\sum_{i=1}^{n} P(s_i)U[H_1(s_i), s_i] \le \sum_{i=1}^{n} P(s_i)U[H_2(s_i), s_i],$$

where for each lottery $L = (\alpha_1, \ldots, \alpha_m)$, $U(L, s_i)$ stands for $\sum_{j=1}^{m} \alpha_j U(z_j, s_i)$.

2. $\hat{f} \succcurlyeq \hat{g}$ iff

$$\sum_{i=1}^{n} \sum_{j=1}^{m} \hat{f}(z_j, s_i)U(z_j, s_i) \le \sum_{i=1}^{n} \sum_{j=1}^{m} \hat{g}(z_j, s_i)U(z_j, s_i).$$

The proof of Theorem 2 makes use of the following theorem from Fishburn (1970, p. 176):

Theorem 3 (Fishburn). Under Axioms 1, 2, and 3, there exist real-valued functions $W_1, \ldots, W_n$ such that $H_1 \prec H_2$ iff

$$\sum_{i=1}^{n} W_i[H_1(s_i)] \le \sum_{i=1}^{n} W_i[H_2(s_i)]. \tag{4}$$

The W_i that satisfy Equation (4) are unique up to a similar positive linear transformation, with W_i constant iff s_i is null.

We provide only a sketch of the proof of Theorem 2. Let $(W_1, \ldots, W_n)$ be the state-dependent utility for horse lotteries guaranteed by Theorem 3, and let $\hat{V}$ be the utility for prize–state lotteries guaranteed by von Neumann and Morgenstern's theorem (1947). All we need to show is that there are $c_1, \ldots, c_n$ and positive $a_1, \ldots, a_n$ such that for each $i = 1, \ldots, n$,

$$W_i(z) = a_i \hat{V}(z, s_i) + c_i, \quad \text{for all } z. \tag{5}$$

If Equation (5) is true, then it follows directly from Equation (4) that $U = \hat{V}$ serves as the state-dependent utility and $P(s_i) = a_i / \Sigma_{k=1}^{n} a_k$ is the probability. The uniqueness follows from the uniqueness of the W_i and $\hat{V}$. To prove Equation (5), let $s = s_j$ for some j and suppose that H_1, H_2, $\hat{f}_1$, $\hat{f}_2$, f_1, and f_2 are as in the statement of Axiom 5. Now consider the set $\mathcal{H}_j$ of all horse lotteries H such that $H(s_i) = H_1(s_i)$ for all $i \neq j$. The stated preferences among this set of horse lotteries satisfy Axioms 1, 2, and 3. Hence there is a utility V_j for this set, and V_j is unique up to positive affine transformation. Clearly, W_j is such a utility, hence we assume that $V_j = W_j$. Next consider the set $\hat{\mathcal{H}}_j$ of all prize–state lotteries $\hat{f}$ that satisfy $\Sigma_{k=1}^{m} \hat{f}(z_k, s_j) = 1$. The stated preferences among elements of $\hat{\mathcal{H}}_j$ also satisfy Axioms 1, 2, and 3. Hence there is a utility $\hat{V}_j$ that is unique up to positive affine transformation. Clearly $\hat{V}$, with domain restricted to $\hat{\mathcal{H}}_j$, is such a utility, hence we will assume that $\hat{V}_j = \hat{V}$. The mapping $T_j : \mathcal{H}_j \to \hat{\mathcal{H}}_j$ defined by $T_j(H)(z, s) = 0$ for all (z, s) with $s \neq s_j$ and $T_j(H) = \alpha_i$ for $z = z_i$ and $s = s_j$, where $H(s_j) = (\alpha_1, \ldots, \alpha_m)$, is one to one and T_j preserves convex combination. It then follows from Axiom 5 that for $H_1, H_2 \in \mathcal{H}_j$, $W_j(H_1) \leq W_j(H_2)$ iff $\hat{V}[T_j(H_1)] \leq \hat{V}[T_j(H_2)]$. Since both $V_j = W_j$ and $\hat{V}_j = \hat{V}$ are unique up to positive affine transformation, we have $W_j = a_j \hat{V} + b_j$ for some positive a_j. This proves Equation (5).

VIII. DISCUSSION

The need for state-dependent utilities arises out of the possibility that what may appear to be a constant prize may not actually have the same value to an agent in all states of nature. Much of probability theory and statistical theory deals solely with probabilities and not with utilities. If probabilities are only unique relative to a specified utility, then the meaning of much of this theory is in doubt. Much of statistical decision theory makes use of utility functions of the form $U(\theta, d)$, where θ is a state of nature and d is a possible decision. The prize awarded when decision d is chosen and the state of nature is θ is not explicitly mentioned. Rather, the utility of the prize is specified without reference to the prize. Although it would appear that $U(\theta, d)$ is a state-dependent utility (as well it might be), one has swept comparisons between states "under the rug." For example, if $U(\theta, d) = -(\theta - d)^2$, one might ask how it was determined that an error of 1 when $\theta = a$ has the same utility as an error of 1 when $\theta = b$.

DeGroot (1970) avoided these problems by assuming that the concept of one event being at least as likely as another is understood without definition. He then proceeded to state axioms implying the existence of a unique subjective probability distribution over states of nature. (For a discussion of attempts to derive quantitative probability from qualitative probability, see Narens 1980.) Further axioms governing preference could then be introduced. These would then lead to a state-dependent utility function. Axioms such as those of Savage (1954), von Neumann and Morgenstern (1947) and Anscombe and Aumann (1963), and deFinetti (1974), which concern only preference among acts like horse lotteries, are not sufficient to guarantee a representation of preference by a unique state-dependent utility and probability. Direct comparisons must be made between lotteries in a specified state of nature and other lotteries in another specified state of nature. These are the prize–state lotteries introduced by Karni (1985). Assuming that preferences among prize–state lotteries are consistent with preferences among horse lotteries, a unique state-dependent utility and probability can be recovered from the preferences.

REFERENCES

Anscombe, F. J., and Aumann, R. J. (1963), "A Definition of Subjective Probability," *Annals of Mathematical Statistics*, 34, 199–205.

Arrow, K. J. (1974), "Optimal Insurance and Generalized Deductibles," *Scandinavian Actuarial Journal*, 1, 1–42.

deFinetti, B. (1974), *The Theory of Probability* (2 vols.), New York: John Wiley.

DeGroot, M. H. (1970), *Optimal Statistical Decisions*, New York: John Wiley.

Fishburn, P. (1970), *Utility Theory for Decision Making*, New York: John Wiley.

Kadane, J. B., and Winkler, R. L. (1988), "Separating Probability Elicitation from Utilities," *Journal of the American Statistical Association*, 83, 357–363. [Chapter 2.1, this volume]

Karni, E. (1985), *Decision Making Under Uncertainty*, Cambridge, MA: Harvard University Press.

Karni, E., Schmeidler, D., and Vind, K. (1983), "On State Dependent Preferences and Subjective Probabilities," *Econometrica*, 51, 1021–1031.

Narens, L. (1980), "On Qualitative Axiomatizations for Probability Theory," *Journal of Philosophical Logic*, 9, 143–151.

Ramsey, F. P. (1926), "Truth and Probability," in *Studies in Subjective Probability*, eds. H. E. Kyburg and H. E. Smokler, Huntington, NY: Krieger, pp. 23–52.
Rubin, H. (1987), "A Weak System of Axioms for 'Rational' Behavior and the Nonseparability of Utility From Prior," *Statistics and Decisions*, 5, 47–58.
Savage, L. J. (1954), *Foundations of Statistics*, New York: John Wiley.
Schick, F. (1986), "Dutch Book and Money Pumps," *Journal of Philosophy*, 83, 112–119.
von Neumann, J., and Morgenstern, O. (1947), *Theory of Games and Economic Behavior*, Princeton, NJ: Princeton University Press.

2.3

Shared Preferences and State-dependent Utilities

MARK J. SCHERVISH, TEDDY SEIDENFELD, AND JOSEPH B. KADANE

ABSTRACT

This investigation combines two questions for expected utility theory:

1. When do the shared preferences among expected utility maximizers conform to the dictates of expected utility?
2. What is the impact on expected utility theory of allowing preferences for prizes to be state-dependent?

Our principal conclusion (Theorem 4) establishes very restrictive necessary and sufficient conditions for the existence of a Pareto, Bayesian compromise of preferences between two Bayesian agents, even when utilities are permitted to be state-dependent and identifiable. This finding extends our earlier result (Theorem 2, 1989a), which applies provided that all utilities are state-independent. A subsidiary theme is a decision theoretic analysis of common rules for "pooling" expert probabilities.

Against the backdrop of "horse lottery" theory (Anscombe and Aumann 1963) and subject to a weak Pareto rule, we show, generally, that there is no Bayesian compromise between two Bayesian agents even when state-dependent utilities

This research was supported by National Science Foundation grants DMS-871770, DMS-8705646, DMS-8805676, SES-8900025, and DMS-9005858, Office of Naval Research contracts N00014-88-K-0013 and N00014-89-J-1851, and the Buhl Foundation. The authors wish to thank the Departmental Editor and referees for helpful suggestions. They would also like to thank Giovanni Parmigiani for helpful comments on an earlier draft.

Reprinted by permission, "Shared Preferences and State-Dependent Utilities," Mark J. Schervish, Teddy Seidenfeld, and Joseph B. Kadane, *Management Science*, Volume 37, Number 12, December 1991. Copyright 1991, The Institute of Management Sciences (currently INFORMS), 2 Charles Street, Suite 300, Providence, Rhode Island 02904 USA.

are entertained in an *identifiable* way. The word "identifiable" is important because if state-dependence is permitted merely by dropping the Anscombe-Aumann axiom (Axiom 4 here) for "state-independence," though a continuum of possible Bayesian compromises emerges, also it leads to an extreme underdetermination of an agent's personal probability and utility given the agent's preferences. Instead, when state-dependence is monitored through (our version of) the approach of Karni, Schmeidler, and Vind (1983), the general impossibility of a Bayesian, Pareto compromise in preferences reappears.

I. INTRODUCTION

This essay combines two questions for expected utility theory:

1. When do the shared preferences among expected utility maximizers conform to the dictates of expected utility?
2. What is the impact on expected utility theory of allowing preferences for prizes to be state-dependent?

Against the backdrop of the "horse lottery" theory of Anscombe and Aumann (1963) and subject to a weak Pareto rule, we show that, in general, there is no Bayesian compromise between two Bayesian agents even when state-dependent utilities are entertained in an *identifiable* way. In order to see the relevance in pairing these two issues, first consider the importance of each question alone.

Regarding shared preferences, it is natural to inquire whether justifications of expected utility theory can be extended from a single agent to a cooperative group in such a way as to preserve those preferences common to them all. For an illustration, consider two coherent decision makers, call them Dick and Jane, who wish to act in unison – with binding agreements possible – in a fashion that their collective choices conform to axiomatic canons of expected utility theory. This is, suppose

i. whenever both Dick and Jane (separately) think that option A_2 is strictly better than option A_1, then A_1 cannot be their cooperative choice if A_2 is available – a (weak) Pareto condition on cooperative, group preference; and suppose
ii. their collective choices are coherent.

Then the first question amounts to asking when there is a Pareto compromise of individual preferences that satisfies the axiomatic constraints which constitute "coherence."

Seidenfeld, Kadane, and Schervish (1989a) examine this question for the horse lottery theory of rationality proposed by Anscombe and Aumann (1963). Their theory, like many others (such as Savage 1954), axiomatizes rational preferences for acts defined as functions from states to outcomes. Anscombe and Aumann's theory is distinguished by its use of simple von Neumann-Morgenstern lotteries for outcomes. (A *von Neumann-Morgenstern lottery*, denoted by L, is a probability distribution over a set $\mathcal{G}$ of *prizes*. A lottery is *simple* if its distribution has finite support. All lotteries discussed in this essay are simple.) Anscombe-Aumann acts, called *horse lotteries* (denoted by H), are functions from states to von Neumann-Morgenstern lotteries. That is, let the set of states be $\mathcal{X}$, a finite set. For each $x \in \mathcal{X}$ and each horse lottery H, $H(x)$ is the von Neumann-Morgenstern lottery which H provides when state x occurs. Thus, in contrast with Savage's theory, horse lottery theory relies on an extraneous account of probability for defining outcomes (and applies just for simple acts: acts which assume only finitely many lottery outcomes).

The use of extraneous probability affords an elegant axiomatization of rational preference for horse lotteries. Let $\prec$ ($\preceq$) denote, respectively, strict (weak) preference between acts. For $0 \leq \alpha \leq 1$, let $\alpha H_1 + (1 - \alpha)H_2$ denote the horse lottery H such that $H(x) = \alpha H_1(x) + (1 - \alpha)H_2(x)$, as in the theory of von Neumann and Morgenstern (1947). Last, let H_{L_i} denote the "constant" horse lottery that awards the lottery L_i in every state. Four axioms summarize Anscombe-Aumann theory for nontrivial preferences. (Note that the first three axioms comprise the von Neumann-Morgenstern Utility theory.)

Axiom 1. Preference is a weak order. *That is, $\preceq$ is reflexive and transitive, with every pair of acts compared.*

Axiom 2. Independence. *For every H_1, H_2, H_3, and every $0 < \alpha \leq 1$,*

$$H_1 \preceq H_2 \quad \textit{if and only if} \quad \alpha H_1 + (1-\alpha)H_3 \preceq \alpha H_2 + (1-\alpha)H_3.$$

Axiom 3. Archimedes. *If $H_1 \prec H_2 \prec H_3$, there exist $0 < \alpha, \beta < 1$, such that*

$$\alpha H_1 + (1-\alpha)H_3 \prec H_2 \prec \beta H_1 + (1-\beta)H_3.$$

The next axiom refers to *nonnull* states.

Definition 1. A state x^* is *null* for an agent if he/she is indifferent between each pair of acts that have the same outcomes (state-by-state) on the remaining states $x \neq x^*$. That is, if x^* is null, it does not matter to the agent's assessment of acts what outcome results when x^* occurs. A state is *nonnull* if it is not null.

Axiom 4. State-independent preference for lotteries. *Given two von Neumann-Morgenstern lotteries L_1 and L_2, let H_1 and H_2 be two horse lotteries which differ only in that, for some nonnull state x^*, $H_1(x^*) = L_1$ and $H_2(x^*) = L_2(H_1(x) = H_2(x)$ for $x \neq x^*$). (That is, the two acts, H_1 and H_2 are (state by state) identical except for state x^*.) Then*

$$H_{L1} \preceq H_{L2} \quad \textit{if and only if} \quad H_1 \preceq H_2.$$

In words, Axiom 4 requires that the agent's preference for outcomes (where outcomes can be thought of also as constant acts) replicate under each state for acts that are "called off," except in that state. Throughout this essay, we will assume that preferences are not trivial. That is, each agent holds some strict preferences. This corresponds to Savage's postulate P5. The four horse lottery axioms have natural counterparts within Savage's theory for simple acts, P1–P6: Axioms 1 and 2 correspond, respectively, to Savage's postulates P1 and P2. The Archimedean Axiom 3 is contained within Savage's technical P6. Last, Savage's P3 (and part of P4) serve the same purposes as Axiom 4.

The next theorem introduces a utility function to be thought of as a function U from the set of prizes $\mathcal{G}$ to the real numbers. If L is a von Neumann-Morgenstern lottery corresponding to the simple probability distribution Q over $\mathcal{G}$, then we write $U(L)$ to stand for $\Sigma_{\mathcal{G}}\, U(g)Q(g)$.

Theorem 1 (Anscombe and Aumann 1963). *Axioms* 1–4 *are satisfied if and only if there exist a (state-independent) utility U (unique up to positive affine transformation) over prizes and a unique personal probability P over states (with $P(s) > 0$ if and only if s is nonnull), satisfying, for every H_1 and H_2,*

$$H_1 \preceq H_2 \quad \textit{if and only if} \quad \sum_j P(x_j)U(L_{1j}) \leq \sum_j P(x_j)U(L_{2j}).$$

That is, rational preference according to Anscombe-Aumann theory is equivalent to expected utility theory with a personal probability P over states and a utility U over prizes.

The central result of Seidenfeld, Kadane, and Schervish (1989a) about shared preferences for two agents, Dick and Jane, whose preferences satisfy these four axioms, is as follows. Let $\preceq_D$ ($\preceq_J$) be, respectively, Dick's (Jane's) preference. By the previous theorem, each preference order is summarized by the probability/utility pair (P_D, U_D) or (P_J, U_J). Suppose these two decision makers have different personal degrees of belief, $P_D \neq P_J$, and different utility functions for prizes, $U_D \neq U_J$. And suppose there are two prizes which they rank order the same. (See footnote 11 in Seidenfeld, Kadane, and Schervish 1989a for a discussion of the significance of this assumption.) Let $\prec$ be the strict partial order created with the (weak) Pareto condition, discussed above. That is, define $H_1 \prec H_2$ if and only if both $H_1 \prec_D H_2$ and $H_1 \prec_J H_2$, i.e., if and only if Dick and Jane each prefers H_2 over H_1.

Theorem 2 (Seidenfeld, Kadane, and Schervish 1989a). *The partial order $\prec$ agrees with no coherent preference $\preceq$ except for the two agents' preferences. That is, except for $\preceq_D$ and $\preceq_J$, no preferences satisfy the Anscombe-Aumann theory while preserving the strict preferences captured by $\prec$.*

In other words, there is no *coherent* (Pareto) compromise of preferences available to Dick and Jane. See Seidenfeld, Kadane, and Schervish (1989a) for discussions of the relation between Theorem 2 and the celebrated Impossibility Theorem of Arrow (1951), and for the relation to important papers by Hylland and Zeckhauser (1979) and Hammond (1981). The negative conclusion of Theorem 2 is reminiscent, also, of recent syndicate-theoretic work by Pratt and Zeckhauser (1989). That theory puts very restrictive conditions on the extent to which agents in a coherent syndicate may hold different beliefs or utilities. The principal contrast with Theorem 2, however, is that the syndicate is formulated with individual veto rights. Each member of the syndicate may exercise a veto whenever an option is judged by that agent to be inferior to the "status quo." Theorem 2 operates under the weak Pareto rule, without an additional veto authority for individuals.

The import of the negative result in Theorem 2 depends, of course, upon the adequacy of horse lottery theory as an account of expected utility theory. For instance, it is easy to show (see footnote 12 of Seidenfeld, Kadane, and Schervish 1989a) that if Axiom 4 is dropped, there is a continuum of Pareto compromises satisfying the first three horse lottery axioms. Thus, we arrive at the second of the two questions posed in the opening paragraph: What is the significance of Axiom 4 for expected utility theory?

We begin our answer by reviewing a formally trivial point. Let the preference order $\preceq$ be represented by expected utility, using a (possibly) *state-dependent* utility. That is, suppose

$$H_1 \preceq H_2 \quad \text{if and only if} \quad \sum_j P(x_j)U_j(H_1(x_j)) \leq \sum_j P(x_j)U_j(H_2(x_j)),$$

where it may be that, e.g., $U_2(g) \neq U_3(g)$ for some prizes g. It may be, contrary to Axiom 4, that the agent's valuation for a particular prize depends upon the state in which it occurs.

Next, let P^* be a probability which is mutually absolutely continuous with P, i.e., P and P^* agree on the "null" states of 0 probability. Define the state-dependent utility U_j^*, by

$$U_j^*(g) = U_j(g)P(x_j)/P^*(x_j). \tag{1}$$

It follows immediately that

$$H_1 \preceq H_2 \quad \text{if and only if} \quad \sum_j P^*(x_j)U_j^*(H_1(x_j)) \leq \sum_j P^*(x_j)U_j^*(H_2(x_j)).$$

Thus, using state-dependent utilities, personal probability is *wholly undefined* (up to null states) when probability and utility are reduced to a preference $\preceq$ over acts. Given only the preference structure $\preceq$, its expected utility representation by a probability/utility pair is maximally underdetermined. (See Schervish, Seidenfeld, and Kadane 1990 for a discussion of state-dependence as it applies to the theories of Anscombe and Aumann 1963 and Savage 1954.)

What is called for, then, is an expected utility theory that does not require Axiom 4 but instead introduces state-dependence in an identifiable way. To duplicate the intent of Theorem 1, we need extra "data" about preferences in order to carry out the measurement of probability and utility, where the latter may be state-dependent. Both of these goals are met in proposals by Karni, Schmeidler, and Vind (1983), elaborated by Karni (1985). In §7 of Schervish, Seidenfeld, and Kadane (1990), we simplify their construction.

The underlying theme is straightforward. The extra data on preferences, needed for investigating state-dependent utilities, involve a comparison of outcomes across states. One way to obtain these data (as suggested by Karni, Schmeidler, and Vind 1983) calls for a von Neumann-Morgenstern construction over prize-state outcomes. That is, the agent is asked to rank order (new kinds of hypothetical) lotteries over outcomes s, but where the state (x_j) is a component of that outcome. A *prize-state lottery* is just a stipulated probability distribution over the set of prize-state pairs (just as von Neumann-Morgenstern lotteries are stipulated probability distributions over the set of prizes). We will denote prize-state lotteries $\hat{L}$ where $\hat{L}(g, x)$ is the stipulated probability of prize-state pair (g, x).

The upshot of this approach is that the agent provides *two* preference orders: $\preceq$ over the original horse lottery acts and $\sqsubseteq$ over prize-state lotteries. That is, unfortunately, the lotteries used to indicate state-dependent preferences are not merged with the horse lotteries. They are not part of a single preference order. The second order ($\sqsubseteq$) reveals the agent's utility U for outcomes, which may be state-dependent. That utility is then used to restrict the potential expected-utility representations for the first order ($\preceq$). The analysis is facilitated by a requirement that, for each nonnull state x, the agent's preferences for prize-state lotteries involving only state x must agree with his/her "called-off" preferences for horse lotteries that agree on all but state x. This "consistency" condition between the two preference schemes, $\preceq$ and $\sqsubseteq$, is captured in Axiom 5, from Schervish, Seidenfeld, and Kadane (1990). The version stated here allows for situations with arbitrarily many states.

Axiom 5. Consistency. *Suppose the following conditions hold:*

- *x^* is a nonnull state.*
- $\hat{L}_1$ and $\hat{L}_2$ are prize-state lotteries whose probability distributions Q_1 and Q_2 assign probability 1 to the set of pairs $\mathcal{G} \times \{x^*\} = \{(g, x^*) : g \in \mathcal{G}\}$ (that is, with x^* fixed).
- *L_1 and L_2 are the von Neumann-Morgenstern lotteries whose simple probability distributions over $\mathcal{G}$ respectively agree with Q_1 and Q_2 when $\mathcal{G}$ is thought of as equivalent to $\mathcal{G} \times \{x^*\}$.*
- *H_1 and H_2 are horse lotteries which satisfy $H_1(x^*) = L_1$, $H_2(x^*) = L_2$ and $H_1(y) = H_2(y)$ for all states $y \neq x^*$.*

Then $\hat{L}_1 \sqsubseteq \hat{L}_2$ if and only if $H_1 \preceq H_2$.

The question we answer in section II is this: When state-dependent preferences are entertained (as we argue they ought to be), under what conditions will there be an expected-utility model for the shared preferences of two agents? That is, when can two agents acting cooperatively find a coherent Pareto compromise of their two preference schemes, $\preceq_i$ and $\sqsubseteq_i$, $i = 1, 2$? We show that the "impossibility" reported in Theorem 2 obtains (with some minor qualifications).

We note, in passing, that our proof of Theorem 2 (from Seidenfeld, Kadane, and Schervish 1989a) does not facilitate a reduction of the state-dependent case to the state-independent case. We see the idea for the reduction prompted by Theorem 13.2, p. 177 of Fishburn (1970). Provided that there are (at least) two constant acts (of unequal value and with state-independent values) to serve as the 0 and 1 utility benchmarks across states, Fishburn's result shows that the conclusion of Theorem 1 may be obtained even when other prizes are available only in designated states. Then, except for the two constant acts (0 and 1), we may as well say that each state has its own set of prizes, disjoint from every other state. The upshot is a version of Theorem 1 in which (except for the two constant acts) utility is vacuously state-independent. The reduction from state-dependence to state-independence occurs by declaring that (apart from 0 and 1) outcomes do not reappear in different states. However, we cannot apply the proof of Theorem 2 under this modification of Theorem 1 because the proof of Theorem 2 uses the structural assumption that each prize is available in each state. (This assumption is also found in the theory of Savage 1954.)

Example 3, discussed in section III, addresses problems of "pooling" opinions. There are well-studied proposals for combining the probabilities taken from several "experts" to form a single distribution that stands for the collective whole. (See Genest and Zidek 1986 for an excellent review.) However, the issue of pooling is a special case of the decision theoretic problem (discussed above), since conditional probability is a special case of state-dependent utility. That is, for a restricted set of acts, e.g., for called-off bets, the agent's utility U_j (given state x_j) is his/her conditional probability given x_j. Thus, we investigate the adequacy of pooling rules from the standpoint of the decisions they induce. Specifically, which pooling rules satisfy the Pareto condition?

II. EXISTENCE OF COMPROMISES

Let the set of possible states of nature be $\mathcal{X}$, a finite set, and let the set of possible prizes be $\mathcal{G}$. We will not assume that the same prizes are necessarily available in every state. Hence, for each state $x \in \mathcal{X}$, we let $\mathcal{G}_x$ be the set of prizes that are available in state x. We will let X stand for the random (unknown) state which eventually occurs. Let S stand for the set of prize-state pairs $s = (g, x)$. Consider two Bayesian agents. Let agent i (for $i = 1, 2$) have subjective probability P_i (with expectations denoted E_i) and state-dependent utility U_i, where $U_i(g, x)$ denotes the utility to agent i of reward g if state x occurs. We will denote by H a general act such that for each $x \in \mathcal{X}$, $H(x)$ is a simple von Neumann-Morgenstern lottery over $\mathcal{G}_x$. We suppose that agent i ranks each act H according to the value of $E_i[U_i(H(X), X)]$. We will also suppose that agent i has ranked all prize-state lotteries using U_i. That is, if Q is a stipulated probability over the set S which specifies prize-state lottery $\hat{L}$, agent i ranks $\hat{L}$ according to the value of $U_i(\hat{L}) = \Sigma_S U_i(s)Q(s)$. Of course, we assume that the rankings of each agent satisfy Axiom 5. All probabilities in this essay are countably additive. The proofs of the new results given in this section appear in the appendix.

We now ask what are the possible representations for a Bayesian compromise between two such agents if we require that the compromise satisfy a weak Pareto condition both for acts and for prize-state lotteries.

Definition 2. By a *ranking of acts*, we mean a preference relation which satisfies Axiom 1 (weak order). If l is a function from acts to real numbers, we will say that *l ranks acts* as follows: l ranks act T as weakly (strictly) preferred to act V if and only if $l(T) \geq (>) l(V)$. We say that a ranking of acts satisfies the *weak Pareto condition* with respect to two agents if, whenever $E_i[U_i(G(X), X)] < E_i[U_i(R(X), X)]$ for two acts G and R and for both $i = 1$ and $i = 2$, R ranks higher than G. Similarly, a ranking of prize-state lotteries satisfies the *weak Pareto condition* with respect to two agents if, whenever $U_i(\hat{L}_1) < U_i(\hat{L}_2)$ for two prize-state lotteries $\hat{L}_1$ and $\hat{L}_2$ and for both $i = 1$ and $i = 2$, $\hat{L}_2$ ranks higher than $\hat{L}_1$. We say that a ranking of acts and a ranking of prize-state lotteries are *Bayesian* if they each satisfy Axioms 1–3 and together they satisfy Axiom 5.

It is clear that every probability/utility pair (P, U) provides a ranking of acts by expected utility as follows. Let E mean expectation. Then set $l(T) = E(U(T(X), X))$. This leads to the following definition.

Definition 3. Let P' and P'' be probabilities with corresponding expectations denoted E' and E''. Let U' and U'' be utilities. We say that $E'U'$ *ranks all acts the same as* $E''U''$ if, for every pair of acts T and V, $E'[U'(T(X), X)] \leq E'[U'(V(X), X)]$ if and only if $E''[U''(T(X), X] \leq E''[U''(V(X), X)]$.

Similarly, every utility U provides a ranking of prize-state lotteries by the values of $U(\hat{L})$.

Our results will concern two Bayesian agents who rank acts according to expected utility and rank prize-state lotteries by utility. That is, we assume that two agents have probabilities P_1 and P_2, respectively, and utilities U_1 and U_2, respectively. We will let expectations be denoted E_1 and E_2, respectively. First, we recall a result of Harsanyi (1955) (see also Fishburn 1984), which says that a ranking of acts satisfies the weak Pareto condition with respect to two agents if and only if it ranks all acts the same as a convex combination of the two expected utilities of the agents.

Theorem 3 (Harsanyi 1955). *A ranking of acts satisfies the weak Pareto condition with respect to our two agents if and only if it ranks acts the same way as*

$$l_\alpha(T) = \alpha E_1(U_1(T(X), X)) + (1-\alpha)E_2(U_2(T(X), X)), \tag{2}$$

for some $0 \leq \alpha \leq 1$.

A corollary to this theorem contains an important ingredient of our results.

Corollary 1. *For* $i = 1, 2$ *and each* $x \in X$, *let* $f_i(x) = P_i(X = x)$. *Let the probability* P *(with corresponding expectation* E*) be defined by* $P = 0.5(P_1 + P_2)$. *Then* $f(x) = P(X = x) = 0.5[f_1(x) + f_2(x)]$. *For each* α, *the utility*

$$U_\alpha(g, x) = \alpha \frac{f_1(x)}{f(x)} U_1(g, x) + (1-\alpha)\frac{f_2(x)}{f(x)} U_2(g, x) \tag{3}$$

has the property that EU_α *ranks all acts the same as does* l_α *from* (2).

In words, a ranking of acts satisfies the weak Pareto condition with respect to our two agents if and only if it ranks all acts the same as EU_α for some α.

Next, we derive a similar result for prize-state lotteries. The following lemma is essentially the same as Theorem 3, except that it refers to prize-state lotteries. (Its proof will not be given because it is identical to that of Theorem 3.)

Lemma 1. *A utility U ranks prize-state lotteries in such a way that it satisfies the weak Pareto condition with respect to our two agents if and only if there exist* $a > 0$, b, *and* $0 \leq \beta \leq 1$ *such that* $aU + b = \beta U_1 + (1 - \beta)U_2$.

We will refer to the convex combination $\beta U_1 + (1 - \beta)U_2$ as $U^{(\beta)}$. Theorem 2 of Schervish, Seidenfeld, and Kadane (1990) says that, in order for the ranking of prize-state lotteries given by $U^{(\beta)}$ to be consistent (Axiom 5) with a ranking of acts by expected utility, there must exist a probability P_* (with expectation E_*) such that the ranking of acts is given by $E_*U^{(\beta)}$.

We are now prepared to put these results together. The question of what nonautocratic Bayesian Pareto compromises exist for both acts and prize-state lotteries becomes the question of what probabilities P_* (with expectation E_*) and which β and α exist such that $E_*U^{(\beta)}$ ranks acts the same as EU_α. We propose to answer this question by fixing $0 < \beta < 1$ and determining under what conditions there exist α and P_* such that $E_*U^{(\beta)}$ ranks acts the same as EU_α.

The next theorem contains the answer to our main question. We first offer some intuition as to the meaning of the conditions of the theorem. (The notation in this discussion includes notation introduced in the theorem.) The set B_1 can be thought of as the set of states such that the conditional probabilities given B_1 are equal. That is, $P_1(A|B_1) = P_2(A|B_1)$. We know when two agents agree on the probabilities of events, there are nonautocratic Pareto compromises available. The set B_2 is the set of states x such that the two utility functions $U_1(\cdot, x)$ and $U_2(\cdot, x)$ are essentially the same (when $c(x)$ and $d(x)$ have the same sign). We know that when agents agree on the utilities of all lotteries, nonautocratic Pareto compromises are available. By assuming that $P(B_1 \cup B_2) = 1$, we assume that almost surely one of the two cases just described occurs (i.e., for each state, the agents either agree on the probabilities or they agree on the utilities). When $c(x)$ and $d(x)$ have opposite signs, then the two utilities are in complete opposition. It is known that nonautocratic Pareto compromises exist in this case also. (See footnote 11 in Seidenfeld, Kadane, and Schervish

1989a.) Note that the function r in Theorem 4 is defined in terms of the functions c and d, which may not be unique. For example, if there exists a set of x values in B_2 such that both $U_1(g, x)$ and $U_2(g, x)$ are constant in g then both $c(x)$ and $d(x)$ are arbitrary for such x values. This means that there might be many functions r of the form specified. Each such function will be called a *version* of r. The reader should also note that, if $x \in B_1 \cap B_2$, then both forms of $r(x)$ in Theorem 4 are the same.

Theorem 4. *Let* $0 < \beta < 1$. *Define*

$$B_1 = \{x : \alpha(1-\beta)f_1(x) = \beta(1-\alpha)f_2(x)\},$$

$$\begin{aligned} B_2 = \{x : \exists b(x), c(x), d(x) \text{ such that } \\ d(x)U_2(g,x) = c(x)U_1(g,x) + b(x) \forall\, g \in \mathcal{G}_x, \\ \text{with not both } c(x) = 0 \text{ and } d(x) = 0\}, \end{aligned}$$

$$r(x) = \begin{cases} \dfrac{\alpha}{\beta} f_1(x) & \text{if } x \in B_1, \\ \dfrac{d(x)\alpha f_1(x) + c(x)(1-\alpha)f_2(x)}{d(x)\beta + c(x)(1-\beta)} & \text{if } x \in B_2 \setminus B_1. \end{cases}$$

Then $E_*U^{(\beta)}$ *ranks all acts the same as* EU_α *if and only if the following conditions hold*:

1. $P(B_1 \cup B_2) = 1$.
2. *There exists a version of r such that* $P(\{x : r(x) \geq 0\}) = 1$, *and* $f_*(x) = r(x)/a$ *is the mass function of* P_*, *where* $a = \Sigma_{y \in X} r(y)$.

Speaking very loosely, Theorem 4 says that there exists a Bayesian Pareto compromise between two agents which applies to both acts and prize-state lotteries if and only if, for each state, either the agents agree on the probability of the state or they agree on the utility in that state. The proof of this theorem appears in the appendix. We illustrate the theorem with several examples in section III.

III. EXAMPLES

Example 1 is the special case handled by Seidenfeld, Kadane, and Schervish (1989a).

Example 1. Suppose that $\mathcal{X} = \{0, 1\}$ and $\mathcal{G}_x = \{0, r, 1\}$ for all x. Suppose that for each i, $P_i(0) = p_i$, $U_i(0, x) = 0$ for all x, and $U_i(1, x) = 1$ for all x. Let $U_1(r, x) = r_1$ and $U_2(r, x) = r_2$. In this way, utilities are state-independent. Suppose that $p_1 \neq p_2$ and $r_1 \neq r_2$. The set B_2 is clearly empty. The set B_1 can contain at most one of the two x values, and then only for exceptional values of α and β. The result is that $P(B_1 \cup B_2) < 1$. We already know that only autocratic Pareto compromises are available in this case, and Theorem 4 confirms this.

Example 2 shows how all of the conditions of Theorem 4 can be met with both B_1 and B_2 nonempty.

Example 2. Let $\mathcal{X} = \{1, 2, 3, 4, 5\}$ and $\mathcal{G}_x = \{g_0, g_1, g_2\}$ for all x. Let the state-dependent utilities be

$$
\begin{aligned}
&U_1(g_j, 1) = j, \quad U_2(g_j, 1) = 2j, \\
&U_1(g_j, 2) = j, \quad U_2(g_j, 2) = \tfrac{1}{2}j, \\
&U_1(g_j, 3) = j, \quad U_2(g_j, 3) = j, \\
&U_1(g_j, 4) = j, \quad U_2(g_j, 4) = 2 - j, \\
&U_1(g_j, 5) = j, \quad U_2(g_j, 5) = \begin{cases} 0 & \text{if } j = 0, \\ 2 & \text{if } j = 1, \\ 1 & \text{if } j = 2. \end{cases}
\end{aligned}
$$

The two probabilities are $P_2(x) = 0.2$ for all x and $P_1(x) = 0.31$ for $x \in \{1, 2, 3\}$, $P_1(4) = 0.02$, $P_1(5) = 0.05$. With $\alpha = \frac{2}{3}$ and $\beta = \frac{1}{3}$, we see that $B_1 = \{5\}$ and $B_2 = \{1, 2, 3, 4\}$. The values of $c(x)$ and $d(x)$ are

$$
c(x) = \begin{cases} 2 & \text{if } x = 1, \\ 1 & \text{if } x = 2, \\ \text{arbitrary} & \text{if } x = 3, \\ 1 & \text{if } x = 4, \end{cases}
$$

$$
d(x) = \begin{cases} 1 & \text{if } x = 1, \\ 2 & \text{if } x = 2, \\ 0 & \text{if } x = 3, \\ -1 & \text{if } x = 4. \end{cases}
$$

We calculate $r(x)$ as

$$r(x)=\begin{cases}0.204 & \text{if } x=1,\\ 0.36 & \text{if } x=2,\\ 0.1 & \text{if } x=3,\\ 0.16 & \text{if } x=4,\\ 0.1 & \text{if } x=5.\end{cases}$$

The sum of these values is a = 0.924. It follows that the mass function of P_* is

$$f_*(x)=\begin{cases}0.2208 & \text{if } x=1,\\ 0.3896 & \text{if } x=2,\\ 0.1082 & \text{if } x=3,\\ 0.1732 & \text{if } x=4,\\ 0.1082 & \text{if } x=5.\end{cases}$$

To verify that everything worked out, we tabulate both $U^{(\beta)}(g, x)$ and $U_\alpha(g, x)$ in the format $(U(g_1, x), U(g_2, x), U(g_3, x))$:

x	$af_*(x)U^{(\beta)}(\cdot, x)$	$U_\alpha(\cdot, x)$
1	$(0, 0.34, 0.68)$	$(0, 0.34, 0.68)$
2	$(0, 0.24, 0.48)$	$(0, 0.24, 0.48)$
3	$(0.033\bar{3}, 0.1, 0.166\bar{6})$	$(0.206\bar{6}, 0.273\bar{3}, 0.34)$
4	$(0.213\bar{3}, 0.16, 0.106\bar{6})$	$(0.133\bar{3}, 0.08, 0.26\bar{6})$
5	$(0, 0.166\bar{6}, 0.133\bar{3})$	$(0, 0.166\bar{6}, 0.133\bar{3})$

We see that the difference between $af_*(x)U^{(\beta)}(g, x)$ and $U_\alpha(g, x)$ is constant in g (although not constant in x). Hence, $E_*U^{(\beta)}$ ranks all acts the same as EU_α.

Example 3 shows how the results of this essay are relevant to considerations of "externally Bayesian" pooling operators.

Example 3. Suppose that interest lies in the joint distribution of two random quantities (X, G) lying in a finite space $\mathcal{X} \times \mathcal{G}$ as well as in the conditional distribution of G given X. Suppose we have two probability functions P_1 and P_2 over $\mathcal{X} \times \mathcal{G}$. For each $0 \le \beta \le 1$, let $P_\beta = \beta P_1 + (1 - \beta)P_2$, which is called the *linear opinion pool.* Let $P_i^{G|X}$ denote the conditional mass function of G given X derived from P_i, and let P_i^X denote the marginal mass function of X. Similarly, let $P_\beta^{G|X}$ and P_β^X denote the conditional and marginal mass functions derived from P_β. The linear opinion pool is *externally Bayesian* if

$$P_\beta^{G|X} = \beta P_1^{G|X} + (1-\beta) P_2^{G|X}. \tag{4}$$

That is to say, if we pool the joint distributions and then condition on X, we get the same result as if we condition each distribution on X and then pool.

An interesting question arises as to whether a linear opinion pool can be externally Bayesian. Theorem 4 provides an answer. Let the set of prizes be $\mathcal{G}$ and let the set of states be $\mathcal{X}$. Define $U_i(g, x) = P_i^{G|X}(g|x)$, that is, let the conditional mass function of G given $X = x$ play the role of the state-dependent utility function. Also, let the marginal distribution of X play the role of the probability. The right-hand side of (4) is just $U^{(\beta)}$. The left-hand side of (4) can be written as

$$\begin{aligned} P_\beta^{G|X} &= \frac{\beta P_1^{G|X} P_1^X + (1-\beta) P_2^{G|X} P_2^X}{\beta P_1^X + (1-\beta) P_2^X} \\ &= \frac{U_\beta}{P_*}, \text{ where} \\ P_* &= \beta P_1^X + (1-\beta) P_2^X. \end{aligned} \tag{5}$$

It follows that (4) holds if and only if $U_\beta = P_* U^{(\beta)}$. Using this correspondence, an act H would correspond to a stipulated probability distribution over $\mathcal{G}$ for each $x \in \mathcal{X}$, say $P_H(\cdot|x)$. Then

$$E_i U_i(H, X) = \sum_{\mathcal{X}} \sum_{\mathcal{G}} U_i(g, x) P_i^X(x) P_H(g|x),$$

and, if we let E_* be the expectation corresponding to P_*,

$$E_* U^{(\beta)}(H, X) = \sum_{\mathcal{X}} \sum_{\mathcal{G}} P_* U^{(\beta)}(g, x)(x) P_H(g|x).$$

It follows that (4) holds if and only if $E_* U^{(\beta)}$ ranks all acts the same as $\beta E_1 U_1 + (1-\beta) E_2 U_2$. This, in turn, holds if and only if the conditions of Theorem 4 hold. Condition 1 says that both P_1^X and P_2^X must assign probability 1 to the union of the two sets B_1 and B_2. Since $\alpha = \beta$ in this case, we can write

$$B_1 = \{x : P_1^X(x) = P_2^X(x)\}.$$

Condition 2 says that P_* must equal P_i^X/a on B_1. But (5) implies that $a = 1$. On $B_2 \backslash B_1$, $P_*(x)$ must be a convex combination of $P_1^X(x)$ and

$P_2^X(x)$, but it will be a different convex combination than (5) unless $b(x) = 0$ and $c(x) = d(x) = 1$. It also follows that

$$B_2 = \{x : P_1^{G|X}(g|x) = P_2^{G|X}(g|x), \text{ for all } g\}.$$

In summary, (4) holds if and only if, for almost all x, either $P_1^X(x) = P_2^X(x)$ or $P_1^{G|X}(\cdot|x) = P_2^{G|X}(\cdot|x)$.

IV. CONCLUSION

Theorem 4 shows that, for two Bayesian agents, nonautocratic (weak) Pareto compromises exist only under very restrictive conditions. This result extends Theorem 2 by allowing for state-dependent utilities. Our results suggest that, even with state-dependent utilities, there is little hope that two Bayesians can arrive at a Bayesian compromise which satisfies the (weak) Pareto rule.

We see three ways for avoiding the unpleasant conclusion that Bayesians cannot find cooperative Bayesian compromises:

1. Savage (1954, §13.5), in his discussion of the group minimax-regret rule, concludes that the standards of rational group behavior need not be the same as the standards of his theory of rational individual behavior. Specifically, he offers that the rational individual, but not the group, ought to be committed to the principle that preference is a weak-order (his postulate P1).
2. Levi (1982) argues for a unified account of individual and (cooperative) group decision making – without the assumption that preference induces a weak-order. But, in rebuttal to our Theorem 2, he finds that the (weak) Pareto condition is unwarranted (see Levi 1990). Roughly put, he defends a logically weaker rule, which he calls "Robust Pareto," wherein the (weak) Pareto condition applies only when, e.g., each agent's preference for, say, H_2 over H_1, is invariant over an interchange of their probabilities (for states), or is invariant over an interchange of their utilities (for outcomes). His position is that the preference relation, as captured by expected utility inequalities, is a derivative notion to be supported by reasons, expressed in terms of probabilities and utilities. Unless the agents can find common reasons for their common preferences, such preferences are *not* to count in a Pareto compromise. The Robust Pareto rule is intended to capture just those cases where common preferences are supported by some common reasons. Under this

modification of Pareto, using Levi's Robust Pareto rule, there exists a convex family of Bayesian compromises for each pair of Bayesian agents. Hence, in Levi's theory, there is no result analogous to our Theorems 2 and 4.

3. In Seidenfeld, Kadane, and Schervish (1989b) and Seidenfeld, Schervish, and Kadane (1990), we explore representations for a theory of preference, where preference is a strict partial order, using sets of probability/utility pairs. This provides a unified standard of rational behavior across individuals and cooperative groups, and yet maintains the (weak) Pareto principle. We hope that approach will afford a viable solution to the challenge of rational group behavior.

APPENDIX. PROOF OF THEOREM 4

The proof of Theorem 4 will proceed through several lemmas. We will use the following notation throughout this appendix. Probabilities on $\mathcal{X}$ will be denoted by P with superscripts or subscripts and the corresponding mass functions and expectations will be denoted by the letters f and E, respectively, with the same superscripts or subscripts. For example, $f''(x) = P''(X = x)$ and $E''(h(X)) = \Sigma_{x\in\mathcal{X}} h(x)f''(x)$.

Lemma 2. *Suppose $E'U'$ and $E''U''$ rank all acts the same. Then there exist $a > 0$ and b, such that, for every act G,*

$$E'[aU'(G(X), X) + b] = E''[U''(G(X), X)]. \tag{6}$$

Proof. First, suppose that there exist two acts T and R such that $E'[U'(R(X), X)] < E'[U'(T(X), X)]$ and $E''[U''(R(X), X)] < E''[U''(T(X), X)]$. Then set

$$a = \frac{E''[U''(T(X), X)] - E''[U''(R(X), X)]}{E'[U'(T(X), X)] - E'[U'(R(X), X)]},$$

$$b = E''[U''(R(X), X)] - aE'[U'(R(X), X)].$$

It follows that both of the next equations hold

$$E'[aU'(R(X), X) + b] = E''[U''(R(X), X)],$$

$$E'[aU'(T(X), X) + b] = E''[U''(T(X), X)].$$

For each act G, there exists $0 \le \alpha \le 1$ such that one of the following is true:

$$\alpha T + (1-\alpha)R \approx G,$$
$$\alpha G + (1-\alpha)R \approx T,$$
$$\alpha T + (1-\alpha)G \approx R,$$

where $\approx$ refers to the common preference ranking of $E'U'$ and $E''U''$. In the first case, both

$$E'[U'(G(X), X)] = \alpha E'[U'(T(X), X)] + (1-\alpha)E'[U'(R(X), X)],$$
$$E''[U''(G(X), X)] = \alpha E''[U''(T(X), X)] + (1-\alpha)E''[U''(R(X), X)],$$

and we see that (6) holds. The proof is similar in the other two cases.

Finally, suppose that both $E'U'$ and $E''U''$ rank all acts as equivalent. Then, for all acts G, $E'[U'(G(X), X)] = c'$, say, and $E''[U''(G(X), X)] = c''$, say. Let $a = 1$ and $b = c'' - c'$ to complete the proof. □

Under the conditions of Lemma 2, the following lemma allows us to replace the two different probabilities with a single probability. (The proof is trivial and not given.)

Lemma 3. *Suppose that* $E'[U'(G(X), X)] = E''[U''(G(X), X)]$ *for every act G. Let P be a probability (with mass function f) such that* $f(x) = 0$ *implies* $f'(x) = 0$ *and* $f''(x) = 0$. *Then*

$$E\left[\frac{f'(X)}{f(X)}U'(G(X), X)\right] = E\left[\frac{f''(X)}{f(X)}U''(G(X), X)\right]$$

for every act G.

The following lemma says that, under the conditions of Lemmas 2 and 3, for each x, the two utilities U' and U'', as functions of g, must be related by an affine transformation.

Lemma 4. *If* $E[U'(G(X), X)] = E[U''(G(X), X)]$ *for every act G, then* $P\{x: U'(g, x) - U''(g, x)$ *is constant in* $g\} = 1$.

Proof. Let

$$C_+ = \{x : \exists g(x), h(x) \in \mathcal{G}_x \text{ such that } U'(g(x), x) - U''(g(x), x) > U'(h(x), x) - U''(h(x), x)\},$$

$$C_- = \{x : \exists g(x), h(x) \in \mathcal{G}_x \text{ such that } U'(g(x), x) - U''(g(x), x) < U'(h(x), x) - U''(h(x), x)\}.$$

We need to prove that $P(C_+ \cup C_-) = 0$. Define two acts

$$G(x) = \begin{cases} g(x) & \text{if } x \in C_+, \\ \text{arbitrary} & \text{if } x \notin C_+, \end{cases}$$

$$H(x) = \begin{cases} h(x) & \text{if } x \in C_+, \\ G(x) & \text{if } x \notin C_+. \end{cases}$$

For $x \in C_+$, set

$$r(x) = U'(g(x), x) - U''(g(x), x) - U'(h(x), x) + U''(h(x), x),$$

which is strictly positive for all $x \in C_+$. Set $r(x) = 0$ for $x \notin C_+$. It is easy to see that

$$0 = E[U'(G(X), X)] - E[U''(G(X), X)] - E[U'(H(X), X)] \\ + E[U''(H(X), X)] = E[r(X)],$$

which implies that $P(C_+) = 0$. A similar proof shows $P(C_-) = 0$. □

We are now in position to prove Corollary 1.

Proof of Corollary 1. For every act G,

$$\alpha E_1[U_1(G(X), X)] + (1-\alpha)E_2[U_2(G(X), X)] \\ = E[\alpha f_1(X)U_1(G(X), X)] + (1-\alpha)f_2(X)U_2(G(X), X)]. \; \square$$

In the remainder of the theorems, U_α will always mean (3), and P will mean $0.5(P_1 + P_2)$.

Lemma 5. *Suppose that $E'U'$ ranks all acts the same as EU_α. The following are true:*

- *There exists P'' such that $E''U'$ ranks acts the same as $E'U'$ and $f(x) = 0$ implies $f''(x) = 0$.*
- *There exists $a > 0$ such that*

$$P\left(x : a\frac{f''(x)}{f(x)}U'(g,x) - U_\alpha(g,x) \text{ is constant in } g\right) = 1.$$

For each x, call the constant value $v_a(x)$.

Proof. If $f(x) = 0$ implies $f'(x) = 0$, we need only prove the second part of the lemma. Let A be the set of all x^* such that $f(x^*) = 0$ but $f'(x^*) > 0$. It must be that

$$P'(x \in A : U'(g, x) \text{ is constant in } g) = P'(A),$$

or else $E'U'$ would be able to distinguish acts that differed only on A, while EU_α would not. Suppose that $c(x)$ is the function such that $P'(\{x : U'(g, x) = c(x) \text{ for all } g\}) = P'(A)$. There are two cases to consider.

1. First, if $P'(A) = 1$, then $E'U'$ must rank all acts as equivalent because $E'(U'(G(X), X)) = E'(c(X))$ for every act G. In this case, set $P'' = P$. Then $E''U'$ will also rank all acts as equivalent.
2. Second, suppose $0 < P'(A) < 1$. For each event B define $P''(B) = P'(B|A^C)$. Now $f(x) = 0$ implies $f''(x) = 0$, and for each act G

$$E'U'(G, X) = P'(A^C)E''U'(G, X) + E'(c(X)).$$

It is now clear that $E''U'$ ranks all acts the same as $E'U'$.

For the second part of the lemma, use Lemma 2 to find $a > 0$ and b such that, for every act G,

$$E[U_\alpha(G(X), X)] = E''[aU'(G(X), X) + b] = E\left[a\frac{f''(X)}{f(X)}U(G(X), X) + b\right].$$

Next, use Lemma 4 to conclude that

$$P\left(x : a\frac{f''(X)}{f(X)}U(g, x) + b - U_\alpha(g, x) \text{ is constant in } g\right) = 1. \quad \square$$

In light of Lemma 5, we will assume that each P_* that we consider satisfies "$f(x) = 0$ implies $f_*(x) = 0$."

Lemma 6. *$E_*U^{(\beta)}$ ranks all acts the same as EU_α if and only if there exists $a > 0$ such that*

$$h_a(g, x) = [a\beta f_*(x) - \alpha f_1(x)]U_1(g, x) + [a(1-\beta)f_*(x) - (1-\alpha)f_2(x)]U_2(g, x) \quad (7)$$

is constant in g a.s.$[P]$. Call this constant value $h_a(x)$.

Proof. For the "if" part, we note that, for every act G,

$$aE_*[U^{(\beta)}(G(X), X)] - E[U_\alpha(G(X), X)] = E\left[\frac{h_a(X)}{f(X)}\right],$$

which is the same for all acts because we assume (7). It follows that $E_*U^{(\beta)}$ ranks acts the same as EU_α. For the "only if" part, apply Lemma 5. □

We are now in position to prove the "if" part of Theorem 4.

Proof (of the "if" part of Theorem 4). Assume that conditions 1 and 2 of Theorem 4 hold. For all $x \in B$,

$$a\beta f_*(x) = \alpha f_1(x),$$
$$a(1-\beta)f_*(x) = (1-\alpha)f_2(x).$$

It follows that $h_a(g, x)$ from (7) equals 0 for all $x \in B_1$. For $x \in B_2$ and such that $d(x) \neq 0$,

$$\begin{aligned} h_a(g,x) &= [\beta r(x) - \alpha f_1(x)]U_1(g,x) + [(1-\beta)r(x) - (1-\alpha)f_2(x)] \\ &\quad \times \left[\frac{c(x)U_1(g,x)+b(x)}{d(x)}\right] \\ &= \frac{b(x)}{d(x)}[(1-\beta)r(x) - (1-\alpha)f_2(x)] \\ &= \frac{b(x)}{\beta d(x) + (1-\beta)c(x)}[\alpha(1-\beta)f_1(x) - (1-\alpha)\beta f_2(x)], \end{aligned}$$

which is constant in g. Similarly, if $c(x) \neq 0$,

$$\begin{aligned} h_a(g,x) &= \left[\beta\frac{d(x)\alpha f_1(x) + c(x)(1-\alpha)f_2(x)}{d(x)\beta + c(x)(1-\beta)} - \alpha f_1(x)\right] \\ &\quad \left[\frac{d(x)U_2(g,x) - b(x)}{c(x)}\right] \\ &\quad + \left[(1-\beta)\frac{d(x)\alpha f_1(x) + c(x)(1-\alpha)f_2(x)}{d(x)\beta + c(x)(1-\beta)} - (1-\alpha)f_2(x)\right]U_2(g,x) \\ &= -\frac{b(x)}{c(x)}[\beta r(x) - \alpha f_1(x)] \\ &= \frac{b(x)}{\beta d(x) + (1-\beta)c(x)}[\alpha(1-\beta)f_1(x) - (1-\alpha)\beta f_2(x)], \end{aligned}$$

which is the same as when $d(x) \neq 0$. So, the conditions of Lemma 6 (the "if" part) are met and we conclude that $E_*U^{(\beta)}$ ranks all acts the same as EU_α. □

Lemma 7. *Let B_1 and B_2 be as in Theorem 4, and suppose that $P(B_1 \cup B_2) = 1$. Suppose also that there exist P_* and α such that $E_*U^{(\beta)}$ ranks all acts the same as EU_α. Then $f_* = dP_*/dP$ must have the form given in Theorem 4, namely $r(x)/a$, for some version of r.*

Proof. First, we look at $x \in B_2$. Lemma 6 says that $h_a(g, x)$ must be constant in g. We deal with four cases here.

Case (i). Both $U_1(g, x)$ and $U_2(g, x)$ are constant in g. Then $h_a(g, x)$ is constant in g, and both $c(x)$ and $d(x)$ can be arbitrary (so long as both are not 0). This means that $r(x)$ is arbitrary, and every $f_*(x)$ has the desired form.

Case (ii). $U_1(g, x)$ is constant in g, but $U_2(g, x)$ is not. In this case $d(x) = 0$ is necessary, and $r(x) = (1 - \alpha)f_2(x)/(1 - \beta)$. It is also true in this case that $h_a(g, x)$ equals a constant plus $[af_*(x)(1 - \beta) - (1 - \alpha)f_2(x)]U_2(g, x)$. For this to be constant, it is necessary that $f_*(x) = r(x)/a$.

Case (iii). $U_2(g, x)$ is constant in g, but $U_1(g, x)$ is not. This is virtually the same as the previous case.

Case (iv). Neither $U_1(g, x)$ nor $U_2(g, x)$ is constant in g. In this case, neither $c(x) = 0$ nor $d(x) = 0$ is possible. It also follows that $c(x)$ and $d(x)$ are unique up to multiplication by nonzero constant (for each x). To see this, suppose that for all g, $c_j(x)U_1(g, x) + b_j(x) = d_j(x)U_2(g, x)$ for $j = 1, 2$. Set $c^*(x) = c_2(x)/c_1(x)$. Then

$$c^*(x)b_1(x) - b_2(x) = [c^*(x)d_1(x) - d_2(x)]U_2(g, x).$$

It is clear that $c^*(x)d_1(x) = d_2(x)$, and we have proven uniqueness. It follows that the form of $r(x)$ is unique in this case. Now, write $h_a(g, x)$ as

$$[\alpha f_*(x)\beta - \alpha f_1(x)]U_1(g, x) + [af_*(x) - (1-\alpha)f_2(x)]\left[\frac{c(x)U_1(g, x) - b(x)}{d(x)}\right]$$

$$= U_1(g, x)\left[af_*(x)\beta - \alpha f_1(x) + \frac{c(x)}{d(x)}af_*(x)(1-\beta)\right.$$

$$\left. - \frac{c(x)}{d(x)}(1-\alpha)f_2(x)\right] - \frac{b(x)}{d(x)}[af_*(x) - (1-\alpha)f_2(x)].$$

This is constant in g if and only if $f_*(x) = r(x)/a$.

Next, look at $x \in B_1 \backslash B_2$. The only way for $h_a(g, x)$ to be constant in g is for the coefficients of both $U_1(g, x)$ and $U_2(g, x)$ to be zero. The reason is that, if not, x would be in B_2. So, $af_*(x)\beta = \alpha f_1(x)$ and we see that $f_*(x) = r(x)/a$. □

Lemma 8. *Let B_1 and B_2 be as in Theorem 4. Suppose that P_* and α exist such that $E_*U^{(\beta)}$ ranks all acts the same as EU_α. Then $P(B_1 \cup B_2) = 1$.*

Proof. Suppose that such a P_* and α exist. Let $a > 0$ and b be as guaranteed by Lemma 2. Let $B_3 = (B_1 \cup B_2)^C$. Then, for every $x \in B_3$ and for every $d(x)$ and $c(x)$, there exist $g_1(x)$, $g_2(x) \in \mathcal{G}_x$ such that

$$d(x)U_2(g_1(x), x) - c(x)U_1(g_1(x), x) > d(x)U_2(g_2(x), x) - c(x)U_1(g_2(x), x), \quad (8)$$

and $\alpha(1-\beta)f_1(x) \neq \beta(1-\alpha)f_2(x)$. So, for each $x \in B_3$, define

$$c(x) = af_*(x)\beta - \alpha f_1(x),$$
$$d(x) = af_*(x)(1-\beta) - (1-\alpha)f_2(x).$$

Both $c(x) = 0$ and $d(x) = 0$ simultaneously is not possible, since this would imply that $x \in B_1$. For these choices of $c(x)$ and $d(x)$, define $g_1(x)$ and $g_2(x)$ to satisfy (8). Now, define two acts

$$G_1(x) = \begin{cases} g_1(x) & \text{if } x \in B_3, \\ \text{arbitrary} & \text{if } x \notin B_3, \end{cases}$$

$$G_2(x) = \begin{cases} g_2(x) & \text{if } x \in B_3, \\ G_1(x) & \text{if } x \notin C. \end{cases}$$

We now have that

$$\begin{aligned} 0 &= E_*[aU^{(\beta)}(G_1(X), X)] - E[U_\alpha(G_1(X), X)] \\ &\quad -\{E_*[aU^{(\beta)}(G_2(X), X)] - E[U_\alpha(G_2(X), X)]\} \\ &= \sum_{x \in B_3} [af_*(x)\beta - \alpha f_1(x)][U_1(g_1(x), x) - U_1(g_2(x), x)] \\ &\quad + [af_*(x)(1-\beta) - (1-\alpha)f_2(x)][U_2(g_1(x), x) - U_2(g_2(x), x)] \\ &= \sum_{x \in B_3} [c(x)U_1(g_1(x), x) - d(x)U_2(g_1(x), x)] \\ &\quad - [c(x)U_1(g_2(x), x) - d(x)U_2(g_2(x), x)] \end{aligned}$$

By (8), each term in this last sum is positive, hence B_3 must be empty. □

Finally, we are ready to prove the "only if" part of Theorem 4.

Proof (of the "only if" part of Theorem 4). Suppose that there exists a P_* as in the statement of the theorem. We will now prove that conditions 1 and 2 hold. Lemma 8 says that if such a P_* exists, then condition 1 holds. Lemma 7 says that if such a P_* exists, then condition 2 holds. □

REFERENCES

Anscombe, F. J. and R. J. Aumann, "A Definition of Subjective Probability," *Ann. Math. Statist.*, 34 (1963) 199–205.

Arrow, K., *Social Choice and Individual Values*, Wiley, New York, 1951.

Fishburn, P. C., *Utility Theory for Decision Making*, Wiley, New York, 1970.

——, "On Harsanyi's Utilitarian Cardinal Welfare Theorem," *Theory and Decision*, 17 (1984), 21–28.

Genest, C. and J. Zidek, "Combining Probability Distributions," *Statist. Sci.*, 1 (1986), 114–148.

Hammond, P., "Ex-ante and Ex-post Welfare Optimality under Uncertainty," *Econometrica*, 48 (1981), 235–250.

Harsanyi, J. C., "Cardinal Welfare, Individualistic Ethics, and Interpersonal Comparisons of Utility," *J. Political Economy*, 63 (1955), 309–321.

Hylland, A. and R. Zeckhauser, "The Impossibility of Bayesian Group Decisions with Separate Aggregation of Belief and Values," *Econometrica*, 47 (1979), 1321–1336.

Karni, E., *Decision Making under Uncertainty*, Harvard University Press, Cambridge, 1985.

——, D. Schmeidler and K. Vind, "On State Dependent Preferences and Subjective Probabilities," *Econometrica*, 51 (1983), 1021–1031.

Levi, I., "Conflict and Social Agency," *J. Philosophy*, 84 (1982), 231–247.

——, "Pareto Unanimity and Consensus," *J. Philosophy*, 87 (1990), 481–492.

Pratt, J. W. and R. Zeckhauser, "The Impact of Risk Sharing on Efficient Decision," *J. Risk and Uncertainty*, 2 (1989), 219–234.

Savage, L. J., *The Foundations of Statistics*, John Wiley, New York, 1954.

Schervish, M. J., T. Seidenfeld and J. B. Kadane, "State-dependent Utilities," *J. Amer. Statist. Assoc.*, 85 (1990), 840–847. [Chapter 2.2, this volume]

Seidenfeld, T., J. B. Kadane and M. J. Schervish, "On the Shared Preferences of Two Bayesian Decision Makers," *J. Philosophy*, 86 (1989), 225–244. [Chapter 1.1, this volume]

——, —— and ——, "A Representation of Partially Ordered Preferences," *Ann. Statistics*, 23 (1995), 2168–2217. [Chapter 1.3, this volume]

——, M. J. Schervish and J. B. Kadane, "Decisions Without Ordering," In W. Sieg (Ed.), *Acting and Reflecting*, Kluwer Academic, Dordrecht, 1990, 143–170. [Chapter 1.2, this volume]
von Neumann, J. and O. Morgenstern, *Theory of Games and Economic Behavior* (2nd Ed.), Princeton University Press, Princeton, NJ, 1947.

2.4

A Conflict Between Finite Additivity and Avoiding Dutch Book

TEDDY SEIDENFELD AND
MARK J. SCHERVISH

ABSTRACT

For Savage (1954) as for deFinetti (1974), the existence of subjective (personal) probability is a consequence of the normative theory of preference. (DeFinetti achieves the reduction of belief to desire with his generalized Dutch-Book argument for *previsions*.) Both Savage and deFinetti rebel against legislating countable additivity for subjective probability. They require merely that probability be finitely additive. Simultaneously, they insist that their theories of preference are weak, accommodating all but self-defeating desires. In this chapter we dispute these claims by showing that the following three cannot simultaneously hold:

i. Coherent belief is reducible to rational preference, i.e. the generalized Dutch-Book argument fixes standards of coherence.
ii. Finitely additive probability is coherent.
iii. Admissible preference structures may be free of *consequences*, i.e. they may lack prizes whose values are robust against all contingencies.

I. INTRODUCTION

One of the most important results of the subjectivist theories of Savage and deFinetti is the thesis that, normatively, preference circumscribes

We would like to thank P. C. Fishburn, Jay Kadane, Isaac Levi, Patrick Maher and a referee for their helpful comments. Also, we have benefited from discussions about consequences with E. F. McClennen.

Reprinted with permission from *Philosophy of Science*, 50 (1983): 398–412, published by the University of Chicago Press.

belief. Specifically, these authors argue that the theory of subjective probability is reducible to the theory of reasonable preference, i.e. coherent belief is a consequence of rational desire. In Savage's (1954) axiomatic treatment of preference, the existence of a quantitative subjective probability is assured once the postulates governing preference are granted. In deFinetti's (1974) discussion of *prevision*, avoidance of a (uniform) loss for certain is thought to guarantee agreement with the requirements of subjective probability (sometimes called the avoidance of "Dutch Book").

Obviously, the significance of these results depends upon the avowed liberality regarding the range of preferences and beliefs the theories are said to tolerate. Both Savage and deFinetti are explicit in their opposition to the stipulation of countable additivity for probability, and, of course, each insists that the constraints imposed on reasonable preference are weak, permitting all but self-defeating desires. It is our purpose in this essay to challenge these claims. We aim to show that the reduction of belief to preference cannot be carried off as Savage and deFinetti suggest without contracting the range of admissible states of preference and belief. In particular, we argue that the purported reduction fails unless

i. subjective probability is countably additive, or
ii. each agent is required to acknowledge the existence of a rich supply of *consequences*, i.e. prizes whose values are robust against the contingencies of nature.

As we see in section III, Savage recognized that consequences serve as expedients in his theory for constructing "constant acts" and should not be essential to subjectivism. We find no convenient stockpile of consequences. In fact, it seems reasonable to deny that there are consequences in practical decisions. Thus, our position is that, lacking consequences, expected utility theory must treat subjective probability distributions as *extraneous* (Fishburn 1970, §12.2 and chapter 13). Otherwise, probabilities which are not countably additive cannot be sanctioned. In light of our findings in section IV, the problem is deeply rooted indeed. The expected utility hypothesis fails for acts with denumerably many outcomes, when probability (extraneous or otherwise) is merely finitely additive and consequences are absent.

Fortunately, Savage's theory is axiomatized so that the first six of his seven postulates deal with the structure of preference for *gambles*, i.e. acts which produce only finitely many different outcomes almost surely. It is the seventh postulate, P7, which carries the extension of expected utility theory to acts in general. Thus, our focus in section II is on the final axiom. We examine several conjectures about the conditions under which P7 remains independent of P1–P6 and demonstrate that the independence is *not* a matter of the additivity of the probability. Hence, one may satisfy P1–P6 with a countably additive probability but violate the expected utility hypothesis for acts that are not gambles. In other words, compliance with P1–P6 fails to guarantee the expected utility hypothesis for random variables in general, even on the condition that probability (based on P1–P6) is countably additive. Readers unfamiliar with Savage's theory may wish to skip section II on a first reading.

In section III, we analyze P7 and show that its role in extending utility to acts in general trades on an undesirable feature consequences are conceded to require. For instance, when P7 is reformulated to avoid this feature of consequences, the resulting theory precludes all but countably additive subjective probability. In our discussion of deFinetti's argument against "Dutch Book" in section IV, we grant his working hypothesis that there is a linear utility function for outcomes (as when he assumes dollars linear in utility), then show that his standards for coherence of previsions prohibit merely finitely additive probability. In parallel with Savage's theory, if consequences are introduced and coherence is confined to previsions involving consequences exclusively, then, as desired, finite additivity is all that follows from avoidance of "Dutch Book". Thus, the dilemma is between mandating consequences and denying the admissibility of merely finitely additive distributions.

II. ON THE INDEPENDENCE OF P7

In his classic *The Foundations of Statistics* (1954), L. J. Savage constructs a theory of utility, axiomatized in seven postulates. The first six of Savage's axioms yield a theory of expected utility for *gambles*, i.e. acts which produce at most finitely many consequences almost surely. The seventh postulate (P7) extends the theory to acts in general. Immediately following the introduction of P7, Savage demonstrates its independence from the first six with the aid of a finitely, but not countably additive probability. He concludes the demonstration with this terse remark:

Finite, as opposed to countable, additivity seems to be essential to this example; perhaps, if the theory were worked out in a countably additive spirit from the start, little or no counterpart of P7 would be necessary (1954, p. 78).[1]

Our purpose in this section is to demonstrate that the conjecture implicit in the above remark is not accurate. That is, we will produce examples involving only countably additive probabilities for which P1–P6 are satisfied but P7 is not. This means, on the condition that the expected utility hypothesis is valid for acts in general, some replacement for P7 is necessary even if the theory is worked out in a countably additive spirit.

We will assume that the reader either is familiar with Savage's postulate system or else has a copy of Savage (1954) readily available. Additionally, we recommend Fishburn (1970, Ch. 14) and Fishburn (1981) for helpful discussions of Savage's theory. In any event, P1–P4 are stated in the proof of Lemma 1 below, a lemma useful for the investigation of conditions under which P7 remains independent of P1–P6. The remaining three postulates, P5–P7, are stated following Lemma 1.

Savage's postulates concern *states* (elements of a set $\boldsymbol{S}$), *events* (subsets of $\boldsymbol{S}$), *consequences* (elements of a set $\boldsymbol{F}$), *acts* (functions from $\boldsymbol{S}$ to $\boldsymbol{F}$), and a relation between acts $\leq$ (read "is not preferred to"). If $\boldsymbol{f} \leq \boldsymbol{g}$ and $\boldsymbol{g} \leq \boldsymbol{f}$, we say $\boldsymbol{g}$ and $\boldsymbol{f}$ are *equivalent*. If $\boldsymbol{f}$ is an act, the consequence of $\boldsymbol{f}$ occurring in the state s is denoted $f(\boldsymbol{s})$. To avoid additional notation and with only slight encumbrance on the reader, we often identify a consequence with the act which produces that consequence in all states, that is, the constant act. If B is an event, we will use the notation I_B to denote the indicator of B, that is, the function which is 1 if B occurs and 0 if not. Following Savage, we will denote the complement of B, $\sim B$.

Lemma 1. Let $\boldsymbol{S}$ be a measurable set, and let F be a subset of the real numbers containing zero and closed under division (by nonzero elements) and multiplication. Assume all acts are measurable functions from $\boldsymbol{S}$ to $\boldsymbol{F}$, and assume that whenever $\boldsymbol{f}$ and $\boldsymbol{g}$ are acts, and B is an event, $\boldsymbol{f}I_B + \boldsymbol{g}I_{\sim B}$ is an act. Assume all constant functions are acts, and denote the act which is constantly 0 as $\boldsymbol{0}$. Let W be a mapping from the set of all acts to the finite real numbers which statisfies $W(\boldsymbol{0}) = 0$,

$$W(\boldsymbol{f}+\boldsymbol{g}) = W(\boldsymbol{f}) + W(\boldsymbol{g}), \tag{2.1}$$

whenever $\boldsymbol{fg} = \mathbf{0}$, and

$$cW(fI_B) = fW(cI_B), \tag{2.2}$$

for all events B and all consequences $f, c \in \boldsymbol{F}$. Define $\boldsymbol{f} \leq \boldsymbol{g}$ if and only if $W(\boldsymbol{f}) \leq W(\boldsymbol{g})$. Then P1–P4 are satisfied.

Proof

P1: The relation $\leq$ is a simple ordering.
This is trivial and needs no proof.

P2: If $\boldsymbol{f}, \boldsymbol{g}$, and $\boldsymbol{f}', \boldsymbol{g}'$ are acts and B is an event such that:

1. for $s \in \sim B$, $f(s) = g(s)$, and $f'(s) = g'(s)$,
2. for $s \in B$, $f(s) = f'(s)$, and $g(s) = g'(s)$,
3. $\boldsymbol{f} \leq \boldsymbol{g}$;

then $\boldsymbol{f}' \leq \boldsymbol{g}'$.

The conditions of P2 say that $\boldsymbol{f}I_{\sim B} = \boldsymbol{g}I_{\sim B}, \boldsymbol{f}'I_{\sim B} = \boldsymbol{g}'I_{\sim B}, \boldsymbol{f}I_B = \boldsymbol{f}'I_B$, and $\boldsymbol{g}I_B = \boldsymbol{g}'I_B$. Since $\boldsymbol{h} = \boldsymbol{h}I_B + \boldsymbol{h}I_{\sim B}$, for every act $\boldsymbol{h}$, and $I_B I_{\sim B} = \mathbf{0}$, it follows from (2.1) that $\boldsymbol{f}' \leq \boldsymbol{g}'$ if $\boldsymbol{f} \leq \boldsymbol{g}$.

P3: If $\boldsymbol{f} \equiv g, \boldsymbol{f}' \equiv g'$, and B is not null; then $\boldsymbol{f} \leq \boldsymbol{f}'$ given B, if and only if $g \leq g'$ (as constant acts).

Savage defines "$\boldsymbol{f} \leq \boldsymbol{f}'$ given B" to mean $\boldsymbol{g} \leq \boldsymbol{g}'$ for every pair of acts $\boldsymbol{g}, \boldsymbol{g}'$ satisfying $\boldsymbol{g}I_B = \boldsymbol{f}I_B$, $\boldsymbol{g}'I_B = \boldsymbol{f}'I_B$, and $\boldsymbol{g}I_{\sim B} = \boldsymbol{g}'I_B$. Under (2.1) and the conditions of P3, this can only happen if $g \leq g'$ (as constant acts).

P4: If f, f', g, g' are consequences, A, B are events, and $\boldsymbol{f}_A, \boldsymbol{f}_B, \boldsymbol{g}_A, \boldsymbol{g}_B$ are acts such that

1. $f' < f, g' < g$ (as constant acts),
2. $f_A(\boldsymbol{s}) = f, g_A(\boldsymbol{s}) = g$ for $s \in A$,
3. $f_A(\boldsymbol{s}) = f', g_A(\boldsymbol{s}) = g'$ for $s \in \sim A$,
4. $f_B(\boldsymbol{s}) = f, g_B(\boldsymbol{s}) = g$ for $s \in B$,
5. $f_B(\boldsymbol{s}) = f', g_B(\boldsymbol{s}) = g'$ for $s \in \sim B$,
6. $\boldsymbol{f}_A \leq \boldsymbol{f}_B$,

then $\boldsymbol{g}_A \leq \boldsymbol{g}_B$.

The conditions of P4 say that $\boldsymbol{f}_A = fI_A + f'I_{\sim A}, \boldsymbol{f}_B = fI_B + f'I_{\sim B}, \boldsymbol{g}_A = gI_A + g'I_{\sim A}$, and $\boldsymbol{g}_B = gI_B + g'I_{\sim B}$. Condition 6 together with (2.1) and (2.2) yields

$$(f - f')\{W(cI_A) - W(cI_B)\} \leq 0,$$

for some positive $c \in \boldsymbol{F}$. Condition 1 implies that $(f - f')$ and $(g - g')$ have the same sign. Hence

$$(g - g')\{W(cI_A) - W(cI_B)\} \leq 0. \tag{2.3}$$

It follows from (2.2) and (2.3) that $\boldsymbol{g}_A \leq \boldsymbol{g}_B$. □

The final three postulates are:

P5: There is at least one pair of consequences f, f' such that $f' < f$ (as constant acts).

P6: If $\boldsymbol{g} < \boldsymbol{h}$, and f is any consequence; then there exists a (finite) partition of $\boldsymbol{S}$ such that, if $\boldsymbol{g}$ or $\boldsymbol{h}$ is so modified on any one element of the partition as to take the value f at every s there, other values being undisturbed; then the modified $\boldsymbol{g}$ remains strictly not preferred to $\boldsymbol{h}$, or $\boldsymbol{g}$ remains strictly not preferred to the modified $\boldsymbol{h}$, as the case may require.

P7: If $\boldsymbol{f} \leq (\geq)\ g(s)$ given B for every $s \in B$, then $\boldsymbol{f} \leq (\geq)\ \boldsymbol{g}$ given B.[2]

The following lemma is stated without proof because it is so straightforward. We then proceed to Savage's example.

Lemma 2. P5 will be satisfied if $\boldsymbol{W}$ assigns different values to at least two different constant acts. Under the conditions of Lemma 1, P6 will be satisfied if for each act $\boldsymbol{g}$, each consequence f, and each $\varepsilon > 0$ there exists a finite partition $B_1, \ldots, B_n$ such that

$$|W(gI_{B_i}) - W(fI_{B_i})| < \varepsilon,$$

for all i.

Example 2.1. (Savage 1954.) Let $\boldsymbol{S}$ be the set of positive integers and $\boldsymbol{F}$ the interval (0.0, 1.0). Let P be any finitely additive probability on $\boldsymbol{S}$ which assigns probability 0 to each integer, assigns probability $\frac{1}{2}$ to the even integers, and admits a (finite) partition of $\boldsymbol{S}$ into events of arbitrarily small probability. Any limit point (as $n \to \infty$) of the sequence of discrete uniform distributions over the first n integers will do. Define

$$W(\boldsymbol{f}) = \int_S f(s)dP(s) + \lim_{\varepsilon \to 0} P\{f(s) \geq 1 - \varepsilon\}.$$

It is easy to see that W satisfies the conditions of Lemmas 1 and 2. Note that if $\boldsymbol{f}$ is a gamble, i.e. having finitely many consequences almost surely,

$$\lim_{\varepsilon\to 0} P\{f(s) \geq 1-\varepsilon\} = 0.$$

Thus, for gamble $\boldsymbol{f}$, $W(\boldsymbol{f})$ is a utility, with $W(\boldsymbol{f}) = f$ for a constant act $\boldsymbol{f} \equiv f$. To see that P7 is violated, let $\boldsymbol{f}$ equal $1 - 1/n$ for even n and 0 for odd n, and let $g(s)$ equal the larger of $\frac{3}{4}$ and $f(s)$. We now have $W(\boldsymbol{f}) = 1$, $W(\boldsymbol{g}) = \frac{11}{8}$, and $W(g(s)) < 1$, for all $s \in \boldsymbol{S}$. So $\boldsymbol{f} \geq g(s)$ given S for all s, but $\boldsymbol{f} < \boldsymbol{g}$.

The following is a similar example which uses countably additive probabilities.

Example 2.2. Let $\boldsymbol{S} = \boldsymbol{F}$ be the interval of real numbers (0.0, 1.0), and let P be uniform probability on Lebesgue measurable subsets of $\boldsymbol{S}$. Let all measurable functions from $\boldsymbol{S}$ to $\boldsymbol{F}$ be acts, and define $a(\boldsymbol{f}) = \inf\{P(E)$: $f(s)$ assumes only finitely many values on $\sim E\}$. Note that $a(\boldsymbol{f} + \boldsymbol{g}) = a(\boldsymbol{f}) + a(\boldsymbol{g})$ whenever $\boldsymbol{fg} = \boldsymbol{0}$. For each act $\boldsymbol{f}$ define

$$W(\boldsymbol{f}) = \int_{[0.1]} f(s)dP(s) + a(\boldsymbol{f}).$$

It is easy to see that W satisfies the conditions of Lemmas 1 and 2. If $\boldsymbol{f}$ is a gamble, $a(\boldsymbol{f}) = 0$, hence $W(\boldsymbol{f})$ is a utility. To see that P7 is violated, let $g(s) = s$ for all s except $s = 1$, and let $g(1) = 0$. Let $f(s) = 1$, for all s. Then $W(\boldsymbol{g}) = 1.5$, $W(\boldsymbol{f}) = 1$, $W(g(s)) = s$ for all s except $s = 1$, and $W(g(1)) = 0$. So $\boldsymbol{f} > g(s)$ for all s, while $\boldsymbol{f} < \boldsymbol{g}$.

The feature that drives Example 2.2 is the fact that the "worth" W of an act is increased from its expected value by the extent to which the act produces uncountably many consequences. Savage proves that (given P1–P6) P7 holds for gambles (effectively Theorem 2.7.3 of Savage 1954). His example (2.1, above) shows that P7 need not hold for acts that assume countably many consequences. The following example shows that this remains the case even when the probability is countably additive (unlike Example 2.1).

Example 2.3. Let $\boldsymbol{S}$ be the half-open interval (0.0, 1.0), and $\boldsymbol{F}$ the rational numbers in $\boldsymbol{S}$. Let P be uniform probability on Lebesgue measurable subsets of $\boldsymbol{S}$. Let all measurable functions $\boldsymbol{f}$ from $\boldsymbol{S}$ to $\boldsymbol{F}$ satisfying

$$a(\boldsymbol{f}) = \lim_{i\to\infty} P\{f(s) \geq 1-2^{-i}\}2^{i} < \infty$$

be acts (all subsets of $\boldsymbol{F}$ being measurable). Define

$$W(f) = \int_S f(s)dP(s) + a(f).$$

Once again, W satisfies the conditions of Lemmas 1 and 2. If f is a gamble, $a(f) = 0$, hence $W(f)$ is a utility. To see that P7 is violated, let $f(s) = 1 - \frac{1}{2^k}$ and let $g(s)$ equal $1 - \frac{1}{2^{k+1}}$ for $1 - \frac{1}{2^{k+1}} > s \geq 1 - \frac{1}{2^k}$. Then $W(f) = \frac{1}{3} + 1 = \frac{4}{3} > W(g(s))$, for each $s \in S$. Yet $W(g) = \frac{2}{3} + 2 = \frac{8}{3}$, so that $g > f$.

What Examples 2.2 and 2.3 illustrate is the independence of the relationship between P1–P6 and P7 from the degree of additivity which personal probability possesses. What we hope to show in the next section is that this independence follows from the special role that consequences play in P7. If we deny the existence of consequences, attempts to reformulate P7 lead to the exclusion of merely finitely additive probabilities.[3]

III. DOMINANCE AND CONGLOMERABILITY OF PROBABILITY

Savage's seventh postulate contrasts acts in general through a comparison of one with the consequences of the other (on some non-null event). This is possible because for each consequence, say $g(s)$, there is an act, the constant act $g^* \equiv g(s)$, which serves naturally as the counterpart for the consequence. However, the reader is reminded that constant acts (consequences) have, in virtue of P3, rather distinguished properties. To wit: as stipulated by P3, the relative values of consequences are unaffected given non-null events, i.e. their values are invariant under different states. Thus, a consequence must behave like a prize whose value is robust against whatever (non-null) information we might acquire.[4]

In practical terms, P3 prohibits approximating a consequence by an award of, e.g. stock options where the relative attractiveness of two stocks may be a function of the state of the economy. Are there good candidates for consequences? As Savage argues in his typically even-handed style,

> what are often thought of as consequences (that is, sure experiences of the deciding person) in isolated decision situations typically are in reality highly uncertain. Indeed, in the final analysis, a consequence is an idealization that can perhaps never be well approximated. I therefore suggest that we must expect acts with actually uncertain consequences to play the role of sure consequences in typical isolated decision situations (1954, p. 84).

If we concede that consequences, in the sense required by P3, are not the entities typically viewed as outcomes in familiar decisions, what can we offer in place of P7 if we ignore the relativization to contrast by consequences? We can recast the question this way. Let $\pi^B = \{h_1, \ldots\}$ be a, possibly infinite, partition of the event B by non-null elements h_i. Can we make sense of a comparison, given B, between act $\boldsymbol{f}$ and each outcome $\boldsymbol{g}I_{h_i}$ even though $\boldsymbol{g}I_{h_i}$ is not a consequence in the fashion of P3? (Unfortunately, we are forced to consider non-null h_i exclusively because in Savage's theory all acts are equivalent given a null event. That is, Savage's program cannot generate probability conditional upon an event of 0 probability.)

Suppose we attempt to avoid consequences entirely. Instead of comparing acts $\boldsymbol{f}$ and $\boldsymbol{g}$ through consequences, we might contrast the outcomes of $\boldsymbol{f}$ and $\boldsymbol{g}$ given h_i ($i = 1, \ldots$) directly. Thus, we have

P8: If $\boldsymbol{f} \leq (\geq)\, \boldsymbol{g}$ given h_i for all i, then $\boldsymbol{f} \leq (\geq)\, \boldsymbol{g}$ given B.

Of course, P8 does not suffice as a replacement for P7 if the goal is to extend expected utility theory to acts in general. To wit: the preference structures $W(\cdot)$ of Examples 2.2 and 2.3 satisfy P1–P6 and P8, yet $W(\cdot)$ does not admit a ranking of acts by their expected utility of consequences. Thus, a substitute for P7 must do more (or other) than P8 to reach the goal of the expected utility hypothesis.

Ideally, we would formulate a rule like P8 without the restriction that h_i be non-null, i.e. to permit strict preferences conditional upon an event of zero probability. To repeat, this move is not available to us within Savage's theory because of his definition of null events. Nonetheless, we find it productive to analyze the relation between P8 and P1–P7. Our investigation uncovers a hidden tie to countable additivity, a tie that, perhaps, underpins Savage's (1954, p. 78) statement about P7.

P8 trades for its plausibility on a dominance principle, extended to infinite partitions by non-null events. That is, the tacit assumption behind P8 is this: since $\boldsymbol{f}$ is not preferred to $\boldsymbol{g}$ on each element of π^B, $\boldsymbol{f}$ should not be preferred to $\boldsymbol{g}$ given B.

Surprisingly, P8 is inconsistent with P1–P7 unless finitely, but not countably additive probability is precluded. That is, P8 fails, though P1–P7 do not, whenever the agent's subjective probability $P(\cdot)$ is not conglomerable in π^B, or equivalently (Dubins 1975) whenever $P(\cdot)$ is not disintegrable in π^B. Without loss of generality, hereafter we assume $B = S$, the sure event.[5]

Non-conglomerability (deFinetti 1972, §5.30) of a probability $P(\cdot)$ occurs in a partition $\pi = \{h_1, \ldots\}$ if, for some event E and constants k_1 and k_2, $k_1 \leq P(E|h_i) \leq k_2$ for each $h_i \in \pi$, yet $P(E) < k_1$ or $P(E) > k_2$. For example (due to Dubins, see deFinetti 1972, p. 205), let the sure-event be the union of the events (i, j) where $i = 1, 2, \ldots$ is an integer and $j = 0$ or 1. Let I stand for the first coordinate, and let J stand for the second coordinate. Let $\pi = \{h_i | h_i = \{(i, 0), (i, 1)\}\}$ be a partition. Consider a finitely, but not countably additive probability $P(\cdot)$ such that $P(J = 0) = P(J = 1) = \frac{1}{2}$, and $P(I = i | J = j) = 2^{-(i+j)}$. There are many such $P(\cdot)$. Each has probability "adherent" along the sequence of $(i, 1)$ events. That is, $P(J = 1) = P(\cup_i\{(i, 1)\}) = \frac{1}{2}$, but $\Sigma_i P(i, 1) = \frac{1}{4}$. By finite additivity, $P(h_i) = 3 \times 2^{-(i+2)}$, so each element of π is non-null. Bayes' Theorem entails that $P(J = j | h_i) = (2 - j)/3$. Hence, for each element of π, $P(J = 1 | h_i) = \frac{1}{3}$; however $P(J = 1) = \frac{1}{2}$. $P(\cdot)$ is not conglomerable in π.

Failure of P8 follows directly from non-conglomerablity of $P(\cdot)$ in π. To show that $P(\cdot)$ does not satisfy P8, let the dollar symbol $ denote utiles, and assume that prizes worth any desired number of utiles exist. (These prizes need not be consequences in Savage's sense.) Consider the acts

f: an even odds bet on E, the event that $J = 1$, with stake of $2,

that is, the agent places $1 on E against an opponent who places $1 on $\sim E$, and

g: the bet on $\sim E\{J = 0\}$ at odds of 3:2 with a stake of $2,

that is, the agent places $1.20 on $\sim E$ against an opponent who places $0.80 on E. Since $P(E) = \frac{1}{2}$, $\mathbf{g} < \mathbf{f}$. However, since $P(E | h_i) = \frac{1}{3}$, for each h_i, $i = 1, \ldots$, $\mathbf{f} < \mathbf{g}$ given h_i for all i. This contradicts P8.

Hence, P1–P7 do not entail a dominance rule P8 unless conglomerability in denumerable partitions is mandated, in which case countable additivity holds (see Schervish, Seidenfeld, and Kadane 1981). Moreover, P1–P6 and P8 are insufficient to extend utility theory from gambles to acts, even when probability is countably additive (Example 2.2) and the space of consequences is denumerable (Example 2.3). In short, the relation between Savage's "extended sure-thing" principle and dominance does not parallel the relation between the basic "sure-thing" principle (P2) and the dominance-in-finite-partition rule which follows from it (Savage 1954, Theorem 2.7.2).[6]

Our failure to duplicate the force of P7 with a principle of dominance raises the following problem, which we discuss at length in the next section. Savage's theory, like deFinetti's which antedates it, extracts probability from preference. There is no extraneous probability in either program. However, unlike Savage's approach, deFinetti's theory begins with coherence of previsions, and establishes the calculus of probability from that. But coherence of previsions is a requirement of dominance; specifically, deFinetti's concern is that one's previsions be undominated. DeFinetti's definition of coherence requires that dominance hold in finite partitions regardless of how we view the status of outcomes, i.e. for coherent previsions dominance in finite partitions does not demand a contrast by consequences. Our question, then, is this. How does deFinetti's approach apply for previsions not limited to finitely many outcomes, without legislating countable additivity while retaining a principle of coherence?

IV. ON THE AVOIDANCE OF "DUTCH BOOK"

DeFinetti (1974, chapter 3) presents his concept of a prevision and offers an "operational definition" in terms of lotteries.[7] An agent's *prevision* is a function P from real-valued random quantities X to real numbers $x^+ = P(X)$, where x^+ is the agent's "fair price equivalent" for the lottery that yields a prize worth x units when $X = x$. If it can be shown that a person's prevision ought to satisfy the constraints:

i. $P(X + Y) = P(X) + P(Y)$ (additive previsions) and
ii. $\inf X \leq P(X) \leq \sup S$ (previsions lie within the range of values for X),

then (normatively) previsions are finitely additive expectations, and entail finitely additive probability for events as a special case. Specifically, if events E are identified with their indicator functions I_E, then previsions for events are subjective probabilities.

DeFinetti defines a set of previsions to be *coherent* if no finite selection of "fair" lotteries yields a uniformly negative return for each possible outcome of the random variables involved in the lotteries. (See, e.g., Shimony 1955 for a discussion of coherence and "Dutch Book" regarding events.) Specifically, for each random variable X, the agent is required to hold a prevision $P(X) = x^+$ that he feels makes "fair" a lottery yielding a prize worth $c(X - x^+)$ units, where c is some real-valued constant selected by an "opponent". Again, the agent's previsions are coherent if among such "fair" lotteries there is no finite selection of non-zero c's that guarantee a uniform loss regardless of which values the random variables assume.[8]

DeFinetti operationalizes all this by supposing that, for modest quantities of money, people are prepared to use dollar(lire)-unit prizes. That is, with the c's confined to some small interval about 0, deFinetti posits the existence of previsions when lotteries are given in small dollar amounts. The effect of this working hypothesis is to make utility linear in dollars in the region near \$0. For example, a prevision of 0.5 for the event E means that for small dollar stakes the agent is prepared to offer even odds, regardless of whether (by choosing $c < 0$) he bets on E, or (by choosing $c > 0$) he bets against E.

We separate deFinetti's thesis that there exist previsions when prizes are measured in the appropriate scale, i.e. utiles, from his working hypothesis, that small dollar amounts are utiles. Is the thesis neutral regarding disputes in inductive inference? Kyburg (1978) argues in the negative. For our purposes, fortunately, we do not need to enter this debate. We are prepared to grant both the thesis and working hypothesis, since our goal is to show that the reduction of belief to preference does *not* follow from the standard of coherence alone and our criticism is compatible with both assumptions.

However, to grant the working hypothesis is not to concede that dollar prizes (or whatever) are consequences (in Savage's sense). For constructing a lottery for a prevision of X, we need only have available prizes that can be awarded in cx units when $X = x$, subject to the thesis that the prevision is independent of the sign and magnitude of c. The unreasonableness of viewing prizes as Savage-type consequences is as apparent to deFinetti as to Savage. When discussing the dispensability of his working hypothesis (to cover the familiar problem of "risk aversion"), he says

> It would be more appropriate, instead of considering the variable x representing the gain, to take f + x, where f is the individual's 'fortune' (in order to avoid splitting hairs, inappropriate in this context, one could think of the value of his estate). Anyway, it would be convenient to choose a less arbitrary origin to take into account the possibility that judgments may alter because in the meantime variations have occurred in one's fortune, or risks have been taken, and in order not to preclude for oneself the possibility of taking these things into account, should the need arise. Indeed, as a recognition of the fact that

the situation will always involve risks, it would be more appropriate to denote the fortune itself by F (considering it as a random quantity), instead of with f (a definite value) (deFinetti 1974, p. 79).

What deFinetti says of the agent's 'fortune' applies *mutatis mutandis* to the payoff of the lottery. That is, given $X = x$, one augments the agent's capital reserve by an amount, cx units. But the prize itself is no different in kind from the fortune (F), each of which the agent owns outright and neither of which (normatively) need be a full-blooded consequence.

In light of our discussion in section III, it should come as no surprise to learn that unless lottery prizes are consequences, deFinetti's criterion of coherence precludes all but countably additive distributions. As before, failure of conglomerability entails a (uniform) failure of dominance. Coherence, after all, requires avoidance of previsions that induce a failure of dominance with respect to the alternative: no bet. But conglomerability in denumerable partitions is equivalent to countable additivity. Thus, for previsions of random variables assuming more than finitely many outcomes, coherence entails countable additivity. The following (typical) construction illustrates the incoherence of merely finitely additive distributions.

For simplicity, let $P(\cdot)$ be a finitely additive probability assuming infinitely many different values. It follows from Theorem 3.1 of Schervish, Seidenfeld, and Kadane (1984) that there exist an event E, a positive number d, and a partition $\pi = \{h_1 \ldots\}$ such both that $P(h_i) > 0$ and $P(E) - P(E \mid h_i) > d$ for all i. That is, conglomerability fails in π with regard to the event E. Let $x_i = P(E \mid h_i)$ and $k = \sup_i x_i$ so that $P(E) \geq k + d$. Consider, next, a wager W that yields a \$1 (one utile) prize in case E occurs and \$0 otherwise. W is worth at least $k + d$. However, given h_i, W is worth at most k. So, the prevision of W, $P(W) \geq k + d$, and $P(W \mid h_i) < k$ for all i. Define the random variable X so that $X = x_i$ just in case h_i obtains. That is,

$$X = \sum_i x_i I_{h_i}.$$

Then X is the (conditional expected) utility of any lottery whose prizes have (conditional expected) utility given h_i equal to x_i for all i. W is such a lottery. Define the lottery Y by saying that Y awards prize WI_{h_i} if h_i obtains for $i = 1, 2, \ldots$. It is clear that $Y = W$; however, note that the prizes WI_{h_i} awarded by Y are not consequences. Since $0 \leq x_i \leq k$ for all i, the prevision for any lottery whose prizes are worth x_i under h_i

should be between 0 and k. Y is such a lottery, hence y^+ should be between 0 and k. But, $W = Y$ so $-(W - y^+)$ should be considered fair. However, $w^+ \geq k + d$, so $-(W - y^+)$ is worth no more than $-d$, and hence is unfair. On the other hand, if y^+ is chosen equal to w^+, then y^+ is not the expected utility over the partition π.

In summary, we see that deFinetti's criterion of coherence for prevision rules out merely finitely additive distributions unless lotteries are restricted to prizes which are consequences in the sense of Savage. In the construction of the previous paragraph, WI_{h_i} serves as a prize whose value, given h_i, equals x_i; however, WI_{h_i} is not a consequence, since its value is not independent of, e.g., the event E. The dilemma is, of course, that consequences are hard to come by and, it would seem, beyond what is required by consideration of rational preference. Can we not argue, like Savage and deFinetti, that there always are risks? The lesson is clear. We will not have all three of the following:

i. reduction of coherent belief to rational preference;
ii. coherence of finitely additive probability;
iii. admissible preference structures free of consequences.

In light of the highly questionable character of consequences, it seems best to us to dispose of either (i) or (ii). Since many classical statistical procedures require finitely additive "priors" in order to be Bayesian (see e.g. Heath and Sudderth 1978), there is good reason to resist abandoning (ii). However, given (ii) and (iii), the expected utility hypothesis fails (as shown above). Without consequences to fall back upon, non-conglomerability of finitely additive distributions cannot be squared with a requirement of (uniform) dominance for acts. This leaves statistical decision theory devoid of a formidable criterion: *admissibility* (see Savage 1954, p. 114). We do not find this an easy choice to make. Perhaps further discussion will suggest a solution.

APPENDIX

In this appendix we examine several results and suggestions of Fishburn (1970) which are related to our discussion in section II. First, Fishburn (1970, Ch. 10) considers the relationship between countable additivity and a set of postulates not equivalent to Savage's P1–P6. For example, Fishburn's (1970, p. 137) postulate $S4$, that the set of probability measures be closed under countable convex combination, is not implied by Savage's system. In fact, this postulate rules out Example 2.3 but not Example 2.2.

Second, Savage (1954 [second edition 1972], p. 78*n.*) references a suggestion by Fishburn (1970, p. 213, ex. 21) for weakening P7 to accommodate acts in general, subject to the constraint that probabilities are countably additive. Fishburn's suggested version, called P7b, requires that

$$\text{if } f \leq (\geq)\, g(s) \text{ given } A, \text{ for each } s \in A, \text{ then } \boldsymbol{f} \leq (\geq)\, \boldsymbol{g} \text{ given } A,$$
$$\text{for constant act } \boldsymbol{f} = f$$

P7b is less demanding than P7 in that it contrasts acts in general with constant acts solely, and not with other acts in general. However, P7b is *not* sufficient for extending utility theory to the class of acts in general even when probability is countably additive. In Example 2.3, $W(\cdot)$ satisfies P7b though acts are not ranked by W in accord with their expected utilities (as fixed by P1–P6). To see that W of Example 2.3 satisfies P7b it is sufficient to verify that:

(case 1) if $f \leq g(s)$ given A, for each $s \in A$, then $fI_A \leq \int g(s) I_A dP(s)$.

Since $a(\cdot)$ is non-negative, it follows that $W(fI_A) \leq W(gI_A)$;

hence $\boldsymbol{f} \leq \boldsymbol{g}$ given A as required.

(case 2) if $f \geq g(s)$ given A, for all $s \in A$, then as $f < 1$, $a(gI_A) = 0$.

Thus $W(fI_A) \geq \int g(s) I_A dp(s) = W(gI_A)$,

and $\boldsymbol{f} \geq \boldsymbol{g}$ given A as required.

It is to be observed that there can be no counterexample to the conjecture that P7b suffices for extending utility theory to acts in general if $\boldsymbol{F}$ is closed under limits of utility (as fixed by P1–P6). Since P1–P6 entail that consequences have finite utility (Savage 1954, p. 81), the assumption of "closure" yields bounded utilities for gambles. However, subject to "closure" of $\boldsymbol{F}$ under preference, P7b entails P7, given P1–P6, regardless of the additivity of probability. This is seen as follows:

Assume the antecedent of P7, that is $\boldsymbol{f} \leq (\geq)\, g(s)$ given A, for each $s \in A$. By hypothesis of "closure", there exist constant acts $\boldsymbol{g}_*$ and $\boldsymbol{g}^*$ which are the infemum and supremum of the consequences of $\boldsymbol{g}$ for $s \in A$. Clearly $\boldsymbol{f} \leq \boldsymbol{g}_* (\geq \boldsymbol{g}^*)$ given A. Then by P7b, $\boldsymbol{g}_* \leq (\boldsymbol{g}^* \geq)\, \boldsymbol{g}$. By transitivity, $\boldsymbol{f} \leq (\geq)\, \boldsymbol{g}$, as required by P7.

NOTES

1 Savage had misgivings about this comment. In a letter to P. C. Fishburn (dated 30 June 1965) he wrote:

You suggest that I review the last sentence on page 78 of F. of S. [*Foundations of Statistics*] It is hard for me now to feel sure what I meant by that sentence, and I have serious doubts that it is defensible. But what it seems to say is not that something stronger than P7 would be needed in a countably additive context, but rather something weaker might suffice.

And in another letter to Fishburn (dated 9 September 1966):

Once you convince yourself, with Zorn's lemma, that the Blackwell-Girshick theorem cannot be had without some counterpart of P7, you will have shown that the conjecture at the bottom of page 78 of F. of S. is more or less incorrect.

We thank Professor Fishburn for bringing these to our attention.

2 Fishburn (1970, Theorem 14.1) offers a weakened version of P7 which suffices to extend expected utility theory to acts in general. The slight weakening of P7 is achieved by insisting on a *strict* inequality, $f < (>)\ g(s)$, in the antecedent. Our Examples 2.2 and 2.3 apply to that form of P7 as well.

3 Fishburn (1970) has also studied the relationship between P7 and countable additivity. See the Appendix to this essay for a discussion of the connection between his results and those of the present essay.

4 This feature of consequences is separate from the requirement discussed by Fishburn (1970, p. 166 and 1981, p. 162) that each consequence be "relevant", i.e. an outcome for some act, for each state. In other words, the entire class of Savage-type consequences F is needed to exhaust the range of outcomes for each state. This restriction prohibits the strategy of adopting outcomes under fine descriptions for consequences, since for different states the descriptions are contraries, in violation of the clause that each consequence be "relevant" to each state. We do not know of an axiomatic approach that avoids completely the existence of (at least a pair of) Savage-type consequences "relevant" for each state.

5 The convenience, $B = S$, is justified by the result that conglomerability of $P(\cdot)$ fails in a partition π of the sure-event just in case it fails for $P(\cdot|B)$, given some non-null B, in the restricted partition π^B (Kadane, Schervish, and Seidenfeld 1986, §4).

6 Theorem 2.7.2 of Savage (1954) reads as follows: If $\pi^B = \{B_1, \ldots, B_n\}$ is a finite partition of B, and $\boldsymbol{f} \leq \boldsymbol{g}$ given B_1 for each i, then $\boldsymbol{f} \leq \boldsymbol{g}$ given B. The proof of this theorem depends only on P1 and P2.

7 DeFinetti offers two criteria of coherence which, of course, he shows to be equivalent. We make use of his first criterion in the text, but our objection applies also to the second criterion. The second criterion is that the agent is attempting to minimize the value of a proper scoring rule (c.f. Savage 1971).

8 The two clauses: (i) that coherence involves *finitely* many lotteries, and (ii) that merely *uniformly* negative losses be avoided, are each necessary to avoid restricting coherence to countably additive previsions. To see that (i) is necessary, consider a finitely additive probability which assigns 0 probability to each

of a denumerable exhaustive collection of pair-wise disjoint events. By accepting a \$1 wager at odds of 1:0 on each of the denumerably many indicator variables (corresponding to these events), the agent permits an "opponent" a \$1 win for sure. To see that (ii) is necessary, consider the random variable $X = x_i = 1/i$ for $i = 1, 2, \ldots$ with a finitely additive probability that assigns $P(x_i) = 0$ for all i. Then the prevision for X, is $P(X) = 0$. But with $x^+ = 0$, $-(X - x^+) < 0$, but not uniformly less than 0.

REFERENCES

deFinetti, B. (1972), *Probability, Induction and Statistics.* New York: Wiley.

deFinetti, B. (1974), *The Theory of Probability* (2 vols.). New York: Wiley.

Dubins, L. (1975), "Finitely additive conditional probabilities, conglomerability and disintegrations", *Annals of Probability 3*: 89–99.

Fishburn, P. C. (1970), *Utility Theory for Decision Making.* New York: Wiley.

Fishburn, P. C. (1981), "Subjective Expected Utility: A Review of Normative Theories", *Theory and Decision 13*: 139–199.

Heath, D. and Sudderth, W. (1978), "On finitely additive priors, coherence, and extended admissibility", *Annals of Statistics 6*: 333–345.

Kadane, J. B., Schervish, M. J., and Seidenfeld, T. (1986), "Statistical Implications of Finitely Additive Probability", in *Bayesian Inference and Decision Techniques*, ed. P. K. Goel and A. Zellner, pp. 59–76. Amsterdam: Elsevier Science Publishers. [Chapter 2.5, this volume]

Kyburg, H. E. (1978), "Subjective probability: considerations, reflections, and problems", *Journal of Philosophical Logic 7*: 157–180.

Savage, L. J. (1954), *The Foundations of Statistics.* New York: Wiley.

Savage, L. J. (1971), "Elicitation of Personal Probabilities and Expectations", *Journal of the American Statistical Association 66*: 783–801.

Schervish, M. J., Seidenfeld, T., and Kadane, J. B. (1984), "The Extent of Non-conglomerability of Finitely Additive Probabilities", *Zeitschrift für Wahrscheinlichkeitstheorie und Verwandte Gebiete 66*: 205–226.

Shimony, A. (1955), "Coherence and the axioms of confirmation", *Journal of Symbolic Logic 20*: 1–28.

2.5

Statistical Implications of Finitely Additive Probability

JOSEPH B. KADANE, MARK J. SCHERVISH, AND TEDDY SEIDENFELD

I. INTRODUCTION

In his classic monograph on the foundations of the theory of probability, Kolmogorov (1956) introduces a postulate [P6], equivalent to the principle of countable additivity, which he justifies as a mathematical expedient for infinite probability structures. Countable additivity requires that the denumerable union of pairwise disjoint events has probability equal to the sum of the individual probabilities, i.e. if $A = \cup_{i=1}^{\infty} A_i$ $(A_i \cap A_j = \varnothing, i \neq j)$, then $P(A) = \Sigma_{i=1}^{\infty} P(A_i)$. His theory less P6 is hereafter described as the theory of finitely additive probability.

Our dispute with Kolmogorov's characterization of countable additivity as an "expedient" does not stem from the fear of non-measurability arising in routine problems of statistical inference. For our purposes it is enough that probability be defined over a σ-field. The questions we ask are:

1. Does countable additivity build in unexpected statistical consequences?
2. Does Bayesian statistics mandate the added restrictions countable additivity imposes?

We thank Morris DeGroot, Isaac Levi and Herman Rubin for their suggestions and criticisms.

Reprinted from *Bayesian Inference and Decision Techniques* (Amsterdam: Elsevier Science Publishers, 1986), pp. 59–76, with permission of the Editors, Prem K. Goel and Arnold Zellner.

We argue that the answers are yes and no (respectively).

Without countable additivity, finitely additive probability undergoes a failure of "conglomerability" (discussed in section II). Without conglomerability, familiar decision theoretic principles, e.g. admissibility (discussed in section III) and other forms of dominance rules (discussed in section IV) are invalid. What reason is there for taking seriously the finitely additive theory if dropping countable additivity undermines what has become accustomed statistical decision theory?

First, if one adopts the standards of coherence defended by deFinetti (1974), Savage (1974), and Lindley (1981), one's beliefs are modeled by a probability that is finitely but not necessarily countably additive. (See Seidenfeld and Schervish, 1983, for discussion of this view.) Second, as we investigate in section V, there are important connections between finite additivity and Bayesian reconstructions of standard (orthodox) statistical inference – reconstructions using "improper" priors in the fashion of Jeffreys (1961) and Lindley (1970). By interpreting improper distributions as finitely and not countably additive probabilities, the way is paved to resolve a host of anomalies thought by some to be evidence of inconsistency in Bayesian theory. In this we agree with Heath and Sudderth (1978), Hill (1980), and Levi (1980).

II. NON-CONGLOMERABILITY OF MERELY FINITELY ADDITIVE PROBABILITY

Define P to be *conglomerable in* π (see deFinetti, 1972, p. 99) when for every event E such that $P(E|h_i)$ is defined for all i, and for all constants k_1, k_2, if $k_1 \leq P(E|h_i) \leq k_2$ for all $h_i \in \pi$, then $k_1 \leq P(E) \leq k_2$. That is to say, conglomerability asserts that, for each event E, if all the conditional probabilities over a partition π are bounded by two quantities, k_1 and k_2, then the unconditional probability for that event is likewise bounded by these two quantities. DeFinetti draws attention to non-conglomerability of finitely additive probability in denumerable partitions.

Example 2.1 (due to P. Lévy, 1930; see deFinetti, 1972, p. 102). Let $P(\cdot)$ be a finitely additive probability defined over the field of all subsets of the denumerable set of points $\{\langle i, j\rangle: i, j \text{ are positive integers}\}$. Thus, one may think of $P(\cdot)$ defined over subsets of integer coordinate points $\langle i, j\rangle$ in the first quadrant. Constrain $P(\cdot)$ so that for any point $\langle i, j\rangle$,

$P(\langle i, j\rangle) = 0$. Hence, $P(\cdot)$ is not countably additive. Also, for any pair $\langle i, j\rangle$, let $P(\langle i, j\rangle | B) = 0$ if B is an infinite set. Hence, conditional on an infinite set B, $P(\cdot|B)$ is again not countably additive.

Now, using the finitely additive probability $P(\cdot)$ (given above), we see that conglomerability must fail with respect to the event $A = \{\langle i, j\rangle: j \geq i\}$. That is, let A be the region whose lower boundary is the diagonal "$i = j$". Consider the partition $\pi_1 = \{h_i: h_i = \{\langle i, j\rangle: j < \omega\}$ and $i < \omega\}$. That is, π_1 is the partition into "vertical" sections. Since, for each $i < \omega$, h_i is an infinite set, $P(A | h_i) = 1$ (since there are only finitely many points below the point $\langle i, i\rangle$, for each $i < \omega$). Let $\pi_2 = \{h'_j: h'_j = \{\langle i, j\rangle: i < \omega\}$ and $j < \omega\}$. That is, π_2 is the partition into "horizontal" sections. Since, for each $j < \omega$, h'_j is an infinite set, $P(A | h_j) = 0$ (since there are only finitely many points to the left of the point $\langle j, j\rangle$, for each $j < \omega$). Thus, conglomerability must fail for at least one of π_1, π_2 as $P(A)$ must differ from at least one of the two values 0, 1.

In an earlier paper (Schervish, Seidenfeld and Kadane, 1984), we investigate several general questions about the existence and magnitude of failures of deFinetti's conglomerability principle. Using a result due to Dubins (1975) (that conglomerability in a partition is equivalent to "disintegrability" in that partition), we provide least upper bounds on the failures of conglomerability with respect to *denumerable* partitions. We also show that for finitely additive probabilities, non-conglomerability in *denumerable* partitions characterizes those which fail to be countably additive, confirming a statement of deFinetti (1972, p. 99).

III. ADMISSIBILITY

Non-conglomerability of finitely additive probabilities, i.e. the phenomenon that an unconditional probability may lie outside the range of values of conditional probabilities over an exhaustive partition, quickly leads to a violation of a familiar decision-theoretic principle, admissibility.

Let O_1 and O_2 be two options in a decision and let π be a partition into states (independent of the options) such that for each state the same option, say O_1, is strictly preferred to the other option, O_2. Then O_2 is *inadmissible*, i.e. O_1 is strictly preferred to O_2 unconditionally.

However, if we consider a choice between O_1: bet on the event A

(as defined in Example 2.1) at even odds, and O_2: bet on the complementary event $\bar{A}$ at even odds, then for each $h_i \in \pi_1$, O_1 is strictly preferred (in expectation) to O_2; while for each $h'_j \in \pi_2$, O_2 is strictly preferred (in expectation) to O_1. It cannot be both that O_1 is strictly preferred to O_2 *and* that O_2 is strictly preferred to O_1. Therefore, failures of conglomerability entail failures of admissibility. This observation permits a simplification of Arrow's (1972) axiom system for preferences which include both a principle of Monotone Continuity (Villegas, 1964) and an admissibility rule, called Dominance. Since the Dominance principle implies countable additivity, Monotone Continuity is redundant.

The fundamental property of admissible procedures, namely that they alone are Bayesian procedures or limits of them, fails for finitely additive probabilities. There are reasonable acts, Bayesian with respect to a finitely additive opinion, which are inadmissible.

Historically, admissibility took on its greatest importance for statistics when Stein (1955) showed that $\bar{x}$ is inadmissible as an estimate of the mean μ of a normal vector with dimension greater than two. Stein showed that drawing in the components of $\bar{x}$ toward an arbitrary origin was a strict improvement over $\bar{x}$ in the admissibility sense. Lindley (1962) later showed in a simple Bayesian model in which μ itself is considered to be normally distributed with mean μ_0, that $\bar{x}$ should be drawn toward μ_0 by an amount determined by the prior variance of μ around μ_0, and the sampling variance of $\bar{x}$ around μ. These observations in turn led to an increased interest in empirical Bayes methods, using the data to estimate hyperparameters like μ_0. In general, the move away from the automatic use of $\bar{x}$, occasioned by Stein's and Lindley's work, has been a healthy development for statistics, we think. Statisticians have been encouraged by these results to consult their prior beliefs more systematically than they had been before.

A consequence of accepting the coherence of merely finitely additive options is to reduce the importance of the concept of admissibility. In particular, there are finitely but not countably additive, prior distributions for which $\bar{x}$ is an optimal Bayesian act regardless of the dimension of $\bar{x}$. We do not interpret this to mean that now people should go back to using $\bar{x}$ (if they ever stopped) without concern for their prior opinion. Rather, we think that our results re-emphasize the importance of using your opinion within a Bayesian paradigm that does not insist on countably additive "priors".

IV. COHERENCE AND "STRONG INCONSISTENCY" WITH FINITELY ADDITIVE PROBABILITY

Associated with a failure of conglomerability are instances of "strong inconsistencies", as Stone (1976) calls them. These are colorfully illustrated by the following game (Stone, 1981), which is a rewording of Stone's (1976) "Flatland" example.

A regular tetrahedral die is rolled a very large number of times. The faces of the die are labelled: e^+ (positron); e^- (electron); μ^+ (muon); and μ^- (antimuon). A record is kept of the outcomes subject to the constraint that if complementary events occur successively in the record, they "annihilate" each other and the record contracts, without trace of the "annihilation". Thus, the sequences $\ldots e^+e^- \ldots$, $\ldots e^-e^+ \ldots$, $\ldots \mu^+\mu^- \ldots$, and $\ldots \mu^-\mu^+ \ldots$, cannot occur in the record. At some (arbitrary) point in the sequence of rolls the player, who is ignorant of the outcome to date, calls for one last roll after which he is shown the final record. He is then asked to gamble on the outcome of the last toss of the die.

Consider a finitely additive prior over the countable set S of all possible states of the record prior to the final toss defined as follows. First arrange all elements of S into a single sequence, and let B_n be the set consisting of the first n of them. For $B \subseteq S$, let $\lambda_n(B) = \#(B \cap B_n)/n$. Take any limit point of the sequence λ_n as a prior λ. Since every finite subset of S has zero probability under λ, define for any finite set C, $\lambda(A|C) = \#(A \cap C)/\#(C)$. This set of conditional probabilities is consistent in the sense that if A and B are any subsets of S, and C is any finite subset such that $A \cap C \neq \emptyset$, $\lambda(B|A \cap C)\, \lambda(A|C) = \lambda(B \cap A|C)$. If the player is a Bayesian who adopts this prior distribution, then upon seeing the record he will assign equal probability to the four possible outcomes compatible with the record. For example, if the record ends $\ldots e^+\mu^+$, he will assign probability 1/4 to the four possible outcomes of the final toss. But then the player assigns probability 1/4 to each of the following four states of the record as it existed immediately before the final roll of the die:

$\ldots e^+$ (corresponding to a final toss landing μ^+),
$\ldots e^+\mu^+e^+$ (corresponding to a final toss landing e^-),
$\ldots e^+\mu^+e^-$ (corresponding to a final toss landing e^+) and
$\ldots e^+\mu^+\mu^+$ (corresponding to a final toss landing μ^-).

Thus, he assigns probability 3/4 to the event that the final toss resulted in an "annihilation" in the record for the final entry. Moreover, with the prior described above, the player assigns probability 3/4 to the event "annihilation" in the record for the final entry (call this event A) for each observation of a non-vacuous record. If he were to observe a blank record, then for certain the last roll resulted in A. Hence,

$$3/4 \leq P(A|x) \leq 1, \quad \text{for each observation } x. \tag{1}$$

Let us call the state of the record just prior to the final toss, θ, the parameter. Then, on the assumption that the die is fair,

$$0 \leq P(A|\theta) \leq 1/4, \quad \text{for each } \theta. \tag{2}$$

Note that $P(A|\theta) = 0$ only when $\theta = \theta_0$, corresponding to a blank record just prior to the final toss. $P(A|\theta) = 1/4$ for all other parameter states.

If conglomerability applies, then incoherence results as $3/4 \leq P(A)$ by (1) and $P(A) \leq 1/4$ by (2). It is trivial to "make book" in such circumstances, by betting at, say, 1:3 odds both against and for the event A. The player who (incoherently) adopts conglomerability is led to accept these gambles, as shown by the two (inconsistent) inequalities for $P(A)$.

What if the player adopts conglomerability in only one of the two partitions? For instance, Levi (1980, pp. 284–287) requires conglomerability in the $\pi_\theta = \{\theta_i\}$ partition, as a consequence of his theory of "direct inference". It appears that most other writers assume conglomerability in this, the margin of the "parameter". The result is a finitely additive distribution which is coherent, in the sense that no finite collection of bets suffice to "make book". However, on pain of a sure loss, the player with such a coherent, finitely additive distribution will decline the offer of an infinite class of "fair" bets, where he is agreeable to every finite subclass of wagers. This is, according to Stone, a "strong inconsistency".

For example, let conglomerability apply in $\pi_\theta = \{\theta_i, i = 0, \ldots\}$. Then prior to the game the player has odds of 1:3 on A (an "annihilation" on the final roll), as $P(A) = 1/4$ (since $P(\theta_0) = 0$). Prior to the game, the player also holds the infinite collection of conditional odds, given $x = x$ $(i = 0, \ldots)$, of (at least) 3:1 on A, for each possible observation x_i. But the player will not accept all the denumerably many called-off bets (called off in case $x = x_i$ fails to occur) at the conditional odds of 3:1 on A, while also agreeing to wager on A at the unconditional odds of 1:3. To do so would expose him to a sure loss. Of course, the player is willing to accept any finite subset of this infinite set of wagers. No finite subset is sufficient to fix a sure loss. Hence, the player is coherent in deFinetti's sense (1974, ch. 3).

A similar argument applies in case the player holds conglomerability only in $\pi_x = \{x_i\}$, the margin of the "observable". Then $P(A) = 3/4$. The requisite infinite class of "fair" gambles is constructed by considering the conditional odds on the event A, given the state of the record immediately prior to the final toss (θ_i). The player is agreeable to each of the denumerably many called-off bets on A, called-off in case $\theta = \theta_i$ fails, at odds of 1:3, as $P(A|\theta_i) \leq 1/4$ ($i = 0, \ldots$). Once again, the player will not accept all such bets at once. Instances of "strong inconsistencies" are thus not violations of coherence.

Additional cases of "strong inconsistency" (without failure of coherence in deFinetti's sense) can arise when a posterior probability is calculated by Bayes' Theorem for densities with an "improper" prior, subject to the assumption of conglomerability in the margin of the parameter (the unobserved quantity). Heath and Sudderth (1978) illustrate this with their example 5.2. We discuss their theory in the next section.

V. "IMPROPER" PRIORS, COHERENT POSTERIORS, AND ATTEMPTS AT CURTAILING NON-CONGLOMERABILITY

The use of improper distributions to represent ignorance has a long history (see, for example, Jeffreys, 1961 [first edition 1939], Lindley, 1970 [first edition 1965], and Hartigan, 1964). Improper distributions are not probabilities because they assign infinite mass to the universal set, e.g. Lebesgue measure on the real line. An improper distribution may be translated directly into a set of finitely additive probabilities, as does Levi (1980, pp. 125–131), who treats improper distributions as σ-finite representations of finitely additive measures. Alternatively, an improper distribution may arise as a limit of finite measures which can be normalized to be probabilities (see Renyi, 1955).

The limits of the normalized measures often do not exist, however. For example, if the universal set is the real numbers, let λ_n be Lebesgue measure restricted to the interval $[-n, n]$. Each λ_n can be normalized to $\mu_n = (2n)^{-1}\lambda_n$ which is a probability. Whereas λ_n converges to Lebesgue measure on the entire line, the sequence $\{\mu_n\}_{n=1}^{\infty}$ does not converge to a countably additive measure. There exist subsequences of

$\{\mu_n\}$, however, which converge to finitely additive probabilities. This can be seen as follows. Each probability μ_n is a function from a field of events $\mathcal{F}$ into the interval [0, 1]. The collection of all such functions is a compact space (in the product topology). Hence, every sequence $\{\mu_n\}$ in this space has a limit point. Since limits preserve finite additivity, each such limit point μ will be a finitely additive probability.

The same reasoning can be applied to any improper prior to produce a collection of finitely additive probabilities associated with it, as follows. For each improper prior λ there exist finitely additive probabilities μ, each of which is a limit point of a sequence of probabilities $\{\mu_n\}$, and each μ_n is λ restricted to a set of finite measures and normalized to be a probability. This establishes a connection between improper priors and finitely additive probabilities.

Having established the connection, we can ask if the inferences made using improper distributions remain valid in the finitely additive theory. For example, suppose the distribution of $X = (\bar{X}, S)$ given $\theta = (\mu, \sigma)$ is that $\bar{X}$ and S are independent, $\bar{X}$ has a normal $N(\mu, \sigma^2/n)$ distribution and nS^2/σ^2 has a chi-squared distribution with $n - 1$ degrees of freedom. If we pretend that the improper prior $1/\sigma$ is a density for (μ, σ), then use of Bayes' Theorem for densities leads to the conclusion that the posterior distribution of μ given X is such that $(n-1)^{1/2}(\mu - \bar{X})/S$ has a t distribution with $n - 1$ degrees of freedom, a proper distribution.

In their important paper, Heath and Sudderth (1978) take the following approach to inferences involving improper distributions. They define the posterior distribution of a parameter θ given the data X as the conditional distribution necessary to make the joint distribution of (X, θ) conglomerable in *both* the X and θ margins if such a posterior distribution exists. That is to say, if the conditional distribution of X given θ (i.e. the likelihood) is $p(\mathrm{d}x \mid \theta)$ and the prior for θ is $\pi(\mathrm{d}\theta)$, then the marginal for X is $m(\mathrm{d}x) = \int p(\mathrm{d}x \mid \theta)\pi(\mathrm{d}\theta)$ and the posterior $q(\mathrm{d}\theta \mid x)$, if it exists, will satisfy

$$\int\int \phi(x,\theta)q(\mathrm{d}\theta|x)m(\mathrm{d}x) = \int\int \phi(x,\theta)p(\mathrm{d}x|\theta)\pi(\mathrm{d}\theta), \qquad (3)$$

for all bounded measurable ϕ. In the above example concerning (μ, σ), Heath and Sudderth find a class of measures, to which $\mathrm{d}\lambda = \mathrm{d}\mu\mathrm{d}\sigma/\sigma$ belongs, which have the property, among others, that there exists a sequence $\{B_n\}_{n=1}^{\infty}$ of sets with $0 < \lambda(B_n) < \infty$, and $\cup_{n=1}^{\infty} B_n$ equal to the space of all (μ, σ) pairs. They then form the sequence of probabilities

$\lambda_n(\cdot) = \lambda(\cdot \cap B_n)/ \lambda(B_n)$. Each limit point π of this sequence is a distinct finitely additive prior for (μ, σ) which has a "posterior" distribution for μ given X agreeing with the one obtained above by use of Bayes' Theorem for densities applied to the improper prior $d\lambda$.

However, Heath and Sudderth (H–S) use the term "posterior" in an overly restricted sense, in our opinion. The implication of their definition is that if no such q exists, there is no "posterior". The work of Schervish, Seidenfeld and Kadane (1984) shows that under mild conditions, each merely finitely additive probability, even one that satisfies (3), will have countable partitions (margins) in which it is not conglomerable. The requirement that a probability be conglomerable in a pair of given (albeit uncountable) partitions before admitting the existence of a posterior is, perhaps, too harsh. After all, the prior $\pi(d\theta)$, if it is merely finitely additive, will not be conglomerable in all countable partitions which result from coarsening the parameter space. Consider the following example:

Example 5.1. Assume $\theta \in \{0, 1\}$, $X \in \{1, 2, 3, \ldots\}$ with a distribution satisfying $p(x|\theta) = 2^{-(x+\theta)}$ and $\pi(\theta) = \frac{1}{2}$. There are many such finitely additive distributions. Each has some probability adherent along the sequence of (x, θ) values $\{(n, 1)\}_{n=1}^{\infty}$ in the sense of deFinetti (1974, p. 240). It follows that the marginal for X is $m(x) = (3)2^{-(x+2)}$ and Bayes' Theorem gives the posterior of θ given X to be $q(\theta|x) = (2 - \theta)/3$ for all x. Let $\phi(x, \theta)$ equal 1 if $\theta = 1$ and zero otherwise. Then

$$\iint \phi(x,\theta)p(dx\,|\,\theta)\pi(d\theta) = \frac{1}{2}$$

and

$$\iint \phi(x,\theta)q(d\theta|x)m(dx) = \frac{1}{3},$$

where the integrals are defined as in Dunford and Schwartz (1958). Yet it is reasonable to claim that q is the posterior for θ given X once finite additivity is accepted. This example makes clear the need for a less restrictive definition of posterior distribution that will allow inference even when a probability cannot be made conglomerable in a specific partition.

Another problem with requiring probabilities to be conglomerable in a specific partition is illustrated by this example. Schervish,

Seidenfeld and Kadane (1984) show that there are finitely additive probabilities on the space of all (x, θ) pairs which assign probability one to the points $\{(x, 1)\}_{x=1}^{\infty}$, are conglomerable in the x margin, and assign zero probability to each individual point. Let P_1 be such a probability. There is a countably additive probability P_0 which assigns probability 2^{-x} to the point $(x, 0)$ for each x. Of course, this probability is conglomerable in the x margin. But $P = \frac{1}{2}P_0 + \frac{1}{2}P_1$ has exactly the form of the probability described in Example 2.1, where we showed that P is not conglomerable in the x margin. Hence, the collection of probabilities conglomerable in a specified partition is not closed under convex combination, and H–S coherence is not preserved under mixtures of H–S coherent distributions.

Finitely additive distributions that are coherent in the usual sense but not H–S coherent reflect "strong inconsistencies" similar to the anomaly of Stone's example (section IV). Likewise, coherent but H–S incoherent distributions fail Robinson's (1979) criterion that there be no "super-relevant" betting procedures.

In contrast to H–S coherence, where "posteriors" are free of strong inconsistency in the two canonical margins (π_x, and π_θ), Levi's (1980) theory requires little more than coherence in deFinetti's sense, and suggests one view of how to calculate posterior probability with an "improper" prior. Levi's (1980, p. 129) analysis supports the familiar manuever with the Bayes formula if the likelihood function is countably additive.

Whether or not one is prepared to insist on H–S coherence (we are not), it seems reasonable to investigate "conditional properties" of coherent distributions. That is, from a Bayesian point of view the distinction between "absolute" and "conditional" probability is tenuous, at best. What is the effect of strengthening conglomerability to fix conditional probability values?

Conditional Conglomerability

Let F be an event for which $P(\cdot | F)$ is defined, and let $\pi = \{h_i\}$ be a partition. Let $h_{iF} = h_i \cap F$, and assume $P(\cdot | h_{iF})$ is defined for all i. Then P is *conditionally conglomerable in π with respect to (given)* F if for each $E \subseteq F$, and for all constants k_1 and k_2 such that

$$k_1 \leq P(E|h_{iF}) \leq k_2, \quad \text{for all } i; \quad k_1 \leq P(E|F) \leq k_2.$$

Trivially, if conditional conglomerability with respect to all events is satisfied in a specified partition, then conglomerability also holds in that margin. The converse is false, however, as we show using the example constructed by Buehler and Feddersen (1963), which was a rebuttal to arguments of Fisher (1956). Thus, H–S coherence admits distributions which are conditionally H–S incoherent.

Let (x_1, x_2) be i.i.d. $N(\theta, \sigma^2)$, with both parameters "unknown". It is obvious that

$$P(x_{\min} \leq \mu \leq x_{\max} | (\mu, \sigma^2)) = 0.5, \quad \text{for each pair} (\mu, \sigma^2). \tag{4}$$

Let $t = (x_1 + x_2)/(x_1 - x_2)$. Buehler and Feddersen (1963) show that

$$P(x_{\min} \leq \mu \leq x_{\max} | (\mu, \sigma^2), \ |t| \leq 1.5) > 0.518, \tag{5}$$

for each pair (θ, σ^2), despite (4).

Moreover, Heath and Sudderth (1978) show there is a class of finitely additive "prior" probabilities over pairs (μ, σ^2) such that for each "prior" conglomerability is satisfied in *both* partitions $\pi_{(x_1,x_2)}$ and $\pi_{(\mu,\sigma^2)}$. (Note that each partition has cardinality of the continuum.) Also, they show that each finitely additive "prior" probability induces a familiar countably additive "posterior" distribution, where

$$P(x_{\min} \leq \mu \leq x_{\max} ; (x_1, x_2)) = 0.5, \quad \text{for each pair} (x_1, x_2). \tag{6}$$

Now, since t is a function of the (x_1, x_2) pairs, we can partition the event $|t| \leq 1.5$ by the set of pairs (x_1, x_2) for which this inequality holds. Thus, by conditional conglomerability applied to this partition of $|t| \leq 1.5$, (7) is a consequence of (6):

$$P(x_{\min} \leq \mu \leq x_{\max} | \ |t| \leq 1.5) = 0.5. \tag{7}$$

But, also we may partition the event $|t| \leq 1.5$ into the continuum of states $(|t| \leq 1.5, \mu, \sigma^2)$, so that with conditional conglomerability applied to this partition of $|t| \leq 1.5$, (8) is a consequence of (5):

$$P(x_{\min} \leq \mu \leq x_{\max} | \ |t| \leq 1.5) > 0.518. \tag{8}$$

At least one of (7) and (8) must fail, hence conditional conglomerability fails in at least one partition where conglomerability holds (see Seidenfeld, 1979).

The failure of conditional conglomerability in a partition that admits conglomerability is restricted to events that have zero probability. We show this as follows. Let $P(F) \neq 0$, and argue that

$$\begin{aligned}\int_\theta P(E|F,\theta)\mathrm{d}P(\theta|F) &= \int_\theta P(E|F,\theta)\frac{P(F|\theta)\mathrm{d}P(\theta)}{P(F)} \\ &= \int_\theta \frac{P(E\cap F|\theta)\mathrm{d}P(\theta)}{P(F)} \\ &= P(E\cap F)/P(F) \\ &= P(E|F),\end{aligned}$$

where the first equality is by conglomerability in π_θ and Bayes' Theorem, the second and fourth equalities are by the multiplication theorem for conditional probability, and the third equality is again by conglomerability in π_θ. Hence, in the Fisher–Buehler–Feddersen problem, it must be that $P(|t| \leq 1.5) = 0$.

Examples such as the Fisher–Buehler–Feddersen paradox, where conditional conglomerability fails in a margin for which conglomerability holds, illustrate the violation of Robinson's (1979) criterion that there be no "relevant" betting procedures. Thus, H–S coherence is insufficient to preclude "relevant" betting schemes, in Robinson's sense. Also, subject to conditional conglomerability, advantage may switch repeatedly between two gamblers by an iteration of who has last say in determining when a bet is "on". This is illustrated by Fraser's (1977) "balanced procedure".

In contrast to the anomaly of "strong inconsistency" that is a consequence of non-conglomerability, we can identify a "weak inconsistency" of finitely additive distributions that arises from failure of a simple dominance (admissibility) rule even where conglomerability applies. Consider the choice between "no-bet" and a gamble that pays off $1/i$ if θ_i occurs ($i = 1, \ldots$). Clearly, for each value of θ the gamble is strictly preferred to "no-bet". However, if one imposes a merely finitely additive probability distribution over θ, such that $P(\theta_i) = 0$ (all i), $E(\text{no-bet}) = E(\text{gamble}) = 0$. That is, one becomes indifferent between the two options even though simple dominance obtains.

We can express this failure of dominance as a violation of a strengthened version of conglomerability. That is, "weak inconsistency" results from violating the following:

$$\text{If there exists a constant } k, \text{ such that } P(E|\theta_i) \gtrsim k \text{ for all } i, \text{ then } P(E) \gtrsim k. \tag{9}$$

In terms of the above example, if $P(E|\theta_i) = 1/i$ and we receive \$1 if E occurs when we choose to gamble, then for each θ_i we gain (in expectation) $1/i$ if θ_i occurs. However, though $P(E|\theta_i) > 0$ (all i), $P(E) = 0$ under the merely finitely additive prior described above, in violation of (9).

A somewhat different example is as follows. Let $x \sim N(\mu, 1)$. As Heath and Sudderth (1978) show, suitably invariant "priors" over μ exist for which conglomerability is satisfied simultaneously in π_μ and π_x, and where (for each x) $P(\mu|x)$ is $N(x, 1)$. Hence, for each x, $P(x \le \mu|x) = 0.5$. Assume that the prior chosen assigns positive probability to $\{x: x < 0\} = X^-$, i.e. $P(X^-) \neq 0$. (We make this assumption to avoid unnecessary questions about probability conditional upon an event of zero probability.) Since we know conglomerability holds in π_x (and since $P(X^-) \neq 0$) we have that $P(x \le \mu|X^-) = 0.5$. However, we note that, for each μ, $P(x \le \mu|X^-, \mu) > 0.5$. (In fact, $P(x \le \mu|X^-, \mu) = 1$ for all $\mu \ge 0$.) Thus, $P(\cdot|X^-)$ fails (9) in π_μ, though conditional conglomerability, given X^-, holds.

Finally, we come to those "marginalization paradoxes" involving the transformation of continuous random variables. These include all but the first of those of Dawid, Stone and Zidek (1973) and Stone and Dawid (1972). Here, by transformation, e.g. changing from $N(\mu, \sigma)$ to (τ, σ), where $\tau = \mu/\sigma$, we create the situation that:

i. $P(\tau|x) = P(\tau|t)$, for t a function of $x = \{x_1, \ldots, x_n\}$, $x_i \sim N(\mu, \sigma)$;
ii. $P(t|(\mu, \sigma)) = P(t|\tau)$, i.e. t depends solely on τ;
iii. conglomerability holds simultaneously in $\tau_{(\mu,\sigma)}$ and π_x;
iv. yet the probability density $p(t|\tau)$ [from (ii)] does not factor the density $P(\tau|t)$ [from (i)] as a function of τ.

As is argued by Seidenfeld (1981) and Sudderth (1980), no violation of conglomerability is present here. Only (9) is violated. But we saw above that (9) can fail in partitions where conglomerability holds. Where conditional conglomerability obtains, but (9) does not, we are faced with betting schemes that violate Robinson's (1979) criterion of "semi-relevance", though satisfying his requirement that there be no "super-relevant" or "relevant" betting policies.

In conclusion, we have noted several strengthened versions of conglomerability and have described statistical anomalies that reflect violation of each. Since each coherent finitely additive distribution must fail conglomerability in some denumerable partition (unless the distribution is countably additive), our reaction to these anomalies is to see them as further evidence that the attempt to modulate non-conglomerability (as in Robinson's criteria prohibiting betting schemes with "super-relevant" or "relevant" selections) is misguided. If conglomerability is necessary for "consistency", then nothing less than countably additive distributions suffice with denumerable partitions, and even "proper" priors may suffer non-conglomerability in non-denumerable partition, as we discuss in the next section.

VI. COUNTABLE ADDITIVITY AND NON-CONGLOMERABILITY IN NON-DENUMERABLE PARTITIONS

Since non-conglomerability in denumerable partitions characterizes merely finitely additive probability, it a mistake to advocate *both* a finitely additive theory of probability *and* standards of coherence entailing dominance or disintegrability in particular partitions (margins). One may cite non-conglomerability as reason enough to reinstate countable additivity. Non-conglomerability may be too high a price to pay for the convenience of "uniform" distributions.

However, Example 6.1 shows that non-conglomerability, which is characteristic of merely finitely additive probability in denumerable partitions, can occur with countably additive measures in non-denumerable partitions.

Example 6.1. Suppose X_1 and X_2 are independent standard normal random variables except that $X_2 = 0$ is impossible (this does not change the joint distribution function). Since X_1 and X_2 are independent, the distribution of X_2 given $A = \{X_1 = 0\}$ is standard normal. Using the usual transformation of variables technique, the conditional distribution of X_2 given $B = \{X_1/X_2 = 0\}$ is that of $(-1)^Y Z$, where $P[Y = 1] = P[Y = 2] = \frac{1}{2}$, Z has chi-squared distribution with two degrees of freedom, and Y and Z are independent. The two events A and B are identical, however.

The fact that the conditional distribution of X_2 given A differs from that given B is an example of the well-known Borel paradox (Kolmogorov, 1956). This paradox was also discussed by Hill (1980, p. 44). The seeming contradiction is often resolved by claiming that the transformation of variables only yields conditional probability given the sigma field of events determined by the random variable X_1/X_2, not given individual events in the sigma field. This approach is unacceptable from the point of view of the statistician who, when given the information that $A = B$ has occurred, must determine the conditional distribution of X_2. A more reasonable approach is to consider the theory of conditional probability spaces as defined by Dubins (1975), among others. In such a theory $P[X_2 \in E|A]$ means the conditional probability that X_2 is in the set E given the event A and is not relative to a sigma field. This, then, is the meaning of conditional probability one assumes when one conditions on the occurrence of a particular event.

To conclude Example 6.1, suppose one must determine conditional probabilities given all events of the form $A_{a,c} = \{X_1 - cX_2 = a\}$. The distribution of X_2 given $A_{a,c}$ can be calculated from the transformation of variables technique, if one wishes, as

$$\text{Normal}\, N\left(\frac{-ac}{1+c^2}, \frac{1}{1+c^2}\right).$$

In particular, if $a = 0$ the distribution of X_2 is normal with mean zero and variance $(1 + c^2)^{-1}$. Consider the uncountable partition of the sample space via the events $\{A_{0,c}| c \text{ a real number}\} = \theta$. Let Φ denote the standard normal distribution function and $E = \{X_2 > 1, X_1 \neq 0\}$. Then $P[E|A_{0,c}] = 1 - \Phi((1 + c^2)^{1/2}) < 1 - \Phi(1) = P(E)$, for all c. Hence, P is not conglomerable in the partition θ. Theorem A1 in the Appendix shows that non-conglomerability in uncountable margins is quite common for countably additive probabilities with continuous distributions, just as non-conglomerability in countable margins is quite common for merely finitely additive probabilities. In fact, the theorem implies that no matter how one defines conditional probability given $A_{a,c}$ in the above example, conglomerability will fail in some partition.

Hence, it appears that finite additivity cannot be rejected merely on the grounds that it allows failures of conglomerability, unless countable additivity is also to be rejected. Since few, if any, statisticians are willing to reject countable additivity, we suggest that finite additivity be judged on issues other than conglomerability.

VII. CONCLUSION

What attitude does our research lead us to with respect to finite additivity? We are comforted to know that various seeming paradoxes in statistical theory are understood once the lack of conglomerability of merely finitely additive probabilities is accounted for. Thus, marginalization, the Fisher–Buehler–Feddersen paradox, etc. no longer are troublesome as problems illustrating a failure in the foundations of probability and (subjective Bayesian) statistical theory.

Should we then advocate the use of merely finitely additive probabilities as likelihoods and/or prior distributions? So much is unknown, or known only to a very few, about the probability theory of merely finitely additive probabilities that, as of 1985, it is very difficult to make such a judgment. On the one hand, merely finitely additive distributions do allow certain kinds of invariant distributions not allowed under the requirement of countable additivity, such as uniform distributions on countable sets, and translation invariant measures on the real line. For certain purposes these may be useful objects (Jeffreys, 1961; Zellner, 1971; Box and Tiao, 1973). Yet countably additive distributions do offer a wide range of expression of opinion as it is. For the moment, then, we propose the dual strategy of encouraging the development of the probability theory of finite additivity, while leaving open the matter of whether the extra generality permitted by going beyond countably additive opinions is worth the cost.

We conjecture that, ultimately, the combined force of deFinetti's argument that only finitely additive probabilities are required for coherence, and the kind of considerations that led Dubins and Savage (1965) to write their book in a finitely additive framework, will lead to a recognition that finitely additive probability is the proper setting for subjective Bayesian inference. The pace at which this innovation occurs will be governed largely, we think, by the development of the necessary probability theory.

APPENDIX: NON-CONGLOMERABILITY OF COUNTABLY ADDITIVE PROBABILITIES

To continue Example 6.1, consider two independent standard normal random variables X_1 and X_2 with $X_2 = 0$ impossible. Define $Y_c = X_1 - cX_2$ and $Z_a = (X_1 - a)/X_2$. Then $A_{a,c} = \{Y_c = a\} = \{Z_a = c\}$. The conditional density of X_2 given $Y_c = a$, $f_c(X_2|a)$, is proportional to

$$\exp\left\{-\tfrac{1}{2}(1+c^2)\left[x_2 + ca(1+c^2)^{-1}\right]^2\right\},$$

and the conditional density of X_2 given $Z_a = c$, $f^a(x_2|c)$, is proportional to

$$|x_2|\exp\left\{-\tfrac{1}{2}(1+c^2)\left[x_2+ca(1+c^2)^{-1}\right]^2\right\}.$$

The set $\{(a, c, x): f^a(x|c) = f_c(x|a)\}$ has zero three-dimensional Lebesgue measure. If we assume that the conditional density of X_2, given $A_{a,c}$, $f(x_2|a, c)$, is defined for all a and c and is a measurable function of (x_2, a, c), then the conditions of Theorem A1 below are satisfied. The conclusion is that conglomerability fails in some uncountable partition. In fact, the partition will be the one determined by one of the random variables Y_c or one of the Z_a.

Theorem A1. Suppose two random variables (X_1, X_2) have a joint density $f(x_1, x_2)$ which is strictly positive over some measurable set G of positive Lebesgue measure. Let $\mathbb{R}$ denote the real numbers. Suppose there exist two sets of random variables $\{Y_c: c \in \mathbb{R}\}$ and $\{Z_a: a \in \mathbb{R}\}$, all measurable functions of (X_1, X_2), such that $Y_c = a$ if and only if $Z_a = c$, and (Y_c, X_2), and (Z_a, X_2) are each one-to-one functions of (X_1, X_2) for all a and c. Suppose the conditional density of X_2, given $A_{a,c} = \{Y_c = a\} = \{Z_a = c\}$, is $f(x_2|a, c)$ and is a measurable function of (x_2, a, c). Suppose that the transformation of variables technique gives the conditional density of X_2, given $Y_c = a$, as $f_c(x_2|a)$, and given $Z_a = c$, as $f^a(x_2|c)$, with the three-dimensional Lebesgue measure of $\{(a, c, x): f^a(x|c) = f_c(x|a)\}$ equal to zero. Then there exists $\delta > 0$, an event E, and a partition $\pi = \{h_b: b \in \mathbb{R}\}$ such that

$$P(E) - \int_\pi P(E|h_b)dP(h_b) > \delta (or < -\delta) \tag{A1}$$

and conglomerability fails in the partition π.

Before proving Theorem A1, we state and prove the following lemma.

Lemma A2. Under the conditions of Theorem A1, there exists $\varepsilon > 0$, such that for some c the intersection of one of the following two sets with G has positive Lebesgue measure:

$$G_1(c,\varepsilon) = g_c^{-1}\{(a, x_2): f(x_2|a, c) - f_c(x_2|a) > \varepsilon\},$$

$$G_2(c,\varepsilon) = g_c^{-1}\{(a, x_2): f(x_2|a, c) - f_c(x_2|a) > -\varepsilon\}.$$

or for some a, the intersection of one of the following two sets with G has positive Lebesgue measure:

$$G_3(a,\varepsilon)=(g^a)^{-1}\{(c,x_2): f(x_2|a,c)-f^a(x_2|c)>\varepsilon\},$$

$$G_4(c,\varepsilon)=(g^a)^{-1}\{(c,x_2): f(x_2|a,c)-f^a(x_2|c)<-\varepsilon\}.$$

Proof of Lemma A2. Let $(Y_c, X_2)=g_c(X_1, X_2)$ and $(Z_a, X_2)=g^a(X_1, X_2)$ be the one-to-one functions guaranteed by the hypothesis of Theorem A1. Since g^a and g_c are both one-to-one measurable functions, if D is any subset of G, then D has positive Lebesgue measure if and only if $g_c(D)$ has positive measure for all c and $g^a(D)$ has positive measure for all a. Assume that for all $\varepsilon > 0$, and a the intersections of $G_3(a, \varepsilon)$ and $G_4(a, \varepsilon)$ with G have zero Lebesgue measure. It follows that the intersection of G with

$$(g^a)^{-1}\{(c,x_2): f(x_2|a,c)\neq f^a(x_2|c)\}$$

has zero Lebesgue measure for all a. It follows that the Lebesgue measure of

$$g^a(G)\cap\{(c,x_2): f(x_2|a,c)\neq f^a(x_2|c)\}$$

is zero for all a. Integrating the measure of this set with respect to one-dimensional Lebesgue measure da gives that the three-dimensional measure of

$$\left[\bigcup_a\{(a,c,x_2):(c,x_2)\in g^a(G)\}\right]\cap\{(a,c,x_2): f(x_2|a,c)\neq f^a(x_2|c)\} \tag{A2}$$

is zero. Call the set on the left of the intersect symbol in equation (A2) A and call the set on the right B. Next, consider the set

$$A\cap\{(a,c,x_2): f(x_2|a,c)=f_c(x_2|a)\}. \tag{A3}$$

Call the set on the right of the intersect symbol in equation (A3) C. It follows from the hypotheses of Theorem A1 that $B^c \cap C$ has zero measure, where B^c is the complement of B. Write

$$A\cap C=(A\cap C\cap B)\cup(A\cap C\cap B^c).$$

Since $A \cap B$ and $C \cap B^c$ both have zero measure, so does $A \cap C$. Since $g_c(x_1, x_2) = (a, x_2)$ if and only if $g^a(x_1, x_2) = (a, x_2)$, it follows that we can also write A as

$$\bigcup_c\{(a,c,x_2):(a,x_2)\in g_c(G)\}.$$

The three-dimensional measure of $A \cap C$ equals the integral over $\mathbb{R}$ with respect to dc of the two-dimensional measure of

$$g_c(G) \cap \{(a, x_2) : f(x_2|a, c) = f_c(x_2|a)\},$$

which must be zero for almost all c by Tonelli's Theorem. Since $g_c(G)$ has positive measure, the intersection of $g_c(G)$ with either

$$\{(a, x_2) : f(x_2|a, c) > f_c(x_2|a)\}$$

or

$$\{(a, x_2) : f(x_2|a, c) < f_c(x_2|a)\}$$

has positive measure for some c. Hence, there exists $\varepsilon > 0$ and c such that the intersection of G with one of the sets $G_1(c, \varepsilon)$ or $G_2(c, \varepsilon)$ has positive measure. □

Proof of Theorem A1. In the notation of Lemma A2, suppose that the intersection of $G_1 = G_1(c_0, \varepsilon)$ with G has positive Lebesgue measure. The proof is similar for the others. Let $f_{c_0}(a)$ denote the marginal density of Y_{c_0}. Consider the partition $\pi = \{h_a: a \in \mathbb{R}\}$, where $h_a = A_{a,c_0}$. This is the partition determined by Y_{c_0}, hence d$P(h_a) = f_{c_0}(a)$da. If E is any event of the form $\{(X_1, X_2) \in D\}$, then

$$P(E) = \iint_D f(x_1, x_2)\mathrm{d}x_1\,\mathrm{d}x_2 = \iint_{g_{c_0}(D)} f_{c_0}(x_2|a) f_{c_0}(a)\mathrm{d}x_2\,\mathrm{d}a. \qquad \text{(A4)}$$

Let $D = G_1$ and $E = \{(X_1, X_2) \in D\}$. Note that $P(E)$ is positive since the intersection of G and D has positive Lebesgue measure. Let $G_0 = g_{c_0}(G_1)$. Since G_{c_0} is one-to-one, equation (A4) implies:

$$P(E) = \iint_{G_0} f_{c_0}(x_2|a) f_{c_0}(a)\mathrm{d}x_2\mathrm{d}a. \qquad \text{(A5)}$$

From the definitions of G_0 and G_1 and equation (A5), it follows that

$$P(E) < \iint_{G_0} f(x_2|a, c_0) f_{c_0}(a)\mathrm{d}x_2\,\mathrm{d}a - \varepsilon \iint_{G_0} f_{c_0}(a)\mathrm{d}x_2\,\mathrm{d}a. \qquad \text{(A6)}$$

Since $P(E)$ is positive, it follows from equation (A6) that f_{c_0} cannot be zero almost everywhere in G_0, hence the second integral on the right-hand side of equation (A6) is positive. Let $\delta = \varepsilon \iint_{G_0} f_{c_0}(a)\,\mathrm{d}x_2\mathrm{d}a > 0$. If we let $G(a) = \{x_2: (a, x_2) \in G_0\}$, then $E \cap A_{a,c_0}$ equals $\{a\} \times G(a)$ and

$$P(E|A_{a,c_0}) = \int_{G(a)} f(x_2|a, c_0)\mathrm{d}x_2, \qquad \text{(A7)}$$

by the definition of $f(x_2|a, c_0)$. The first integral on the right-hand side of (A6) can be rewritten, using equation (A7), as

$$\int_{\mathbb{R}}\int_{G(a)} f(x_2|a, c_0)f_{c0}(a)\mathrm{d}x_2\mathrm{d}a = \int_{\mathbb{R}} P(E|A_{a,c0})f_{c0}(a)\mathrm{d}a$$
$$= \int_{\pi} P(E|h_a)\mathrm{d}P(h_a). \quad \text{(A8)}$$

Together equation (A6), equation (A8), and the definition of δ imply equation (A1) (with $< -\delta$). In the terminology of Dubins (1975), we have proven that the joint distribution of X_1 and X_2 is not disintegrable in π. Dubins proves that this implies that conglomerability fails in π also. □

To conclude Example 6.1, assume that the conditional density of X_2 given $A_{a,c}$ is given by $f(x_2|a, c)$. As long as this is a measurable function of (x_2, a, c), Theorem A1 shows that conglomerability fails in some partition determined by one of the Y_c or one of the Z_a.

REFERENCES

Arrow, K.J. (1972) "Exposition of the theory of choice under uncertainty", in: C.B. McGuire and R. Radner, eds., *Decision and Organization* (North Holland, Amsterdam).

Box, G.E.P. and G. Tiao (1973) *Bayesian Inference in Statistical Analysis* (Addison-Wesley Publishing Co., Reading).

Buehler, R.J. and A.P. Feddersen (1963) "Note on a conditional property of Student's *t*", *Ann. Math. Stat.* 34, 1098–1100.

Dawid, A.P., M. Stone and J.V. Zidek (1973) "Marginalization paradoxes in Bayesian and structural inference", *J.R.S.S.*, Ser. B 35, 189–233 (with discussion).

deFinetti, B. (1972) *Probability, Induction and Statistics* (Wiley, New York).

deFinetti, B. (1974) *Theory of Probability*, Vol. 1 (Wiley, New York).

Dubins, L. (1975) "Finitely additive conditional probabilities, conglomerability and disintegrations", *Ann. Probability*, 3, 89–99.

Dubins, L. and L.J. Savage (1965) *How to Gamble if You Must: Inequalities for Stochastic Processes* (McGraw-Hill, New York).

Dunford, N. and J.T. Schwartz (1958) *Linear Operations, Part I: General Theory* (Interscience Publishers).

Fisher, R.A. (1956) "On a test of significance in Pearson's Biometrika Tables (No. 11)", *J. Roy. Statist. Soc.*, Ser. B 18, 56–60.

Fraser, D.A.S. (1977) "Confidence, posterior probability, and the Buehler example", *Ann. Stat.*, 5, 892–898.

Hartigan, J. (1964) "Invariant prior distributions", *Ann. Math. Statist.*, 35, 836–845.

Heath, J.D. and W. Sudderth (1978) "On finitely additive priors, coherence, and extended admissibility", *Ann. Stat.*, 6, 333–345.

Hill, B. (1980) "On some statistical paradoxes and non-conglomerability", in: Bernardo, DeGroot, Lindley and Smith, eds., *Bayesian Statistics* (University Press, Valencia, Spain), pp. 39–49 (with discussion).

Jeffreys, H. (1961) *Theory of Probability* (Oxford University Press).

Kolmogorov, A. (1956) *Foundations of the Theory of Probability* (Chelsea, New York).

Levi, I. (1980) *The Enterprise of Knowledge* (The MIT Press, Boston).

Lindley, D.V. (1962) "Discussion of C. Stein's paper, 'Confidence sets for the mean of a multivariate normal distribution'", *J. Roy. Statist. Soc.*, Ser. B 24. 285–287.

Lindley, D.V. (1970) *Introduction to Probability and Statistics* (Cambridge University Press).

Lindley, D.V. (1981) "Scoring rules and the inevitability of probability", unpublished report ORC 81-1. Operations Research Center, University of California at Berkeley.

Renyi, A. (1955) "On a new axiomatic theory of probability", *Acta Math. Acad. Sci. Hung.*, 6, 285–335.

Robinson, G.K. (1979) "Conditional properties of statistical procedures", *Ann. Stat.*, 7, 742–755.

Savage, L.J. (1974) *Foundations of Statistics* (Wiley, New York).

Schervish, M., T. Seidenfeld and J. Kadane (1984) "The extent of non-conglomerability of finitely additive probabilities", *Z. f. Wahrscheinlichkeitstheorie*, 66, 205–226.

Seidenfeld, T. (1979) *Philosophical Problems of Statistical Inference* (D. Reidel, Dordrecht, Holland).

Seidenfeld, T. (1981) "Paradoxes of conglomerability and fiducial inference", in: J. Los and H. Pfeiffer, eds., *Proceedings of the 6th International Congress on Logic, Methodology and Philosophy of Science* (North Holland, Amsterdam).

Seidenfeld, T. and M.J. Schervish (1983) "A conflict between finite additivity and avoiding Dutch Book", *Phil. Sci.*, 50, 398–412. [Chapter 2.4, this volume]

Stein, C. (1955) "Inadmissibility of the usual estimator for the mean of a multivariate normal distribution", in: *Third Berkeley Symposium* (University of California Press, Berkeley), pp. 197–206.

Stone, M. (1976) "Strong inconsistency from uniform priors", *J.A.S.A.*, 71, 114–116 (with discussion, 117–125).

Stone, M. (1981) "Review and analysis of some inconsistencies related to improper priors and finite additivity", in: J. Los and H. Pfeiffer, eds., *Proceedings of the 6th International Congress on Logic, Methodology and Philosophy of Science* (North Holland, Amsterdam).

Stone, M. and A.P. Dawid (1972) "Un-Bayesian implications of improper Bayes inference in routine statistical problems", *Biometrika*, 59, 369–375.

Sudderth, W. (1980) "Finitely additive priors, coherence and the marginalization paradox", *J. Roy. Statist. Soc.*, Ser B 42, 339–341.

Villegas, C. (1964) "On quantitative probability σ-algebras", *Ann. Math. Statist.*, 35, 1787–1796.

Zellner, A. (1971) *An Introduction to Bayesian Inference in Econometrics* (Wiley, New York).

PART 3

Non-cooperative Decision Making, Inference, and Learning with Shared Evidence

3.1

Subjective Probability and the Theory of Games

JOSEPH B. KADANE AND PATRICK D. LARKEY

ABSTRACT

This chapter explores some or the consequences of adopting a modern subjective view of probability for game theory. The consequences are substantial. The subjective view of probability clarifies the important distinction between normative and positive theorizing about behavior in games, a distinction that is often lost in the search for "solution concepts" which largely characterizes game-theory since the work of von Neumann and Morgenstern. Many of the distinctions that appear important in conventional game theory (two-person versus n-person, zero-sum versus variable sum) appear unimportant in the subjective formulation. Other distinctions, such as single play versus repetitive-play games, appear to be more important in the subjective formulation than in the conventional formulation.

> . . . "Probability has often been visualized as a subjective concept more or less in the nature of an estimation. Since we propose to use it in constructing an individual, numerical estimation of utility, the above view of probability would not serve our purpose. The simplest procedure is, therefore, to insist upon the alternative, perfectly well founded interpretation of probability as frequency in the long run."
>
> von Neumann and Morgenstern
>
> [50. p. 19]

Reprinted by permission, "Subjective Probability and the Theory of Games," Joseph B. Kadane and Patrick D. Larkey, *Management Science*, Volume 28, Number 2, February 1982.

The Theory of Games and Economic Behavior (von Neumann and Morgenstern, [50]) has directly and indirectly spawned an enormous body of work. Theories of games are found in several disciplines including mathematics (Lucas [26]), statistics (Blackwell and Girshick [6]), economics (Schotter and Schwodiauer [42]), political science (Riker and Ordeshook [36]) and social psychology (Miller and Steinfatt [29]). Game theory has become a field in its own right with journals devoted primarily to the topic and academics pursuing careers as "game theorists."

Normative and positive work on game theory has been motivated, in the main, by an interest in rigorously understanding human behavior in situations of strategic interaction, particularly conflict situations. The prospective applications of this understanding are in advising parties to conflict situations and in predicting the outcomes of conflicts. The work on game theory has been largely directed at constructing formal models and deducing, within given game structures (number of players, number of plays, form of payoffs, etc.) and given assumptions about human behavior (usually a set of axioms or postulates for "rationality"), what the outcome of the games' play will be. These deduced outcomes are called "solution concepts."

In some of the disciplinary applications, games are explicitly constructed to serve as metaphors or analogies for naturally occurring conflict phenomena. In economics, game theoretic models have been used as metaphors for the market situations of duopoly, oligopoly, and perfect competition. Political scientists have used game theoretic frameworks to represent a variety of voting and other institutional decision contexts. In mathematics there is less concern with what the formal game theoretic models might represent empirically; game theory has proved to be a very rich source of intrinsically interesting mathematical problems. Some statistical decision theorists have adapted the minimax criterion from game theory to analyze statistical problems as if they were games against "nature."

There are other descendants of von Neumann and Morgenstern, although the lineage is less obvious. Von Neumann and Morgenstern revived interest in cardinal utility theory and the principle of maximizing expected utility as a canon of rational behavior. However, their version of probability was "objective," probabilities derived from relative "frequencies in the long run." Thus, they proposed finding expected utility by integrating subjective utility with respect to objective probability. This line was explored and extended by Abraham Wald, particularly in *Statistical Decision Functions* [51], and further explored by Savage in his *Foundations of Statistics* [40]. Savage's work began as a defense of the von Neumann-Morgenstern-Wald minimax approach; he concluded by shifting to a subjective view of probability, upholding the principle of subjective utility integrated with respect to subjective probability. This work was the starting point for the modern Bayesian view of statistics and econometrics, as exemplified by the work reviewed in Lindley [24], and the essays in Fienberg and Zellner, eds. [14], and Zellner, ed. [54].

It is a curiosity of intellectual history that these two lines of inquiry have had so little to do with one another despite their common heritage. There has been some slight contact (Harsanyi [15]), (Aumann and Maschler [3]), (Harsanyi and Selten [18]), (Mertens and Zamir [28]), (Zamir [53]), (Ponssard and Zamir [32]), (Ponssard [31]), (Kohlberg [22]), (Sorin [46]), having to do with games in which some of the payoffs are unknown.

The primary purpose of this essay is to explore some of the consequences for game theory of taking the modern subjective view of probability in the sense of Savage and the writers who have followed in that tradition.[1] The essay is offered in the spirit of a query to game theorists: Why has game theory followed the course of development that it has? As individuals interested in the development of useful prescriptive and predictive theories of individual and collective decision-making behavior, we have often been perplexed by the emphases in game theoretic research. In particular, we do not understand the search for solution concepts that do not depend on the beliefs of each player about the others' likely actions and yet are so compelling that they will become the obvious standard of play for all those who encounter them. The essay is exploratory. We sketch some of the more obvious implications of a subjective probability view for game theory.

Harsanyi [17], [15] was among the first to characterize game players as "Bayesians." For a variety of reasons including the perceived mathematical difficulties from "an infinite regress in reciprocal expectations on the part of players" and the perceived necessity of deriving determinate solution concepts, Harsanyi introduces special assumptions (e.g., a "basic probability distribution" that influences players' priors) that compromise a purely subjective view of probability. Later, Harsanyi [16] suggested a "tracing procedure" for analyzing games that, while Bayesian in some mechanical respects, does not use subjective probability. The procedure looks for criteria by which a player's prior must be of a certain form, thus adopting the "necessitarian" conception of probability (Jeffreys [20]) rather than the subjectivist position. The work of Boge and Eisele [7] is also relevant. They assume that the opponent's utility function is unknown and that opponents will also act in accordance with Bayes' principle. These special assumptions lead to some very difficult mathematics and they do not explore the more general implications of the subjective probability view for game theory. The most closely related work in point of view we have been able to discover is Sanghvi and Sobel [38], [39], who look for the theoretical case in which the players will act in such a way as to leave unchanged each player's probability distribution on the other's action. Of course, this requires many plays of the same game and special assumptions. See also Eliashberg [11], [12].

Basic to the Savage tradition (and the allied work of deFinetti [9]), a decision-maker has a subjective probability opinion with respect to all of the unknown contingencies affecting his payoffs. In particular in a simultaneous-move two-person game, the player whom we are advising is assumed to have an opinion about the major contingency faced, namely what the opposing player is likely to do. If I think my opponent will choose strategy i ($i = 1, \ldots I$) with probability p_i, I will choose any strategy j maximizing $\Sigma_{i=1}^{I} p_i u_{ij}$, where u_{ij} is the utility to me of the situation in which my opponent has chosen i and I have chosen j. Since the opponent's utilities are important only in that they affect my views $\{p_i\}$ of what my opponent may do, the distinction between zero and non-zero sum games is much less important in this theory than in von Neumann and Morgenstern's formulation. Dominance, weak or strong, is obviously still important in this theory; if a strategy j exists that maximizes u_{ij} for all i, I will choose j, whatever subjective probabilities, p_i, I assign.

What role would the minimax principle for a zero-sum game play in such a theory? Suppose for example that our opponent announced his intention, and committed himself in an unbreakable contract, to use the minimax strategy. Then there would be, in all games without dominant strategies for both players, several choices j each of which would yield an expected utility equal to the value of the game, and whose mixture with appropriate weights would be my minimax strategy. Choosing any one of these, or the minimax or any other mixture of them would be equally good, and would yield the value of the game to me. Thus, the minimax strategy is not ruled out by the subjective approach, but does not here have the strong probative force given it by von Neumann and Morgenstern.

Minimax theory is, of course, incomplete, in that it does not suggest what I should do if I believe that my opponent is not playing the minimax strategy. The experimental evidence (Rapoport and Orwant [35]; Lieberman [23]) suggests that minimax players are the exception. And if my opponent is not playing the minimax strategy, there will be, in most games, a strategy that I can follow which is superior to minimax.

That minimax strategies are a special case is an illustration of an important general connection between the subjective theory of games and the von Neumann-Morgenstern view, namely *solution concepts are a basis for particular prior distributions.*[2] This helps to explain the difficulty in non-zero sum, N-person game theory of finding an adequate solution concept: no single prior distribution is likely to be adequate to all players and all situations in such games.

Multiplayer games present no essentially new conceptual difficulties. To decide on an optimal strategy, a player needs to know his probability, $P_{i_1,i_2 \ldots i_k}$, that player 1 will choose i_1, player 2 strategy i_2, etc. Then the optimal strategy is to choose a strategy j that maximizes $\Sigma P_{i_1,i_2 \ldots i_k} u_{i_1,i_2 \ldots i_k,j}$ where the summation extends over all strategies available to the other players. Such games may, however, pose important new information collection and processing difficulties. It becomes more costly, if not infeasible, to acquire firm priors on the behavior of many players.[3]

Similarly games studied by an outside observer are different from the point of view of a player. Thus, an outside observer will have a probability distribution on the priors of the players of the game. The analysis from the viewpoint of an observer will be somewhat similar to the analysis of Lindley, Tversky and Brown [25] on the elicitation problem from the viewpoint of an external person.

A possible problem with the theory advocated here is the infinite regress.[4] If he thinks I think he'll do x, then he'll do y. If he thinks I think he thinks I think he'll do y, etc. It is true that a subjectivist Bayesian will have an opinion not only on his opponent's behavior, but also on his opponent's belief about his own behavior, his opponent's belief about his belief about his opponent's behavior, etc. (He also has opinions about the phase of the moon, tomorrow's weather and the winner of the next Superbowl.) However, in a single-play game, all aspects of his opinion except his opinon about his opponent's behavior are irrelevant, and can be ignored in the analysis by integrating them out of the joint opinion.

Multiple-play games do introduce a new complication in the Bayesian theory. To help fix ideas, we will take a two game sequence in which game 2 is played, and then game 1. (We use reverse numbering because backwards induction, an essential part of the reasoning for this case, makes it convenient.) When only game 1 is left, only my opinion about my opponent's action in game 1 is relevant. In thinking about my action in game 2, I must take into account not only my opinion about what my opponent is likely to do in game 2, but also my opinion about the effect my strategy in game 2 may have on my opponent's strategy in game 1, conditional on each of the actions I may take in game 2. It is this feature that makes multiple-play games very interesting from the subjectivist Bayesian perspective. This also suggests that certain multiple-play games, such as the prisoner's dilemma, may be illuminated by this approach, since it is precisely the effect of today's decision on the actions of others, tomorrow, that has caused so much debate (see Rapoport and Chammah [34] and Axelrod [5]).

It is notable that, although Bayesian theory is basically prescriptive (Savage [40] and deFinetti [9]), predictive theories are not neglected. Recently a line of psychological investigation has suggested that human beings find it quite difficult to meet the coherence requirements of various rational decision prescriptions, including Bayesian theory (for reviews see Tversky and Kahneman [47], [48]; Hogarth [19]; Slovic *et al.* [45]; and Nisbett *et al.* [30]). These observations in turn have led to questions about whether the normative approach of standard Bayesian theory, or an alternative model of human cognition that fits the facts better, is more deserving of scientific attention. As one of the players, or as an advisor to one of the players, the Bayesian axioms of Savage are prescriptively compelling and consequently we would seek to play in accordance with them. This requires the best predictive theory we can find of the likely actions of the other player(s). Thus, both the prescriptive and predictive theories appear to have natural roles; neither need or should be chosen to the exclusion of the other. It may or may not be the case that the best prescriptive theory and the best predictive theory are one and the same in any given instance. This is an empirical question. As competitors, we seek to profit from whatever peculiarities we find in the play of our opponent(s).

Finally, there is the question of where all these prior distributions "come from." The experimental literature on elicitation of prior beliefs in general is indeed meager. One line, pursued especially by Savage [41], tries to obtain elicitations of priors (and utilities) as optimal behavior under certain specified conditions. This literature is very close to material on proper scoring rules (Brier [8]). A second line is more heuristic, trying to find questions easily answered by the subject and still transformable into the information sought. A recent paper along these lines is Kadane *et al.* [21]. Finally, there is the paper of Lindley, Tversky and Brown [25] which takes a Bayesian approach to elicitation itself, although with very special models. None of this literature on elicitation has, to our knowledge, dealt with beliefs in a game context. This is a potentially fruitful area for further research.

In this essay we have explored the consequences for game theory of adopting a subjective view of probability. The consequences are large. Distinctions that appear to be important in von Neumann and Morgenstern (two-person vs. $n > 2$ person, zero-sum vs. variable sum) appear not to be critical in this formulation. However, the distinction between single-play and repeated-play games seems more important than in the original von Neumann and Morgenstern work.

The Bayesian view of games clarifies the proper, respective roles of prescriptive and predictive theory. The view also raises some fundamental questions about the value of pursuing solutions to games, solutions that presume symmetrical behavior in the two-player case and homogeneous behavior for all players in the multi-player case.

> . . . the achievement of determinate solutions for two person, non-zero-sum games through the estimation of subjective probabilities requires the introduction of an assumption to the effect that the individual employs some specified rules of thumb in assigning probabilities to the choices of the other player. But this is not a very satisfactory position to adopt within the framework of the theory of games. Logically speaking, there is an infinite variety of rules of thumb that could be used in assigning subjective probabilities, and game theory offers no persuasive reason to select any one of these rules over the others. This problem can be handled by introducing new assumptions (or empirical premises) about such things as the personality traits of the players. But such a course would carry the analyst far outside the basic structure of the theory of games, requiring a fundamental revision of the basic perspective of game theory. (Young [52, pp. 28–29])

From the subjectivist Bayesian perspective, game theorists are already "(employing) some specified rules-of-thumb in assigning probabilities to the choices of the other player." Assuming that your opponent will play a minimax strategy which you attempt to construct from his perspective given information in any particular game about his payoffs and preferences is an example of such a "rule-of-thumb." At best, this rule-of-thumb is a partial basis for forming your prior about your opponent's likely behavior in certain simple game situations. It is not a logically compelling prescription for your own play (Ellsberg [13]). And it is not a very efficient predictive theory for most games (Rapoport and Orwant [35]).

Perhaps it is time to reunite the two streams of work descended from von Neumann and Morgenstern [50], prescriptive theories of individual decisionmaking and theories of strategically interactive decisions, and to look to other disciplines such as cognitive psychology for predictive theories of decisional behavior.

NOTES

1 Morris H. DeGroot [10, p. 4] has summarized the modern view as follows:

> According to the subjective, or personal, interpretation of probability, the probability that a person assigns to a possible outcome or some process represents his own judgment of the likelihood that the outcome will be obtained. This judgment will be based on that person's beliefs and information about the process. Another person, who may have different beliefs or different information, may assign a different probability to the same outcome.

2 Most "solution concepts" are from the perspective of an external observer of games. For the observer, the solution concept is the basis of a prior about the game's outcome. At the level of the individual player, assumptions about opponents' behavior correspond to the individual's priors. In minimax, the assumption is that the opponent will certainly do his best assuming that I will do my best with full information; each player is assumed to believe that the other is sure to play his minimax strategy.

3 A common analytic response to the informational difficulties arising from the multiplayer circumstance is to treat the other players as an undifferentiated mass behaving in an analytically tractable fashion:

> . . . in market environments it is assumed, at least implicitly, that there are many agents. In such contexts then the reward to a single agent depends not only on his own decision but also on the decisions of other agents. Thus, to predict the decisions which agents will make in a multi-agent environment there is needed some notion of consistency. We emphasize here as did Rothschild [37] that *the things which each agent takes as given in making his own decision must be consistent with maximizing behavior on the part of the other agents*. (Prescott and Townsend [33, p. 2])

This sort of simplification strategy would be useful to a player with no basis for differentiating among other agents and a firm prior on the decision mechanism of other agents (e.g. "maximizing behavior") as an accurate predictive theory of their behavior.

4 Harsanyi [15, p. 163] speculates that:

. . . the basic reason why the theory of games with incomplete information has made so little progress so far lies in the fact that these games give rise, or at least appear to give rise, to an infinite regress in reciprocal expectations on the part of the players.

Also see Riker and Ordeshook [36] and Young [52, p. 28] for other statements on the perceived importance of the "infinite regress problem."

REFERENCES

1. Aumann, R. J., "Mixed and Behavior Strategies in Infinite Extensive Games," *Advances in Game Theory*, Princeton Univ. Press, Princeton, N.J., 1964, pp. 627–650.
2. ——, "Some Thoughts on the Theory of Cooperative Games," *Advances in Game Theory*, Princeton Univ. Press, Princeton, N.J., 1964, pp. 407–442.
3. —— and Maschler, M., "Repeated Games with Incomplete Information, The Zero-Sum Extensive Case," Report to the U.S. Arms Control and Disarmament Agency, Washington D.C. Final Report on Contract ACDA/ST-143, prepared by *Mathematica*, Princeton, 1968, Chapter III, pp. 25–108.
4. —— and Shapley, L., *Values of Non-Atomic Games*, Princeton Univ. Press, Princeton, N.J., 1974.
5. Axelrod, R., "Evolution of Cooperation in the Prisoner's Dilemma," Institute of Public Policy Studies, University of Michigan, 1979.
6. Blackwell, D. and Girshick, M. A., *Theory of Games and Statistical Decisions*, Wiley, New York, 1959.
7. Boge, W. and Eisele, T., "On Solutions of Bayesian Games," *Internat. J. Game Theory*, Vol. 8 (1979), pp. 193–215.
8. Brier, G. W., "Verification of Forecasts Expressed in Terms of Probability," *Monthly Weather Rev.*, Vol. 75 (1950), pp. 1–3.
9. DeFinetti, B., *Theory of Probability* (2 vols.), Wiley, New York, 1974.
10. DeGroot, M. H., *Probability and Statistics*, Addison-Wesley, Reading, Mass., 1975.
11. Eliashberg, J., "Bayesian Analysis of Competitive Decision-Making Situations," unpublished mimeo., 1980.
12. ——, "An Investigation of Competitive Preference Structures and Posterior Performance Through a Bayesian Decision-Theoretic Approach," 1980.
13. Ellsberg, D., "Theory of the Reluctant Duelist," reprinted in [52].
14. Fienberg, S. E. and Zellner, A., eds., *Studies in Bayesian Econometrics and Statistics in Honor of Leonard J. Savage*, North-Holland, Amsterdam, 1975.
15. Harsanyi, J., "Games with Incomplete Information Played by 'Bayesian' Players," *Management Sci.*, Vol. 14, No. 3 (1967), pp. 159–182; Vol. 14, No. 5 (1968), pp. 320–334; Vol. 14, No. 7 (1968), pp. 486–502.
16. ——, "The Tracing Procedure: A Bayesian Approach to Defining a Solution for *n*-person Noncooperative Games," *Internat. J. Game Theory*, Vol. 4, No. 2 (1975), pp. 61–94.
17. ——, "Bargaining in Ignorance of the Opponent's Utility Function," *J. Conflict Resolution*, Vol. 6 (1962), pp. 29–38.
18. Harsanyi, J. and Selten, R., "A Generalized Nash Solution for Two-Person

Bargaining Games with Incomplete Information." *Management Sci.*, Vol. 18. No. 5 (1972), pp. 80–106.
19. Hogarth, R. M. "Cognitive Processes and the Assessment of Subjective Probability Distributions," *J. Amer. Statist. Assoc.*, Vol. 70. No. 350 (1975).
20. Jeffreys, H., *Theory of Probability* (3rd edition), Oxford Univ. Press, London, 1961.
21. Kadane, J. B., Dickey, J. M., Winkler, R. L., Smith, W. S. and Peters, W. C., "Interactive Elicitation of Opinion for a Normal Linear Model," *J. Amer. Statist. Assoc.*, Vol. 75 (1980), pp. 845–854.
22. Kohlberg, E., "Optimal Strategies in Repeated Games with Incomplete Information," *Internat. J. Game Theory*, Vol. 4 (1975), pp. 7–24.
23. Lieberman, B., "Experimental Studies of Conflict in Some Tw and Three-Person Games," in Massarik, F. and Ratoosh, P. (eds.), *Mathematical Explorations in Behavioral Science*, Irwin, Homewood, Ill., 1[illegible].
24. Lindley, D., "Bayesian Statistics: A Review," SIAM, [illegible]hiladelphia, Penn., 1971.
25. Lindley, D., "Tversky, A. and Brown, R., "On the Reconciliation of Probability Assessments," *JRSS*, Series A (1979), Vol. 142, pp. 146–162.
26. Lucas, W. F., "An Overview of the Mathematical Theory of Games," *Management Sci.*, Vol. 18. No. 5 (1972).
27. McKelvey, R. D. and Ordeshook, P. C., "An Experimental Test of Solution Theories for Cooperative Games in Normal Form," manuscript, 1980.
28. Mertens, J. F. and Zamir, S., "The Value of Two-Person Zero-Sum Repeated Games with Lack of Information on Both Sides," *Internat. J. Game Theory*, Vol. 1 (1971), pp. 39–64.
29. Miller, G. R. and Steinfatt, T. M. "Communication in Game Theoretic Models of Conflict," in Miller, G. R. and Simons, H. W. (eds.), *Perspectives on Communication in Social Conflict*, Prentice-Hall, Englewood Cliffs, N.J., 1974.
30. Nisbett, R. E., Crandall, R., Borgida, E. and Reed, H., "Popular Induction: Information is Not Necessarily Informative," in *Cognition and Social Behavior*, Lee W. Gregg, editor, Lawrence Erlbaum Assoc., Hillsdale, N.J., 1976.
31. Ponssard, J. P., "Zero-Sum Games with 'Almost' Perfect Information," *Management Sci.*, Vol. 21, No. 7 (1975), pp. 794–805.
32. —— and Zamir, S., "Zero-Sum Sequential Games with Incomplete Information," *Internat. J. Game Theory*, Vol. 2 (1973), pp. 99–107.
33. Prescott, E. C. and Townsend, R. M., "Equilibrium Under Uncertainty: Multi-Agent Statistical Decision Theory," in [54].
34. Rapoport, A. and Chammah, A. M., *Prisoner's Dilemma*, Ann Arbor, Mich., 1965.
35. —— and Orwant, C., "Experimental Games: A Review," *Behavioral Sci.*, Vol. 7 (1962), pp. 1–37.
36. Riker, W. H. and Ordeshook, P. C., *An Introduction to Positive Political Theory*, Prentice-Hall, Englewood Cliffs, N.J., 1973.
37. Rothschild, M., "Models of Market Organization with Imperfect Information," *J. Political Economy*, Vol. 81 (1973), pp. 1283–1308.
38. Sanghvi, A. P. and Sobel, M. J., "Bayesian Games as Stochastic Processes," *Internat. J. of Game Theory*, Vol. 5, No. 1 (1976), pp. 1–22.

39. —— and ——, "Sequential Games as Stochastic Processes," *Stochastic Processes and Their Applications*, Vol. 6 (1978), pp. 323–336.
40. Savage, L. J., *Foundations of Statistics*, Wiley, New York, 1954.
41. ——, "Elicitation of Personal Probabilities and Expectations," *J. Amer. Statist. Assoc.*, Vol. 66 (1971), pp. 783–801.
42. Schotter, A. and Schwodiauer, G., "Economics and Game Theory: A Survey," *J. Economic Literature* (June 1980).
43. Shubik, J., "Some Experimental Non-Zero Sum Games with Lack of Information About the Rules," *Management Sci.*, Vol. 8, No. 2 (1962), pp. 215–234.
44. Simon, H. A., "A Comparison of Game Theory and Learning Theory," *Models of Man*, H. A. Simon, Wiley, New York, 1957, pp. 274–279.
45. Slovic, P., Fischoff, B. and Lichtenstein, S., "Behavioral Decision Theory," *Annual Rev. Phych.*, Vol. 28 (1977), pp. 1–39.
46. Sorin, S., "A Note on the Value of Zero-Sum Sequential Repeated Games with Incomplete Information," *Internat. J. Game Theory*, Vol. 8, No. 4 (1979), pp. 217–223.
47. Tversky, A. and Kahneman, D., "Judgment Under Uncertainty: Heuristics and Biases," *Science*, Vol. 185 (1974), pp. 1124–1131.
48. —— and ——. "The Framing of Decisions and the Psychology of Choice," *Science*, Vol. 211 (1980), pp. 453–458.
49. von Neumann, J. "On the Theory of Games of Strategy," *Contributions to the Theory of Games*, Princeton Univ. Press, Princeton, N.J., pp. 13–42.
50. von Neumann, J. and Morgenstern, O., *The Theory of Games and Economic Behavior*, Princeton Univ. Press, Princeton, N.J., 1944.
51. Wald, A., *Statistical Decision Functions*, Wiley, New York, 1950.
52. Young, O. R., *Bargaining: Formal Theories of Negotiation*, Univ. of Illinois Press, Chicago. Ill., 1975.
53. Zamir, S., "On the Relation Between Finitely and Infinitely Repeated Games with Incomplete Information," *Internat. J. Game Theory*, Vol. 1, No. 1 (1971), pp. 179–198.
54. Zellner, A., ed., *Bayesian Analysis in Econometrics and Statistics, Essays in Honor of Harold Jeffreys*, North-Holland, Amsterdam, 1980.

3.2

Equilibrium, Common Knowledge, and Optimal Sequential Decisions

JOSEPH B. KADANE AND TEDDY SEIDENFELD

I. INTRODUCTION

In a paper described as an initial exploration of what two Bayesians need to know in order to play a sequential game against each other, DeGroot and Kadane (1983) argue that optimal sequential decisions need not conform to much of what traditional game theory requires of rational play. Specifically:

1. The players' optimal strategies need not form a Nash equilibrium.
2. Nor do the players need to know (or even believe that they know) the optimal choices of their opponent; there is no requirement of "common knowledge," in that sense.

Nonetheless, these authors propose that

3. Reasoning by backward induction succeeds in locating optimal play.

Each of these three claims is a point of active dispute. For example, regarding (1), in an extended defense of a refined equilibrium concept, Harsanyi and Selten (1988) argue that because of common knowledge

We dedicate this paper to the memory of our dear friend and colleague, Morris H. DeGroot.

Research for this work was supported, in part, by NSF Grants DMS-8705646, DMS-8701770, SES-8900025, by ONR Contract N00014-89-J-1851, and by the Buhl Foundation.

From *Knowledge, Belief and Strategic Interaction*, edited by Cristina Bicchieri and Maria Luisa Dalla Chiara (Cambridge: Cambridge University Press, 1992), pp. 27–45.

(of mutual rationality) the agents *ought to* settle on an equilibrium solution – but a refined equilibrium (see also Harsanyi 1989). Aumann (1987), like Harsanyi and Selten, seeks to reconcile Bayesian and game-theoretic rationality but is led to a theory of correlated equilibrium based on an assumption of a "common prior" for different players. Relating to (2), Binmore and Brandeburger (1988), wary of common-knowledge assumptions, seek other grounds for justifying equilibrium solutions. Concerning (3), Bicchieri (1989) questions the validity of backward induction in cases where the agents have too much or too little common knowledge.

The position we take in this chapter is based on the initial exploration of DeGroot and Kadane (1983). We extend the central example of that work to include the Harsanyi–Selten "trembling hand" model of choices. We argue that, in the extended example as in the original one, equilibrium is not a norm for rational play. (This is a position already announced by Kadane and Larkey 1982.) Based on our views about what may serve as states of uncertainty in a common prior across players, we argue that rational play need not result in a correlated equilibrium. And we argue that there is no problem of too much common knowledge. That is, with respect to the standards of expected utility, backward induction is a valid method for arriving at optimal play. In short, we respectfully disagree with each of the authors mentioned above!

The analysis we offer in this essay supports the condition of "rationalizability" for strategies, a view intelligently defended in papers by Bernheim (1984) and Pearce (1984). In addition, a version of the DeGroot–Kadane game that introduces common priors leads us to a conclusion similar to that reported by Rubinstein (1989); namely, approximating common knowledge does not yield strategies that approximate optimal play under common knowledge. Thus, we endorse the attitude expressed by many who emphasize a careful assessment of *what* players in a game know of each other's beliefs and, especially for sequential games, *when* they come to know it.

II. THE DEGROOT–KADANE GAME

The DeGroot–Kadane game is played between two agents by successive moves of a visible pointer located on the real line. Following their (1983) presentation, suppose the game has three moves: First player 1 moves the pointer, then player 2, and last player 1 moves it again. The payoff (a loss given in utiles) to each player is a function of two components: how far the player has moved the pointer and, after the final move, how far the pointer is from that player's designated target value (x for player 1 and y for player 2). Put formally, let s_0 be the initial location of the pointer and let s_i be its location after the three moves (i = 1, 2, 3). Thus, $u = s_1 - s_0$ is player 1's first move, $v = s_2 - s_1$ is player 2's move, and $w = s_3 - s_2$ is player 1's second (and final) move. The payoff (loss) to player 1 is

$$L_1 = q(s_3 - x)^2 + u^2 + w^2 \tag{2.1}$$

and the payoff to player 2 is

$$L_2 = r(s_3 - y)^2 + v^2. \tag{2.2}$$

DeGroot and Kadane examine two version of this game. In both versions, s_0, q, and r are quantities known to both players. However, in the first version each player knows both targets (so that x and y are common knowledge), while in the second version of the game each player knows only his or her respective target and is uncertain about the opponent's target. In both versions of the game the players are (expected) utility maximizers, and each player models the opponent in that way too. Thus, in the language of Pearce (1984) and Bernheim (1984), the players construct rationalizable strategies.

In the simple version, where the targets are common knowledge, DeGroot and Kadane show that optimal play, as identified by backward induction, yields the following strategies:

$$w = \frac{q(x - s_2)}{q + 1} \tag{2.3}$$

$$v = (1 - k)(m - s_1), \tag{2.4}$$

and

$$u = \frac{(q+1)(x - s_0) + r(x - y)}{(q+1)[1 + (q+1)/qk^2]}, \tag{2.5}$$

where

$$m = (q+1)y - qx \quad \text{and} \quad k = \frac{(q+1)^2}{r + (q+1)^2}.$$

Evidently, these strategies are in equilibrium; that is, each constitutes the best reply if the opponent's play is as specified above. Because the solutions are unique, these strategies make for a "strong equilibrium" in Harsanyi's (1977, p. 104) sense. Moreover, the game has no maximin value for either player. That is, given a proposed value $V < 0$, there are "silly" moves the opponent is permitted to make that force the player's loss to exceed V.

Because the players cannot cooperate in this game (there are no binding agreements), it is not surprising that optimal play does not yield Pareto efficiency. For example, as noted by DeGroot and Kadane (1983, Thm. 2-ii, p. 202), if $s_0 < x < y$ then player 1's first move u is negative – *player 1 moves the pointer away from both targets* – if and only if $x - s_0/(y - x) < r/(1 + q)$. Such a move is clearly Pareto inefficient, as it leads to increased losses for each player compared to what they can achieve subject to binding agreements.

III. AN OBJECTION BY BICCHIERI

In response to a challenge raised by Bicchieri (1989) concerning the legitimacy of backward induction, we note that these strategies are optimal under the assumptions of the model for the game. As we understand Bicchieri's worry, it is that hypothetical reasoning used with backward induction does not accurately reflect how players would react were the hypothetical conditions realized. In terms of the simple DeGroot–Kadane game, we believe her objection takes the following form.

The optimality of player 2's move, given by (2.4), depends upon the assumption that player 1 makes the final move according to (2.3). (That choice of w is determined, according to backward induction, by the fact that – regardless of what has happened on the first two moves – player 1 minimizes L_1 by adhering to (2.3).) Then, regardless of player 1's initial move, player 2 does best by conforming to (2.4). However, according to (our interpretation of) Bicchieri's objection, were player 2 to observe that the initial move by player 1 fails to satisfy (2.5), then the assumption that player 1 will satisfy (2.3) also becomes questionable; hence the backward induction reasoning leading to (2.4) is undone by player 2's observation of the initial move by player 1. To conclude (our account of) Bicchieri's analysis: The hypothetical reasoning that (2.4) is best for player 2, *regardless* of what player 1 chooses as an initial move, is not correct; hence, backward induction is not valid.

Our response to this objection is that the backward induction reasoning, as illustrated by the argument that player 2's move should agree with (2.4), does not indicate what player 2 should do if the model fails, as would be indicated by player 1 failing to satisfy (2.5). Rather, backward induction is used by DeGroot and Kadane merely as an algorithm for determining what is optimal *under the model* for their game. Backward induction employs hypothetical reasoning of the form: "If player 1 has moved to s_1 and this accords with the model (i.e., if that is best for player 1), then what is player 2's best move?" Of course, we discover that player 1's final move accords with the model if and only if it satisfies (2.3), player 2's move accords with the model if and only if it satisfies (2.4), and player 1's first move accords with the model if and only if it satisfies (2.5). But we can use backward induction (and hypothetical reasoning) to discover this without requiring the model be consistent with each (physically) possible move that the players are capable of making.

What is "the model" for the game? In the DeGroot–Kadane game it is the combined assumptions that the players know the initial position s_0, targets (x, y), and the permissible moves, they know the loss functions (2.1) and (2.2), and they accept that each player is concerned to minimize his or her loss and knows this of the other. Under the DeGroot–Kadane model, the solution (2.3)–(2.5) is optimal; that is, it maximizes each player's utility and backward induction correctly identifies the solution. Reasoning by backward induction to arrive at the DeGroot–Kadane solution is not to be confused with reasoning how (for example) player 2 would play when the model fails.

Let us illustrate the difference. Suppose, contrary to the DeGroot–Kadane model, that by some action (internal or external to the game) player 1 can cause player 2 to believe that the final move might not conform to (2.3). Under this alternative model, optimal play for player 1 can differ from that prescribed by (2.5).

Consider the game with $q = r = 1$, $s_0 = 0$, $x = 1$, and $y = 2$. Optimal play (under the DeGroot–Kadane model) yields choices $u = 4/33$, $v = 19/33$, and $w = 5/33$, with a payoff (loss) to player 1 of $L_1 = 2/33$ (≈ 0.06) and to player 2 of $L_2 \approx 1.66$. Suppose, in fact, that player 2 adopts the DeGroot–Kadane model for the game and that player 1 knows it. What would result if the first player were to depart from the strategy (2.5), $u = 4/33$, and instead make the surprising move $s_1 = s_0$? What would happen if player 1 chose not to move the pointer on the first round ($u = 0$)?

With (2.5) failed, it would establish conclusively for player 2 that the hypothesized DeGroot–Kadane model is false. How would player 2 react? Might not player 2 come to doubt that player 1 will maximize in choosing w? Might not player 2 come to think player 1 will again refuse to move the pointer and choose $s_3 = s_2$ (corresponding to $w = 0$)? If that is player 2's reaction to the surprise of $u = 0$ (i.e., if then player 2 predicts the choice $w = 0$), the very best player 2 can do is to make the move $v = 1$ and earn the loss $L_2 = 2$. But that yields player 1 the best possible score, $L_1 = 0$, by choosing $w = 0$ – ironically, just as player 2 predicts. Can player 1 anticipate player 2's reaction to the initial move ($u = 0$) and fool player 2 into this false model for the choice of w (which coincidentally happens to yield a correct prediction for the third move if player 2 chooses $v = 1$)? Such a reaction by player 2 improves player 1's payoff over the strategy (2.3)–(2.5). Does this hypothetical reasoning refute the backward induction argument leading to the strategies (2.3)–(2.5)?

We do not think the question of how player 2 would respond to a failure of his model for the game is to be answered by the logic of decisions. How player 2 would react to the move $u = 0$ is instead an empirical matter. Our point here is simple: A correct interpretation of the backward induction argument is to see it as reasoning used to identify optimal play *under the conditions of a model for the game.* Backward induction does not include (or require) the counterfactual reasoning that is needed when a player's model of the game is falsified, so the strategy (2.3)–(2.5) is optimal under the conditions of the DeGroot–Kadane model. In Section IV we illustrate how a small change in the DeGroot–Kadane game leads to a model of rational play consistent with all possible observations.

IV. TREMBLING HANDS IN THE SIMPLE DEGROOT–KADANE GAME

John Harsanyi and Reinhard Selten question the adequacy of Nash's equilibrium concept when applied to the normal-form version of an extensive-form game. They deny the equivalence of normal and extensive game forms. Instead, they advocate a refined equilibrium concept for extensive-form games, based on a "trembling-hand" model of choice. (But even such refined equilibria are subject to criticism, as illustrated by Pearce 1984, p. 1044.) An equilibrium for extensive forms is acceptable, according to their account, provided it is robust over small perturbations in choice. Specifically, in order to avoid "imperfect

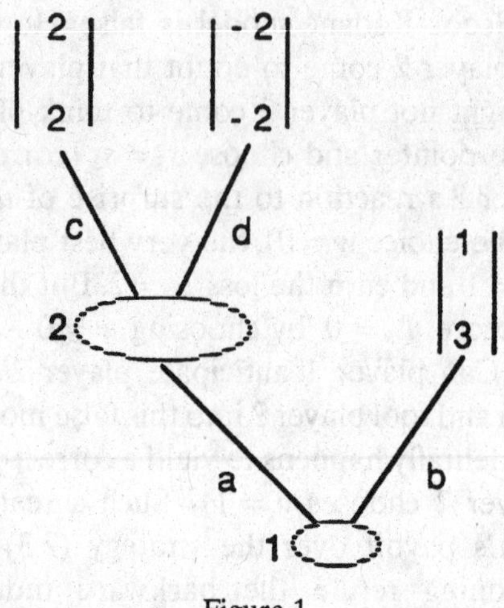

Figure 1

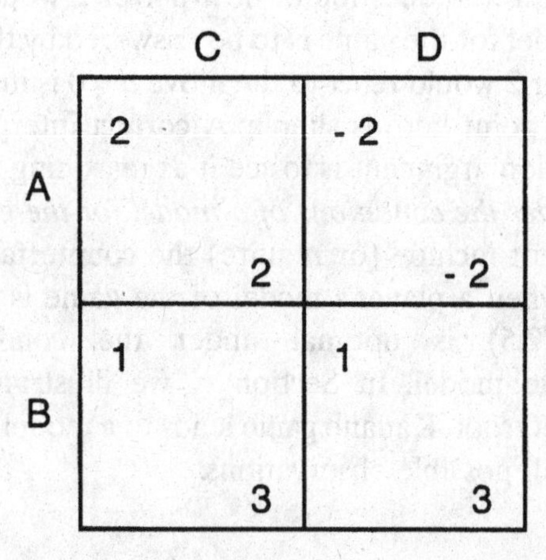

Figure 2

equilibria," they alter the basic moves in a game so that an agent selects one from a set of distributions (on pure options); a player chooses a mixed strategy rather than a pure option.

One of their examples beautifully illustrates the difference between the two kinds of equilibria. Figures 1 and 2 report (respectively) the extensive and normal forms of their game. Figure 3 gives the normal form for the perturbed game, where players may choose one of two mixed strategies in a perturbed extensive-form game (not pictured).

	C^*	D^*
A^*	$2-5\varepsilon+4\varepsilon^2$ $2+3\varepsilon+4\varepsilon^2$	$-2+7\varepsilon-4\varepsilon^2$ $-2+9\varepsilon-4\varepsilon^2$
B^*	$1+\varepsilon-4\varepsilon^2$ $3-\varepsilon-4\varepsilon^2$	$1-3\varepsilon+4\varepsilon^2$ $3-5\varepsilon+4\varepsilon^2$

Figure 3

In their game, each player has two pure strategies. In the extensive form, strategy $\{a, b\}$ for player 1 and – provided 1's information set is reached (provided player 1 chooses a) – strategy $\{c, d\}$ for player 2; in the corresponding normal form, $\{A, B\}$ for player 1 and $\{C, D\}$ for player 2. (Note that the normal form fails to distinguish between the extensive form of Figure 1 and a different game where both play simultaneously, i.e., where player 2's information set does not reflect whether player 1 chooses a or b.) In the perturbed game, the normal-form options given in Figure 3 arise by using a two-point distribution, with probabilities $(1 - \varepsilon)$ and ε assigned to each pure option in the corresponding perturbed extensive form.

Observe that, corresponding to the normal-form Figure 2, there are two equilibria: the pairs $\{A, C\}$ and $\{B, D\}$. However, the latter is "imperfect" in the extensive form of Figure 1, as that requires player 2 to (threaten to) play option d in case choice node **2** is reached. Of course, at node **2**, player 2 maximizes by playing option c instead, and player 1 knows this fact.

In the perturbed versions of the game, this difference between the two solution pairs (which are in equilibrium in the game form of Figure 2) is made evident. In the normal form of Figure 3, only the pair $\{A^*, C^*\}$ is in equilibrium. The $\{B^*, D^*\}$ pair is not in equilibrium because, when player 1 chooses B^*, player 2 improves 2's (expected) payoff by shifting from D^* to C^*; that is, D^* is not player 2's best response to B^*.

The Harsanyi–Selten idea is that imperfect equilibria are deficient because, in extensive game forms, they require a player to choose an outcome that fails to maximize utility. Nonetheless, the suspect choice is justified by Nash's criterion of equilibrium in the corresponding normal form – and it can be viewed as "threat" in the extensive form.

We agree with the Harsanyi–Selten objection to such imperfect equilibria. In the extensive form of their game, player 2 does not maximize utility by choosing option d (if node **2** arises) – d is an idle threat. That move is inconsistent with the assumption that the players are utility maximizers. In contrast, the trembling-hand model of choice eliminates the imperfect equilibrium from the normal form of the game. When the choices are mixed options, the imperfect equilibria fail to be Nash equilibria.

We may incorporate trembling hands in the DeGroot–Kadane game by limiting players to distributions for the location of the pointer, rather than supposing that player moves fix the pointer location exactly. Suppose a player moves by determining the mean of the distribution for the pointer, and suppose that distribution has a fixed and finite variance. Thus, a player may aim as follows:

the player may fix $E_k(s) = k$
by aiming the pointer at location k; (4.1)

and

the distribution has known, finite variance,
$E_k\left[(s-k)^2\right] = c \quad (0 \leq c < \infty)$,
where c does not depend upon k but may reflect the
stage of the game and past locations of the pointer. (4.2)

The DeGroot–Kadane game (of Section II) is a special case, where $c = 0$. The loss functions for the modified game are again given by (2.1) and (2.2), as formulated in terms of the successive, observed pointer locations. We do this in order to preserve complete information in the game. (We understand the sequential game to have *complete information* if the payoffs to each player are a known function of the public outcomes, outcomes that both players observe. This condition does not require that the players' *choices* be known to both, as the following example illustrates.) The feature of complete information would be lost if, instead, losses were defined through the unobserved aiming points.

Aside. The class of distributions specified by (4.1) and (4.2) is more general than Harsanyi–Selten's version of trembling hands. In our version, by contrast, (i) we do not require either that there be a point mass concentrated on a single pure strategy – there does not have to be probability mass assigned to point k; (ii) nor do we suppose that errors are symmetrically distributed across the alternative pure strategies.

Under this modification of the simple version of the DeGroot–Kadane game, backward induction leads to the same choice of aiming points as in the degenerate case, where $c = 0$.

Theorem. *For the (modified) simple version of the DeGroot–Kadane game, with moves specified by* (4.1) *and* (4.2), *the optimal aiming points are given by* (2.3)–(2.5), *which is the same solution as in the original game where* $c = 0$.

Proof. The reason for the theorem is the well-known fact (about point estimates) that the mean of a distribution minimizes expected squared error.

We illustrate the calculation for player 1's final move, on the third round of the game, given the observations of s_0, s_1, and s_2 (respectively, the initial location of the pointer, its location after player 1's first move, and after player 2's move). The prospective loss to player 1 after the final move, given by (2.1), is $L_1 = q(s_3 - x)^2 + u^2 + w^2$. Player 1's selection of moves is the choice of where to aim s_3 ($= s_2 + w$), where $E_k[s_3] = s_2 + E_k[w] = k$. Thus, the expected loss in choosing $E_k(s_3) = k$ is

$$
\begin{aligned}
E_k(L_1) &= E_k\left[q(s_2 + w - x)^2 + u^2 + w^2\right] \\
&= q(s_2 - x)^2 + 2q(s_2 - x)E_k(w) + (q+1)E_k[w^2] + u^2 \\
&= q(s_2 - x)^2 + 2q(s_2 - x)(k - s_2) + (q+1)\left(c + [k - s_2]^2\right) + u^2.
\end{aligned}
$$

Solving the equation $0 = dE_k[L_1]/dk$ (to minimize expected loss) yields

$$
w = (k - s_2) = \frac{q(x - s_2)}{q+1},
$$

as required by (2.3). Equations (2.4) and (2.5) follow in a similar fashion. □

Thus, the DeGroot–Kadane solution (2.3)–(2.5) is also the refined, perfect equilibrium solution advocated by the Harsanyi–Selten theory. (This result obtains because (2.3)–(2.5) are, trivially, limit points of the solutions generated by letting $c \to 0$.) There is, nonetheless, an interesting difference between the two forms of the (perfect information) DeGroot–Kadane game. In the original version, with $c = 0$, the model for the game is consistent with exactly one line of play, as dictated by (2.3)–(2.5). But, in the modified form of the simple game, if the error distribution has full support on the real line (e.g., using a normally distributed shot aimed at its mean), then each (logically) possible combination of locations for s_1, s_2, and s_3 is consistent with the model that both players are utility maximizers. No matter what player 2 sees for the location s_1 – that is, no matter where the pointer stops after player 1's first move – that outcome is consistent with the hypothesis that player 1 chose optimally in accord with (2.5).

Hence, in the trembling-hands version of this simple game ($c > 0$), with suitable error distributions, backward induction reasoning may be used to answer hypothetical questions of the form, "What would player 2 do if u is observed equal to u_1?" In other words, with such trembling-hand moves, Bicchieri's concern is satisfied: Within the model that players are utility maximizers and know this of each other, backward induction accommodates all possible moves. No possible outcome leads to a counterfactual situation.

Of course, some outcomes will be surprising though consistent under the model. Even so, we do not require of the players that they retain their belief in the model, regardless of the observed outcomes. Again, we advocate the strategies (2.3)–(2.5) as optimal under the model. Our view on this matter is no different from our view regarding statistical models generally. Sometimes observations force reevaluation of the statistical model; other times, as with "outliers," it is reasonable to do so even though the data are (formally) consistent with the model.

V. VERSION 2 OF THE DEGROOT–KADANE GAME: TARGETS ARE NOT COMMON KNOWLEDGE

In the first versions of the DeGroot–Kadane game (with and without "trembles"), optimal play according to expected utility theory leads to strategies that are in perfect equilibrium. That consequence does not obtain when the game is modified and a target point is known *only* to the player for whom it is the target – where player 1 knows x but not y and player 2 knows y but not x, and this information difference itself is common knowledge. Let us rehearse the DeGroot–Kadane solutions to the second game (where $c = 0$) to see why optimal play will not form a Nash equilibrium.

According to backward induction, at the final move with u and v given, in order to minimize loss L_1, player 1 takes no interest in player 2's target y. Hence player 1's last move w is determined once again by (2.3):

$$w = \frac{q(x - s_2)}{q + 1}. \tag{5.1}$$

Player 2 does not know x, but player 2 knows that player 1 will choose w to minimize L_1. Thus the argument leading to (2.4) does not apply directly. However, as an expected utility maximizer, player 2 has a personal probability for x (given the datum u) with mean $E_2(x|u)$. In light of the squared-error form of the loss function (2.2) and knowing that player 1 will choose w according to (5.1), player 2 minimizes expected loss by choosing

$$v = (1 - k)[M(u) - s_1], \tag{5.2}$$

where $M(u) = (q + 1)y - qE_2(x|u)$.

How shall player 1 determine the initial move u? Player 1 knows neither y nor the quantity $E_2(x|u)$. But, knowing that player 2 solves (5.2) to find v, player 1 establishes an optimal choice for u in terms of the personal joint probability for these two quantities: y and $E_2(x|u)$. The resulting optimization is given by solving the equation

$$0 = \frac{1}{2}\frac{d}{du}E_1\{K^2(u)\} + \frac{u(q+1)}{q}, \tag{5.3}$$

where $K(u) = (s_0 + u)k - x + (1 - k)M(u)$.

Although these moves are the best responses to what a player believes about the opponent's moves – that is, they maximize the subjective expected utilities of each player (under the common model that they are subjective utility maximizers) – these strategies do not form a Nash equilibrium. The strategies are not in equilibrium for the simple reason that the model does not include the targets as common knowledge. The model does not result in players knowing what the other will do, even under optimal play.

For example, if at stage **2** of the game player 1's rule for choosing w (formula (5.3)) were made known to player 2 by revealing the target x, this would alter player 2's belief set – unless player 2 thinks u reveals where x is or that x is already known, in which case $Var_2[x|u] = 0$ – with the result that player 2's move would change to agree with (2.4). Likewise, if player 1 learns both 2's target y and that player 2 learns x prior to 2's best move at stage **2** of the game, then w is selected according to (2.5), not according to (5.3).

In short, exposing details of the opponent's strategy, as the Nash condition requires for ascertaining whether a replay is also a "best response," radically changes the epistemic conditions of the game. The common-knowledge assumptions leading to strategies (5.1)–(5.3) are not consistent with players verifying that theirs is a best response. The epistemic change required for satisfying the Nash condition is inconsistent with the model for the second version of the game.

This feature of our analysis is not affected by the use of trembling-hand moves. That is, if (as in Section IV) a player moves by choosing an aiming point rather than by fixing the quantity u, v, or w for certain, then optimal play does not form a Nash equilibrium, just as optimal play according to (5.1)–(5.3) does not result in Nash equilibrium. This is shown by the following.

Theorem. *For the (modified) second version of the DeGroot–Kadane game – where target points are not common knowledge – with trembling-hand moves specified by* (4.1) *and* (4.2), *the optimal aiming points are again given by* (5.1)–(5.3).

Proof. As before, the solution arises because the mean squared error of an estimate is minimized at the mean. In particular, player 1's final move (the choice of where to aim s_3, given u and v) is optimized by aiming w so that $0 = dE_k[L_1]/dk$; hence w satisfies (2.3). Likewise, at stage **2**, player 2 knows how the first player will aim the last shot, though player 2 may remain uncertain of player 1's target x. Nonetheless, player 2 has a personal probability for x, given u, whose mean $E_2(x|u)$ enters the optimization just as in the previous version of the game (the version without trembling hands). Because the mean minimizes the expected loss, player 2 chooses the aiming point for v according to (5.2). Equation (5.3), governing the first aiming point, is obtained in the same fashion. □

To repeat the point of this exercise, under a model for the DeGroot–Kadane game where players have common knowledge that they are expected utility maximizers but where they lack common knowledge of their targets, and where moves are subject to trembles, optimal play does not result in a Nash equilibrium.

VI. AUMANN'S CORRELATED EQUILIBRIUM AND THE DEGROOT–KADANE GAME

Aumann (1987) proposes an original unification of the game-theoretic and Bayesian decision-theoretic viewpoints. He identifies the game-theoretic perspective with a generalized account of (Nash) equilibrium, leading to what Aumann terms *correlated equilibrium*. These are best-response strategies that may rely on correlated (rather than independent) joint distributions to form mixed options. That is, the distribution used by player 1 to create a mixed strategy can be correlated with the distribution used by player 2. Aumann's account of Bayesian rationality in games leads to the result that Bayes-rational players will adopt strategies that are in correlated equilibrium. Moreover, each correlated equilibrium can be a model (with specific informational constraints on the individual players) for Bayes-rational play. Hence, there is a reconciliation of the two viewpoints.

We agree with Aumann that Bayesian decision theory should apply to games; the logic of choice is the same whether our uncertainty is about "Nature" or an opponent's moves (see Kadane and Larkey 1982). However, we take issue with (what we understand to be) Aumann's formulation of Bayes rationality in games. He requires a very rich space Ω of states of the world:

> The term "state of the world" implies a definite specification of all parameters that may be the object of uncertainty on the part of any player of [the game] G. In particular, each ω includes a specification of which action is chosen by each player of G at the state ω. Conditional on a given ω, everybody knows everything; but in general, nobody knows which is really the true ω. (1987, p. 6).

Though agents are permitted private information about Ω, Aumann requires that (each) player i's personal probability (over Ω), here denoted by $p_i(\Omega|D_i)$, is a conditional probability that arises from a common prior: $p_i(\Omega) = p(\Omega)$ given i's (perhaps) private data $d_i \in D_i$; however, the prior is the same for each player i. That is, apart from private evidence, the players are required to have the same opinions about the set of states Ω. Because (by the severe assumption that) each stage ω specifies "all parameters that may be the object of uncertainty on the part of any player," Aumann argues that the information sets D_i (though not the private information d_i) also are common knowledge to all players.

We object to Aumann's condition that there be a common prior (across players) in games. He recognizes this challenge in Section 5 of his paper. Concerning ordinary decisions, we believe there is no basis within (say) Savage's decision theory for that assumption, regardless of the detail with which states (of Nature) are defined. Savage's opposition to what he called "necessary" Bayesian theory (Savage 1954, p. 61; 1962, p. 102; 1967) leads us to think he rejected a common-prior requirement even in the structured setting of parametric statistical inference, where likelihoods are specified, a fortiori in less structured game settings where likelihoods are not so determined.

Our second concern is with consequences of demanding that Ω be as detailed as Aumann proposes. In particular, we are uncomfortable with the prospect that agents are required to hold (nontrivial) probabilities over their own current choices. (Again, we observe that Savage's theory is carefully formulated to distinguish between acts and states; states but not acts are assigned personal probability.) We do not see a problem when an agent assigns personal probabilities (more accurately, personal conditional probability) now to future choices, because the agent cannot now make those states true or false. Nor do we find a conceptual problem in assigning a personal probability to past choices, since the agent may have forgotten those past choices. The difficulty with personal probability over one's current choices is that such probabilities do not support the familiar betting-odds interpretation. (See Spohn 1977, Kadane 1985, and Levi 1989 for related discussions.)

The second version of the DeGroot–Kadane game serves to illustrate our position on this issue. Recall that in the second version, players know their respective targets but are uncertain of the other's target, and this informational structure is itself common knowledge. Recall also that it is the uncertainty about the opponent's target that alone differentiates the two versions of the game. In the first version of the game, when the informational structure of the game includes common knowledge of the targets, the optimal strategies (2.3)–(2.5) are common knowledge, too; there is no uncertainty for either player about what he or she will do.

We propose, therefore, to analyze the second version of the game (without trembles) using pairs of targets for the states $\Omega' = \{(x, y): x$ is player 1's target, y is player 2's target}. We introduce a common prior $p(\bullet)$ over these states to allow a comparison with Aumann's theory. As we make clear shortly, the set Ω' is not Aumann's set of states Ω for this game. (Also, to agree with Aumann's presentation, we are prepared to use the game's normal form. That is, we see the selection of "states" as the relevant issue here, not the collapse of extensive to normal form.)

Suppose the two players begin their analysis with a common prior over Ω'; that is, they do not yet know their targets, yet they share the following background information: It is given that both players are utility maximizers, that their respective loss functions are L_1 and L_2, that the initial pointer location is s_0, and that all this is common knowledge. For simplicity, before learning their targets, assume the players have a (common) joint distribution $p(x, y)$ which is bivariate normal (μ, Σ), with known means $\mu = (x_0, y_0)$, with known and equal variances $(\sigma_x = \sigma_y = \sigma)$, and with (x, y) independent $(\rho_{xy} = 0)$.

Then, after player 1 learns x, 1's probability for player 2's target, $p_1(y|x) = p_1(y)$, is normal $N(y_0, \sigma^2)$, since x and y are independent. Likewise, after learning y, player 2 has uncertainty about x, denoted by $p_2(x|y) = p_2(x)$, which is normal $N(x_0, \sigma^2)$. These distributions are common knowledge. In particular, prior to any moves, player 1 knows that player 2's expected value for x is x_0; that is, player 1 knows $E_2(x|y) = E_2(x) = x_0$ and player 2 knows $E_1(y|x) = E_1(y) = y_0$.

Despite the common prior, this common knowledge does not induce a correlated equilibrium with respect to Ω'. That the addition of a common prior for Ω', even one that makes (x, y) independent variables, fails to yield a correlated equilibrium is explained by tracing the impact of the common prior on the solutions (5.1)–(5.3). With respect to player 1's choice of a final move w, the prior $p(x, y)$ is irrelevant because x is known and y plays no role in minimizing L_1 through the choice of s_3. At the second move, when player 2 is contemplating the choice of v, what is relevant is the quantity $E_2(x|u, y)$. But the common prior $p(x, y)$ does not determine this expectation! It fails to do so since it leaves open what might be player 2's beliefs about player 1's choice of u. That is, all of $p_2(u)$, $p_2(u|y)$, and $p_2(u|x, y)$ are underdetermined by the common prior on Ω'.

For instance, both players know that

$$p_2(x|u, y) = p_2(u|x, y) \cdot p_2(x|y) / p_2(u|y).$$

Also, it is common knowledge that $p_2(x|y) = p_2(x)$, where x is a normal $N(x_0, \sigma^2)$ distribution. But the common prior in (x, y) does not fix the ratio $p_2(u|x, y)/p_2(u|y)$, which is known to player 2 only. The terms $p_2(u|x, y)$ and $p_2(u|y)$ cannot be derived using Bayes's theorem merely by giving player 2 privileged information about (x, y). Specifically, the probability $p_2(u|x, y)$ should not be confused with the (point mass) solution for u, given by (2.5) from the first version of the DeGroot–Kadane game (where both targets are common knowledge and player 2 knows u for certain). It is important to correctly interpret the compound conditioning event in $p_2(u|x, y)$. That conditioning event specifies both targets, but it leaves x known to player 1 only. It is important to distinguish two cases:

1. Conditioning on the event (x, y) when these are common knowledge, as in the first version of the DeGroot–Kadane game, leading to (2.3)–(2.5).
2. Conditioning on the event (x, y), when target x is known to player 1 alone and y is known to player 2 alone.

When (x, y) are not common knowledge, as in the second version of the game, it is the second of these two cases that the players face when evaluating the term $p_2(u|x, y)$. For some discussion on the range of values $p_2(u|x, y)$ can take (all of which are unknown to player 1), see Corollaries 1 and 2 in DeGroot and Kadane (1983, p. 206).

Thus, we see the impact of Aumann's selection of fine-grained states Ω on his result equating Bayes rationality (subject to a common prior) with correlated equilibria. For Aumann's theorem to apply, the agents must include player 1's choice of u, as well as the targets x and y, in the states of uncertainty. Then, with a common prior over the refined states $\Omega = \{(u, x, y)\}$, the problematic term $E_2(x|u, y)$ becomes common knowledge. However, to demand a common prior over Ω mandates two conditions that we find unwarranted for rational play in this game. Aumann's analysis mandates: (i) that player 2's beliefs about player 1's choice u are transparent to player 1; and (ii) that player 1 holds nontrivial probabilities about 1's own actions. What is the basis for demanding condition (i)? What is the interpretation, from player 1's perspective, of assigning (nontrivial) probabilities to the choice u?

VII. ON RUBINSTEIN'S THESIS: ALMOST COMMON KNOWLEDGE IS NOT GOOD ENOUGH

The preceding section explored a consequence of imposing a common prior distribution $p(x, y)$ on the set of target states Ω' for the second version of the DeGroot–Kadane game. The upshot of that analysis is that a common prior on Ω' is insufficient for defining player 2's choice of move v, since it leaves open player 2's conditional distribution for player 1's move u, given targets x and y. Thus, the common prior on Ω' also leaves open player 1's first move u, since that depends upon player 1's expectation of player 2's expectation of u, and so forth.

This argument is valid for each value $\sigma^2 > 0$. (Recall that σ^2 is the common variance for the targets.) However, the first version of the game, with targets (x_0, y_0) common knowledge, corresponds to the limiting distribution $\sigma^2 = 0$. Thus, the first version of the DeGroot–Kadane game is not necessarily the limit of the second-version games with common priors, where $\sigma^2 \to 0$. That is, the optimal strategy (2.3)–(2.5) for the first game (where targets are common knowledge) may not equal the limit (as $\sigma^2 \to 0$) of optimal strategies for the second version of the game, constrained by a common prior.

Let us illustrate the point. To simplify the formulas, take $q = r = 1$ and $s_0 = 0$. With the variance $\sigma^2 > 0$ given, denote with subscripts u_σ and v_σ the choices for the first two moves. And, with some slight abuse of notation, use the subscripted u_0 and v_0 to denote the limit of these moves as $\sigma \to 0$. Suppose player 2 reasons as follows.

In the first version of the DeGroot–Kadane game, with targets common knowledge, according to (2.5) my opponent's first move u is linear in the targets x and y. That is, were our targets known, player 1 would choose

$$u = (3x - y)/8.25. \tag{7.1}$$

So, I'll take my expectation for x to be linear in u_σ and y:

$$E_2(x|u_\sigma, y) = a_\sigma + b_\sigma u_\sigma + c_\sigma y. \tag{7.2}$$

Then my move v_σ satisfies

$$v_\sigma = 0.2[2y - a_\sigma - (1 + b_\sigma)u_\sigma - c_\sigma y]. \tag{7.3}$$

The move v_σ, (7.3), contrasts with player 2's choice (from 2.4) of

$$v = 0.2[2y - u - x] \tag{7.4}$$

for the case where targets are common knowledge. Recall that x and y are uncorrelated. Therefore, in order to make (7.3) equal (7.4) as $\sigma \to 0$ (i.e., for $v_0 = v$), it is necessary and also sufficient that $a_\sigma \to 0$, $b_\sigma \to 1$, $c_\sigma \to 0$, and $u_\sigma \to x$.

Now, in case player 1 knows that player 2 has the linear expectation (7.2) (without in general knowing the coefficients a_σ, b_σ, and c_σ), DeGroot and Kadane (1983, p. 206) have shown that player 1's optimal choice of u_σ, given x, satisfies

$$u_\sigma = \frac{E_1\{(0.2b_\sigma - 0.8)\cdot(0.2[(2-c_\sigma)y - a_\sigma] - x)\}}{2 + E_1\{(0.2b_\sigma - 0.8)^2\}}. \tag{7.5}$$

For the limiting values of the coefficients necessary to make (7.3) and (7.4) agree, this yields

$$u_0 = E_1\{(0.6x - 0.24y)/2.36\}. \tag{7.6}$$

However, $u_0 \neq u$, (7.6) does not agree with (7.1) (which is player 1's choice for u when targets are common knowledge), and neither does $u_0 = x$, as is necessary for (7.3) to agree with (7.4).

In short, the limit of optimal play (with $\sigma \to 0$) here does not correspond to the optimal play at $\sigma = 0$. The singularity (at $\sigma^2 = 0$) occurs because merely shrinking the variance ($\sigma^2 \to 0$) of the prior distributions for the targets does not suffice also for shrinking player 2's conditional probability $p_2(u|x, y)$ to the point mass for u concentrated at the solution (2.5). It fails to do so because, in part, the correct interpretation of this conditional probability in the second version of our game is not to be confused with the first-version interpretation, which corresponds to common knowledge of the targets.

Though the limit ($\sigma^2 \to 0$) of the common priors is common knowledge of the targets, the limit of the optimal strategies based on these common priors need not be the optimal strategy based on common knowledge of the targets. Rubinstein is correct: Almost common knowledge is not good enough!

Remark. By supplying the two players with additional, common evidence about the targets (x, y), we can implement the dynamics of a common "posterior" with shrinking variance. For example, if both players observe n pairs (x'_i, y'_i) ($i = 1, \ldots, n$) of i.i.d. bivariate normal variates, with (unknown) means ($\mu_{x'} = x$, $\mu_{y'} = y$), known (equal) variances σ'^2, and zero correlation ($\rho_{x'y'} = 0$), then their common posterior distribution for the targets will be as independent bivariate normal variates with a (common) variance that shrinks to 0 as the sample size n grows without bound.

VIII. CONCLUSION

We have used a relatively simple sequential game between two utility maximizers to emphasize that optimal play among Bayesians (who model each other as such) does not put their strategies into equilibrium. Even with a common prior over the uncertain components of the game (which itself is common knowledge), that is, even with a common prior over the target points, optimal play does not require a correlated equilibrium. The optimal extensive-form strategies are rationalizable in the sense of Pearce (1984), as the reasons for (5.1)–(5.3) make clear. That is, those strategies are derived by backward induction using the common knowledge that the opponents are utility maximizers.

It is right to develop Bayesian game theory. A decision against an opponent, rather than against Nature, does not require novel principles. However, especially in sequential games, the challenge of doing Bayesian game theory against a Bayesian opponent is considerable. It has many facets: Not only must players represent their uncertainties about ordinary events (which, in the DeGroot–Kadane game, corresponds to the players' beliefs about each other's target), but each player must be prepared to formalize how his own actions will affect the other's subsequent choices. In order to do that while respecting the model of common knowledge (where each player is an expected utility maximizer), each must think about how the other models himself. That is, I must ponder what the opponent believes about my beliefs, and so on. The complexity of this thought, the depth to which each player must evaluate iterated expectations of beliefs about the other in order to apply backward induction, depends upon the number of turns in the game. Already, in the simple three-move game of this chapter (without common knowledge of the targets), that task is not trivial for player 2.

The subtleties that attend the difference between "common knowledge" and "almost common knowledge" hint at the number of different faces of Bayesian game theory. In this essay we have focused on one game where it is common knowledge that players are rational. Not all games have that form, even when the players are, in fact, all rational. Perhaps this is a direction to look in to gain a better understanding of such tactics as bluffs and feints. We trust the challenges of Bayesian game theory will be met: some through analysis and some through empirical enquiry.

REFERENCES

Aumann, R. J. (1987), "Correlated Equilibrium as an Expression of Bayesian Rationality." *Econometrica* 55: 1–18.

Bernheim, B. D. (1984), "Rationalizable Strategic Behavior." *Econometrica* 52: 1007–28.

Bicchieri, C. (1989), "Backward Induction without Common Knowledge." In A. Fine and J. Leplin (eds.), *PSA 1988*, vol. 2. East Lansing, MI: Philosophy of Science Association. Presented at the workshop on "Knowledge, Belief, and Strategic Interaction" (June 1989). Castiglioncello, Italy.

Binmore, K., and Brandeburger, A. (1988), "Common Knowledge and Game Theory." Technical Report 89-06, Department of Economics, University of Michigan, Ann Arbor.

DeGroot, M. H., and Kadane, J. B. (1983), "Optimal Sequential Decisions in Problems Involving More Than One Decision Maker." In Rizvi, Rustagi, and Siegmund (eds.), *Recent Advances in Statistics*. New York: Academic Press, pp. 197–210.

Harsanyi, J. C. (1977), *Rational Behavior and Bargaining Equilibrium in Games and Social Situations*. Cambridge: Cambridge University Press.

Harsanyi, J. C. (1989), "Game Solutions and the Normal Form." Presented at the workshop on "Knowledge, Belief, and Strategic Interaction" (June 1989). Castiglioncello, Italy.

Harsanyi, J. C., and Selten, R. (1988), *A General Theory of Equilibrium Selection in Games*. Cambridge, MA: MIT Press.

Kadane, J. B. (1985), "Opposition of Interest in Subjective Bayesian Theory." *Management Science* 31: 1586–8.

Kadane, J. B., and Larkey, P. D. (1982), "Subjective Probability and the Theory of Games." *Management Science* 28: 113–20. [Chapter 3.1, this volume]

Levi, I. (1989), "Feasibility." Department of Philosophy, Columbia University, New York.

Pearce, D. G. (1984), "Rationalizable Strategic Behavior and the Problem of Perfection." *Econometrica* 52: 1029–50.

Rubinstein, A. (1989), "The Electronic Mail Game: Strategic Behavior Under 'Almost Common Knowledge'." *American Economic Review* 79: 385–91.

Savage, L. J. (1954), *The Foundations of Statistics*. New York: Wiley.

Savage, L. J. (1962), *The Foundations of Statistical Inference: A Discussion*. London: Methuen.

Savage, L. J. (1967), "Implications of Personal Probability for Induction." *Journal of Philosophy* 64: 593–607.

Spohn, W. (1977), "Where Luce and Krantz Do Really Generalize Savage's Decision Model." *Erkenntnis* 11: 113–34.

3.3

A Fair Minimax Theorem for Two-Person (Zero-Sum) Games Involving Finitely Additive Strategies

MARK J. SCHERVISH AND
TEDDY SEIDENFELD

ABSTRACT

In this chapter we discuss the sensitivity of the minimax theorem to the cardinality of the set of pure strategies. In this light, we examine an infinite game due to Wald and its solutions in the space of finitely additive (f.a.) strategies.

Finitely additive joint distributions depend, in general, upon the order in which expectations are composed out of the players' separate strategies. This is connected to the phenomenon of "non-conglomerability" (so-called by deFinetti), which we illustrate and explain. It is shown that the player with the "inside integral" in a joint f.a. distribution has the advantage.

In reaction to this asymmetry, we propose a family of (weighted) symmetrized joint distributions and show that this approach permits "fair" solutions to fully symmetric games, e.g., Wald's game. We develop a minimax theorem for this family of symmetrized joint distributions using a condition formulated in terms of a pseudo-metric on the space of f.a. strategies. Moreover, the resulting game

We thank our colleagues in Mathematics, Russell Walker and Juan Schaffer, for references concerning Glicksburg's theorem. Research for this essay was partially supported through an N.S.F. Grant DMS 88-05676 and by O.N.R. Contract N00014-88-K0013. Some preliminary result from this essay were presented at the 40th NBER-NSF Seminar on Bayesian Inference and Econometrics in May 1990 at George Washington University.

From *Bayesian Analysis in Statistics and Econometrics*, edited by D. Berry, C. Chaloner, and J. Geweke (New York: John Wiley & Sons, Inc., 1996), pp. 557–568.

This essay contains the proofs of theorems and lemmas omitted, for reasons of space, in the previously published version.

can be solved in the metric completion of this space. The metrical approach to a minimax theorem is contrasted with the more familiar appeal to compactifications, and we explain why the latter appears not to work for our purposes of making symmetric games "fair." We conclude with a brief discussion of three open questions relating to our proposal for f.a. game theory.

INTRODUCTION

In this essay we derive results for finitely additive (mixed) strategies in two-person, zero-sum games with bounded payoffs. We establish a minimax theorem which is novel in that it allows for joint (finitely additive) distributions which make symmetric (bounded) games fair. That is, the minimax value of a fully symmetric game is 0 under our proposal.

In section I we review the sensitivity of the familiar minimax theorem (of von Neumann and Morgenstern, 1947) to the cardinality of the set of pure strategies. That result, which uses mixed strategies taken from the class of countably additive probabilities, does not apply when the set of pure strategies is infinite. A simple game due to Wald (1950), "Pick the Bigger Integer" (Example 1.1), illustrates the problem. (In this game, the payoff is 0 if both players pick the same integer, otherwise the winner receives 1.) When all strategies are countably additive, this game has no value. If, however, only one player is allowed to use a (merely) finitely additive mixed strategy, Wald's game has a value and that player wins. Allowing both players to use finitely additive mixed strategies leads to a value for the game (as shown by Heath and Sudderth, 1972), but it has the unfortunate consequence that the value depends upon the order of integration over the two mixed strategies. We relate this phenomenon, as it appears in "Pick the Bigger Integer," to P. Lévy's 1930 example of what deFinetti (1972) calls "non-conglomerability" of finitely additive probability.

We find that for all games the player with the "inside" integral occupies the favored position (Theorem 2.2). This means that even for games with symmetric payoffs, as in Wald's game, there may be no symmetry reflected in the value of the game if it is solved by using a particular order of integration. Also in section II we investigate joint finitely additive distributions created from the players' two mixed strategies by taking convex combinations of the two "extreme" joint distributions, where the "extreme" versions of a game correspond to fixing the order of integration first one way and then the other. This leads to a parametrized class of joint distributions, r_w, indexed by a weight w, $0 \leq w \leq 1$. (That is, the "extreme" versions of the game correspond to the values $w = 0$ and $w = 1$.) We prove the minimax theorem (Theorem 2.3) for the set of joint distributions r_w under an assumption (Condition $\mathcal{A}$) expressed in terms of metrical properties of the space of mixed strategies. A simple corollary is that for fully symmetric games (where payoffs satisfy $f(s, t) = -f(t, s)$ and with $w = 0.5$) the game is fair, i.e., the value for the game is 0. Also under condition $\mathcal{A}$, Theorem 2.4 establishes the existence of minimax strategies in the (metric) completion of the space of mixed strategies.

We illustrate the phenomenon that the solution to a game may fail to be a mixture of the minimax solutions from the two "extreme" versions of that game (Example 3). There is an important non-convexity of minimax solutions associated with joint distributions formed by convex combinations of the two "extreme" versions of a game.

In section III we provide a brief account of some related literature. (We defer our review of others' work until section III to allow a contrast with the position taken in this report.) In section IV we indicate a connection between our treatment of games and "improper" priors, and our concluding section V addresses several open questions which we find of interest.

I. THE EFFECT OF INFINITELY MANY PURE STRATEGIES ON THE MINIMAX THEOREM

I.1. *Two-Person, Zero-Sum Games with Infinitely Many Pure Strategies*

In a two-person, zero-sum game, player-1 has pure strategies $s \in S$ and player-2 has pure strategies $t \in T$. Let $f(s, t)$ be the real-valued payoff to player-1, so that $-f(s, t)$ is the payoff to player-2, when player-1 uses strategy s and player-2 uses strategy t. Allow the players to have mixed strategies, i.e., player-1 may use a distribution p ($\in P_\sigma$) on S and player-2 may use a distribution q ($\in Q_\sigma$) on T; for P_σ and Q_σ sets of (σ-additive) probabilities. Then, the (expected) value of strategy pair (p, q) is (assuming f is bounded and measurable with respect to the product measure, $p \times q$):

$$E_{p\times q}f(s,t) - E_p[E_q f(s,t)] = E_q[E_p f(s,t)].$$

That is, the joint distribution is the product measure which, by Fubini's theorem, may be written as the double integral in either order.

Let S and T be finite sets;, then (von Neumann and Morgenstern, 1947) the fundamental result of two-person, zero-sum games asserts:

Theorem 1.1 ("Minimax").

$$\sup_{P\sigma} \inf_{Q\sigma} E_{p\times q}f(s,t) = \inf_{Q\sigma} \sup_{P\sigma} E_{p\times q}f(s,t) = V \qquad \text{(i)}$$

That is, *the game has a value V*. Also:

$$\exists p^* \forall q E_{p^*\times q} f(s,t) \geq V \text{ [maximin strategy]} \qquad \text{(ii)}$$

and

$$\exists q^* \forall p E_{p\times q^*} f(s,t) \leq V \text{ [minimax strategy]}$$

Thus, the strategy pair (p^*, q^*) solves the game. □

This minimax result, even the fact that the game has a value, depends upon there being only finitely many pure strategies. Wald (1950) provides an elegant counterexample when the sets S and T are denumerable.

Example 1.1 ("Pick the Bigger Integer"). Let the pure strategies, $S = T = \{1, 2, \ldots,\}$, be the positive integers. Define the payoff function to player-1 (which is the negative of the payoff to player-2) by:

$$\begin{aligned} f(s,t) &= 1, \quad \text{if } s > t \\ &= 0, \quad \text{if } s = t \\ &= -1, \quad \text{if } s < t. \end{aligned}$$

Then,

$$\sup_P \inf_Q E_{p\times q} f(s,t) = -1$$

while

$$\inf_Q \sup_P E_{p\times q} f(s,t) = +1.$$

Proof. For each σ-additive measure μ on the positive integers, given ε $(0 < \varepsilon < 1)$, there is another σ-additive measure ν and integer n_ε where $\nu\{n: n > n_\varepsilon\} > 1 - \varepsilon$ while $\mu\{n: n < n_\varepsilon\} > 1 - \varepsilon$. □

Thus, game-1 is without a value in the class of countably additive strategies.

Suppose we allow *one* player the use of a finitely additive (f.a.) mixed strategy in Wald's game. Then we can show that the game is determined and that player wins.

Definition. Call p a *purely finitely additive* (p.f.a.) probability if, given $\varepsilon > 0$, there is a denumerable partition $\pi = \{h_i: i = 1, \ldots\}$ with $\Sigma_{hi \in \pi} p(h_i) < \varepsilon$. Also, we refer to these as *diffuse* probabilities.

If player-1 adopts a diffuse probability $p_d(n) = 0$ ($n = 1, \ldots$), but player-2 is restricted to the set Q_σ, then $\inf_{Q\sigma} E_{p_d \times q} f(s, t) = +1$. Likewise, when player-2 adopts a diffuse probability $q_d(n) = 0$ ($n = 1, \ldots$), but player-1 is restricted to the set P_σ, then $\sup_{P\sigma} E_{p \times q_d} f(s, t) = -1$.

This notation is warranted because, on countable spaces, if one of the players uses a countably additive mixed strategy, the order of integration is irrelevant. That is,

Lemma 1.1.

$$\forall (q \in Q_\sigma) E_{p_d}[E_q f(s,t)] = E_q[E_{p_d} f(s,t)] = +1$$

Also,

$$\forall (p \in P_\sigma) E_{q_d}[E_p f(s,t)] = E_p[E_{q_d} f(s,t)] = -1.$$

(The proof is straightforward and is omitted.)

On the other hand, suppose we allow both players in Wald's game to use finitely additive mixed strategies (p, q). The solution now depends upon how the joint strategy, "$p \times q$", is defined. In general, with finitely additive distributions p and q, $E_q[E_p f(s, t)] \neq E_p[E_q f(s, t)]$.

Example 1.2 (attributed to P. Lévy by deFinetti, 1972). Consider a diffuse probability r on the set of all pairs $\langle s, t\rangle$, for s and t positive integers, with the following two restrictions:

$$r(\langle s, t\rangle) = 0,$$

that is, r is 0 on finite sets; and

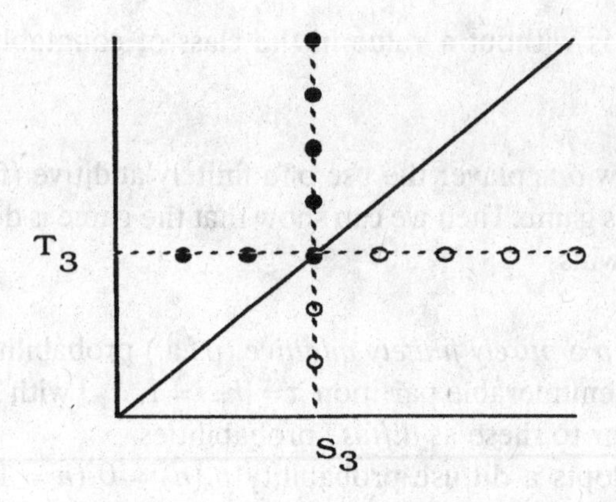

Event E corresponds to pairs <s,t> below the main diagonal

Event E ○

Event E^c ●

Figure 1. Diagram for P. Lévy's example. Only finitely many points on each vertical section lie below the diagonal. Only finitely many points on each horizontal section lie above the diagonal.

$$r(\langle s,t\rangle|F) = 0 \quad \text{if } F \text{ is an infinite set,}$$

that is, conditionally, r is again diffuse given an infinite set F.

Define the events:

$$E = \{\langle s,t\rangle : s > t],$$

$$S_m = \{\langle s,t\rangle : s = m\} \quad (m = 1, \ldots),$$

and

$$T_n = \{\langle s,t\rangle : t = n\} \quad (n = 1, \ldots).$$

Then, $r(E|S_m) = 0$ for $m = 1, \ldots,$ yet $r(E|T_n) = 1$ for n = 1, ... That is, conditioned on each vertical section, the r-probability of an outcome below the main diagonal is 0. However, conditioned on each horizontal section, the r-probability of the same event is 1.

Figure 1 illustrates what is happening.

I.2. *Non-conglomerability of Finitely Additive Probabilities*

Definition (Dubins, 1975). Say that probability p is *conglomerable* in the (denumerable) partition $\pi = \{h_i: i = 1, \ldots\}$ provided that, for each bounded random quantitiy X and $\forall(k_1, k_2)$, if

$$k_1 \leq E(X|h_i) \leq k_2 \quad (i = 1, \ldots)$$

then

$$k_1 \leq E(X) \leq k_2,$$

where $E(\bullet)$ means expectation with respect to p.

Equivalently (Dubins, 1975), p is conglomerable in partition π just in case p is disintegrable in π, and for the special case of an event E (identified with its indicator function)

$$p(E) = \int_{h \in \pi} p(E|h)\,dp(h), \quad \text{for all } E.$$

In Lévy's example, with the two partitions $\pi_1 = \{s_i: i = 1, \ldots\}$ and $\pi_2 = \{t_j: j = 1, \ldots\}$, we see that

$$\int_{s \in \pi_1} p(E|s)\,dp(s) = 0 \quad \text{and} \quad \int_{t \in \pi_2} p(E|t)\,dp(t) = 1.$$

So p fails to be conglomerable in at least one of the two (denumerable) partitions π_1 and π_2.

The lack of conglomerability is endemic to merely finitely additive probabilities. That is, each f.a. probability that is *not* countably additive experiences non-conglomerability in some denumerable partition (Schervish et al., 1984). More precisely, we can say this. Each f.a. probability p has a (unique) decomposition into a convex combination of a σ-additive probability p_σ and a purely finitely additive probability p_d:

$$\forall p \exists! (0 \leq a \leq 1) \quad p = ap_d + (1-a)p_\sigma$$

and p_d [or p_σ] is unique if a [or if $1 - a$] is positive (Yoshida and Hewitt, 1952). The quantity a is the least upper bound on the extent of non-conglomerability of p with respect to events. That is, given a f.a. probability p, for each $\varepsilon > 0$ there is a denumerable partition $\pi = \{h_i: i = 1, \ldots\}$ and event E where,

$$p(E) - p(E|h_i) > a - \varepsilon, \quad \text{for all } i.$$

II.1. *A Proposal for the (Finitely Additive) Joint Distribution "$p \times q$"*

Reconsider Wald's game ("Pick the Bigger Integer"). Observe that in Lévy's example, the event E corresponds to the outcomes where player-1 wins. The non-conglomerability of the probability r (in Lévy's example) illustrates the effect of changing the order of integration in creating a joint distribution on $S \times T$ from the two diffuse "marginal" distributions, p and q. These marginal probabilities correspond to the players' strategies on S and on T (respectively). Let P and Q denote the sets of finitely additive mixed strategies on sets S and T. Thus, when players use (diffuse) purely finitely additive probabilities (p_d, q_d) which assign probability zero to each pure strategy, then

$$\forall (p \in P) E_p[E_{q_d} f(s,t)] = -1 \quad \text{and} \quad \forall (q \in Q) E_q[E_{p_d} f(s,t)] = +1.$$

In particular,

$$E_{p_d}[E_{q_d} f(s,t)] = -1 \neq +1 = E_{q_d}[E_{p_d} f(s,t)].$$

How shall we define the joint distribution that results when player-1 adopts the f.a. strategy p and player-2 adopts strategy q? The condition which motivates our solution, below, is to allow that symmetric bounded games, such as Wald's, admit a solution which makes them fair. That is, when $f(s, t) = -f(t, s)$ for each pure strategy pair (s, t) (and when payoffs are bounded), so the game is symmetric, we require there to be a finitely additive solution to the game that makes it a "draw." We require of such games that they have a value 0.

Proposal. Given f.a. probabilities p and q on S and T (respectively), and given $0 \leq w \leq 1$, adopt the joint probability r_w on the power set $\mathcal{P}[S \times T]$, as follows. For event $E \in \mathcal{P}[S \times T]$,

$$r_w[E] = w\int_T \int_S \chi_E \, dp\, dq + (1-w)\int_S \int_T \chi_E \, dq\, dp,$$

where χ_E is the indicator function for event E. The joint distribution r_w can be thought of as a w-weighted coin-flip between the two joint distributions obtained by fixing the order of integration.

Generally, given p, q and a (bounded) function $f(s, t)$ on $S \times T$, define the E_{r_w}-expectation of f as:

$$E_{r_w}(p,q) = wE_q[E_p f(s,t)] + (1-w)E_p[E_q f(s,t)].$$

The parameter w weights the contribution to the joint expectation E_{r_w} on $S \times T$ of the two "extreme" distributions $E_{r_1} = E_q[E_p f]$ and $E_{r_0} = E_p[E_q f]$. As we explain next (Theorem 2.2), these two are "extreme" in the sense that, in zero-sum games, $w = 1$ favors the first (p-)player, whereas $w = 0$ favors the second (q-)player. Each player prefers the E_{r_w}-game where his expectation has the "inside" integral.

Heath and Sudderth (1972) show that (for games with bounded payoffs) when the joint distribution is determined by fixing the order of integration, then the game has a value. That is, their result is:

Theorem 2.1.

$$V_1 = \sup_P \inf_Q E_q[E_p f(s,t)] = \inf_Q \sup_P E_q[E_p f(s,t)]$$

and

$$V_2 = \sup_P \inf_Q E_p[E_q f(s,t)] = \inf_Q \sup_P E_p[E_q f(s,t)].$$

In our notation, V_1 is the value of the E_{r_1}-game, and V_2 is the value of the E_{r_0}-game. It is a simple corollary to the Heath-Sudderth result that there are minimax strategies which achieve these values.

Corollary 2.1. Corresponding to the E_{r_1}-game there are strategies, (p_1, q_1) such that

$$V_1 = \inf_Q E_{r_1}(p_1, q) = \sup_P E_{r_1}(p, q_1),$$

and corresponding to the E_{r_0}-game there are strategies (p_2, q_2) such that

$$V_2 = \inf_Q E_{r_0}(p_2, q) = \sup_P E_{r_0}(p, q_2). \quad \square$$

This corollary also appears as Theorem 2.1 in Kindler (1983).

Theorem 2.2 ($\forall w$).

$$V_2 \leq \sup_P \inf_Q E_{r_w}(p,q) \leq \inf_Q \sup_P E_{r_w}(p,q) \leq V_1$$

Proof. The players' minimax strategies (each taken from his *dis*-favored game) provide the desired bounds. Corollary 2.1 ensures these strategies exist. Specifically,

$$\sup_P \inf_Q E_{r_w}(p,q) \geq \inf_Q E_{r_w}(p_2,q) \geq w(\inf_Q E_q[E_{p_2} f(s,t)]) \\ +(1-w)\inf_Q E_{p_2}[E_q f(s,t)] \geq V_2.$$

Only the third of these inequalities requires an explanation. It follows from the two facts: (i) $\inf_Q E q_2[E_q f(s, t)] = V_2$, since q_2 solves the E_{r_0} game; and (ii) $\inf_Q E_q[E_{p_2} f(s, t)] \geq V_2$, since, if on the contrary ($\inf_Q E_q[E_{p_2} f(s, t)] < V_2$), then $\exists(t^* \in T)$ such that $E_{p_2} f(s, t^*) < V_2$. But that contradicts p_2 as a minimax solution to the E_{r_0}-game. A similar argument, with "p", "q", and inequalities all interchanged, proves the result about V_1. □

Hence, for all bounded games, $V_2 \leq V_1$, and a player's advantage is with the "inside" integral.

To express our minimax theorems concerning the E_{r_w}-game, we pseudo-metrize the set of strategies.

Definition. Say that two strategies for a player are *equivalent in the E_{r_w}-game* provided they have the same (expected) value against each possible strategy of the opponent. (Denote this relation by $\equiv$, where the game is identified by context.)

That is,

$$(p_1 \equiv p_2) \text{ in game } E_{r_w} \text{ if and only if } (\forall q) E_{r_w}(p_1,q) = E_{r_w}(p_2,q);$$

likewise

$$(q_1 \equiv q_2) \text{ in game } E_{r_w} \text{ if and only if } (\forall p) E_{r_w}(p,q_1) = E_{r_w}(p,q_2).$$

It is obvious that $\equiv$ is an equivalence relation.

Consider the pseudo-metrics ρ_P *on* P-strategies and ρ_Q on Q-strategies defined by

$$\rho_P(p_1,p_2) = \sup_Q |E_{r_w}(p_1,q) - E_{r_w}(p_2,q)|$$

and

$$\rho_Q(q_1,q_2) = \sup_P |E_{r_w}(p,q_1) - E_{r_w}(p,q_2)|.$$

Lemma 2.1. $\rho_P(\rho_Q)$ is a pseudo-metric on the set P (set Q). □

Wald (1950, chapter 2) introduced the same pseudo-metrics to deal with countably additive mixed strategies. The proof that they are pseudo-metrics is the same as in the countably additive case.

Clearly, $\rho_P(p_1, p_2) = 0$ just in case $(p_1 \equiv p_2)$, and similarly with ρ_Q; thus, these two pseudo-metrics are metrics on their respective $\equiv$-equivalence classes of strategies. The pseudo-metrics express by how much two strategies may be separated through the choice of the opponent's reply.

Next, we formulate a sufficient condition for existence of a value for the E_{rw}-game. The condition concerns approximately maximin strategies for one player and corresponding approximately best responses by the other player. Some notation is needed, first, to make these concepts precise.

Suppose $\sup_P \inf_Q E_{rw}(p, q) = a$. Then, given $\varepsilon > 0$, $\forall p \in P$, $\exists q \in Q$ with $E_{rw}(p, q) \leq a + \varepsilon$. Given p, define the quantity:

$$v(p) = \inf_Q E_{rw}(p, q)$$

and the set

$$R^{\varepsilon}(p) = \{q : E_{rw}(p, q) \leq v(p) + \varepsilon\}.$$

$v(p)$ is the (limit of the) *value* of best replies q-player can make against strategy p. $R^{\varepsilon}(p)$ is the *set* of ε-best replies q-player has against the strategy p.

Let

$$B_{\rho_Q}(q^*, \varepsilon) = \{q : \rho_Q(q, q^*) \leq \varepsilon\}, \text{ the set of } q\text{'s near to } q^*.$$

Also, because $\sup_P \inf_Q E_{rw}(p, q) = a$, given $\varepsilon > 0$, $\exists p \in P$, $\forall q \in Q$ such $E_{rw}(p, q) \geq a - \varepsilon$.

Define the set

$$P^{\varepsilon} = \{p : \forall q E_{rw}(p, q) \geq a - \varepsilon\}.$$

P^{ε} is the set of ε-maximin strategies for p-player. Clearly, the sets $R^{\varepsilon}(p)$ and P^{ε} are nonempty and convex. Next, we state the condition under which we prove our minimax theorems. Observe that it is formulated asymmetrically between the two players. We discuss this and other features of the condition below.

Condition $\mathcal{A}$.

$$\exists(k > 0), \forall(\delta > 0), \exists(0 < \varepsilon \leq \delta)\, \exists p \in P^{\varepsilon}\, \exists(n > 0; q_i \in Q, 1 \leq i \leq n)$$
$$\text{with } R^{2\varepsilon}(p) \subseteq \cup_i B_{\rho_Q}(q_i, k\varepsilon).$$

Condition $\mathcal{A}$ requires that, for each small $\varepsilon > 0$, there is some ε-maximin strategy p, each of whose 2ε-best replies is $k\varepsilon$-near to (in the sense of ρ_Q) one of some finitely many q-strategies. In simpler words, there is a "safe" p-strategy $[p \in P^\varepsilon]$ whose set of "best" responses $[R^{2\varepsilon}(p)]$ is covered by a finite number $[n]$ of "small" balls $[B_{\rho_Q}(q_i, k\varepsilon)]$. Within ε-approximations, Condition $\mathcal{A}$ is that there exists some maximin p-strategy, where each good q-reply to p is close to one of some finite collection of q's.

Condition $\mathcal{A}$ is truly asymmetric; it may be satisfied under one order of the players' strategies, but not with roles reversed. For example, consider an extreme version of Wald's game, Example 1.1, with $w = 1$. Recall, $E_{r_1}(p, q) = E_q E_p[f(s, t)]$. The p-player has the advantage (Theorem 2.2) and the game's value is 1. According to the (corollary to the) Heath-Sudderth theorem (Theorem 2.1), minimax strategies exist. For p-player, the minimax strategies all belong to the same ρ_P-equivalence class – any diffuse strategy, p_d, is minimax and only diffuse strategies are minimax for the first player. However, p_d is an "equalizer" strategy: $\forall_q$, $E_{r_1}(p_d, q) = 1$. Hence, all of Q is the set of "best" responses to p_d.

Since $\rho_Q(q, q') \geq 1$ whenever q and q' are different (point-mass) pure strategies, there is no $p \in P^\varepsilon$ satisfying Condition $\mathcal{A}$. The "best" responses to p_d are not contained within finitely many small ρ_Q-neighborhoods. Nonetheless, by considering Condition $\mathcal{A}$ with the players' roles reversed, we discover that, against a diffuse strategy q_d, all good p-responses in the E_{r_1}-game are near to the equivalence class of diffuse strategies, represented by the strategy p_d. That is, with the alternative reading, Condition $\mathcal{A}$ is satisfied using a single ρ_P-neighborhood of p-strategies near to $p_d (k = 2)$.

It is easy to verify that Condition $\mathcal{A}$ obtains in the non-extreme versions of Wald's game: all diffuse strategies (p_d or q_d) lie in the same equivalence class for that strategy space and, in fact, each such strategy is minimax. For $0 < w < 1$ and for sufficiently small ε, all good responses to p_d are close (in the sense of ρ_Q) to q_d. Thus, there is a single neighborhood ($n = 1$, $k = 2$) which satisfies Condition $\mathcal{A}$ and it applies to either player. For non-extreme versions of Wald's game, $\mathcal{A}$ obtains both ways.

Theorem 2.3. Provided Condition $\mathcal{A}$ holds,

$$\sup_P \inf_Q E_{r_w}(p, q) = \inf_Q \sup_P E_{r_w}(p, q).$$

Proof. Clearly, $\sup_P \inf_Q E_{rw}(p, q) \leq \inf_Q \sup_P E_{rw}(p, q)$, regardless of Condition $\mathcal{A}$. We argue for the reverse inequality using two lemmas (which do not require Condition $\mathcal{A}$).

Lemma 2.2. Let $\sup_P inf_Q E_{rw}(p, q) = a$. Then there is a finitely additive measure, μ, on the space of Q-strategies with the property that $\forall p \in P$, $\int_Q E_{rw}(p, q)\, d\mu \leq a$. □

Proof (of lemma). This proof is very similar to that of Heath and Sudderth's (1972, p. 2072) "Theorem 1." Consider the set $\mathcal{B}$ of all bounded functions from Q to the reals, $\mathfrak{R}$. For each $p \in P$, $E_{rw}(p, \bullet) \in \mathcal{B}$. Let $K_1 = \{f \in \mathcal{B}: \forall q, f(q) > a\}$. Let $K_2 = \{E_{rw}(p, \bullet): p \in P\}$. The assumption, $\sup_P \inf_Q E_{rw}(p, q) = a$, implies $K_1 \cap K_2 = \phi$. Clearly, K_1 and K_2 are convex. By Theorem 8 (p. 417 of Dunford and Schwartz, 1958), there exist a scalar c and a non-zero linear function π on $\mathcal{B}$ with the following two properties:

$$\forall p \in P, \forall h \in K_1, \quad \pi[E_{rw}(p, \bullet)] \leq c \leq \pi(h);$$

$$\text{If } \lim_{n\to\infty} h_n = h \text{ uniformly on } Q, \text{ then } \lim_{n\to\infty} \pi(h_n) = \pi(h).$$

Since π is non-zero and linear and since each constant function greater than a belongs to K_1, it follows that $\pi(1) > 0$. Normalize π so that $\pi(1) = 1$. Then it is clear that $c \leq a$ since $\pi(a) = a$ and the constant functions $a = \varepsilon \in K_1$ (whenever $\varepsilon > 0$). It follows that $\pi[E_{rw}(p, \bullet)] \leq a$.

The proof is concluded by showing there exists a finitely additive probability μ on Q such that, $\forall h \in \mathcal{B}$, $\pi(h) = \int_Q h(q)d\mu(q)$. Consider $C \in 2^Q$, a set of q's. Given $\varepsilon > 0$, let $h_\varepsilon(q) = a + \varepsilon + \chi_C(q)$. *So* $h_\varepsilon \in K_1$ and $\pi(\chi_C) \geq 0$ for each C. Define $\mu(C) = \pi(\chi_C)$. Since π is linear, μ is finitely additive; and as $\pi(1) = 1$, μ is a probability. If h is a simple function, by linearity of π, then $\pi(h) = \int_Q h(q)d\mu(q)$. Because every bounded function can be uniformly approximated by simple functions, we obtain the desired representation: $\forall h \in \mathcal{B}$, $\pi(h) = \int_Q h(q)d\mu(q)$. □

Based on Lemma 2.2, define the function $\vartheta(p) = \int_Q E_{rw}(p, q)d\mu$ and denote by $I^\varepsilon(p)$ the set $I^\varepsilon(p) = \{q: |E_{rw}(p, q) - \vartheta(p)| \leq \varepsilon\}$.

That is, $I^\varepsilon(p)$ is the set of q-strategies whose value against p is close to the integral $\vartheta(p)$.

Lemma 2.3. The family $\{I^\varepsilon(p): p \in P, \varepsilon > 0\}$ has the finite intersection property. □

Proof (of lemma). We give the proof in two cases, depending on the size, n, of the finite family $\{I^{\varepsilon_j}(p_j): p_j \in P, \varepsilon_j > 0, j = 1, \ldots, n\}$. First we argue for the elementary case, $n = 1$, which we generalize to the other case, $n > 1$.

Define $L = \inf_{(S\times T)} f(s, t)$, $U = \sup_{(S\times T)} f(s, t)$, and $d = U - L$. Let $N \geq 2d/\varepsilon$. For $i = 1, \ldots, N - 1$, define $g_i(p) = \{q: L + d(i - 1)/N \leq E_{rw}(p, q) < L + di/N\}$, and for $i = N$, let the last inequality be $\leq$. Let $c_{i,p} = \mu(g_i(p))$, where μ is the f.a. measure taken from Lemma 2.2. For each i where $g_i(p) \neq \emptyset$, let $q_i \in g_i(p)$ (for other i, let q_i be arbitrary). Define $q_{\varepsilon,p} = \Sigma_{i=1}^{N} c_{i,p}q_i$. It is evident that $E_{rw}(p, q_{\varepsilon,p}) = \Sigma_{i=1}^{N} c_{i,p}E_{rw}(p, q_i)$. We conclude the case ($n = 1$) by showing that $q_{\varepsilon,p} \in I^{\varepsilon}(p)$.

Define the simple function $h_{p,\varepsilon}: Q \to \Re$ as: $h_{p,\varepsilon}(q) = E_{rw}(p, q_i)$, for $q \in g_i(p)$. It is easy to see that $\forall q$, $|h_{p,\varepsilon}(q) - E_{rw}(p, q)| < \varepsilon$. Hence, $|\int_Q h_{p,\varepsilon}(q)d\mu - \vartheta(p)| < \varepsilon$. Since $h_{p,\varepsilon}(q)$ is a simple function, constant on all $q \in g_i(p)$, we have that $\int_Q h_{p,\varepsilon}(q)d\mu = \Sigma_{i=1}^{N} c_{i,p}E_{rw}(p, q_i)$, which proves the point.

For $n > 1$, we simplify by noting that $I^{\varepsilon'}(p) \supseteq I^{\varepsilon}(p)$ whenever $\varepsilon' \geq \varepsilon$. That is, without loss of generality we take $\varepsilon = \min\{\varepsilon_i\}$ and prove finite intersection property for $\{I^{\varepsilon}(p_j)\}$. Consider the common refinement of the partitions generated by the Nn sets, $g_i(p_j)$. For each $(i_1, \ldots, i_n) \in \{1, \ldots, N\}^n$, let

$$g_{i_1,\ldots,i_n} = \bigcap_{j=1}^{n} g_{i_j}(p_j)$$

As before, let $q_{i_1,\ldots,i_n} \in g_{i_1,\ldots,i_n}$, when the latter is non-empty; otherwise, let $q_{i_1,\ldots,i_n}$ be arbitrary. Let $c_{i_1,\ldots,i_n} = \mu(g_{i_1,\ldots,i_n})$ and let Σ'_j denote the $n - 1$ fold summation over all the indices other than the jth index. Set $c^j_{i_j} = \Sigma'_j c_{i_1,\ldots,i_n}$. Last, define $q^j_{i_j} = (1/c^j_{i_j})\Sigma'_j c_{i_1,\ldots,i_n} q_{i_1,\ldots,i_n}$. Since $q_{i_1,\ldots,i_n} \in g_{i_j}(p_j)$ for each j and i_j, it follows from the convexity of $g_{i_j}(p_j)$ that $q^j_{i_j} \in g_{i_j}(p_j)$. Therefore, select

$$q^* = \sum_{i_j=1}^{N} c^j_{i_j} q^j_{i_j} = \sum_{i_1=1}^{N} \cdots \sum_{i_n=1}^{N} c_{i_1,\ldots,i_n} q_{i_1,\ldots,i_n}$$

We then have that $q^* \in I^{\varepsilon}(p_j)$, for $j = 1, \ldots, n$. □

Thus, Lemma 2.3 asserts that the intersection of finitely many $I^{\varepsilon}(p)$ sets is not empty,

$$\emptyset \neq \bigcap_j I^{\varepsilon_j}(p_j) \quad (j = 1, \ldots, m).$$

Lemma 2.4. Given a p-strategy which is ε-maximin, $p \in P^\varepsilon$, then $R^{2\varepsilon}(p) \supset I^\varepsilon(p)$. □

Proof (of lemma). To see that $R^{2\varepsilon}(p) \supseteq I^\varepsilon(p)$ when $p \in P^\varepsilon$, note that $\forall\, q \in I^\varepsilon(p)$, $a - \varepsilon \leq E_{r_w}(p, q) \leq a + \varepsilon$. The first inequality arises from the fact that, since $p \in P^\varepsilon, \forall q \in Q, E_{r_w}(p,q) \geq a - \varepsilon$. The second inequality obtains because, $\forall p, \vartheta(p) \leq a$; hence, $\forall q \in I^\varepsilon(p), E_{r_w}(p, q) \leq a + \varepsilon$. Last, observe that for $p \in P^\varepsilon, R^{2\varepsilon}(p) \supseteq \{q{:}a - \varepsilon \leq E_{r_w}(p, q) \leq a + \varepsilon\}$. □

To complete the theorem we argue indirectly that $\sup_P \inf_Q E_{r_w}(p, q) \geq \inf_Q \sup_P E_{r_w}(p, q)$. Assume (on the contrary) that $\inf_Q \sup_P E_{r_w}(p, q) = b > a$. Choose $0 < \varepsilon < (b - a)/(k + 2)$. By Condition $\mathcal{A}$, there is a p-strategy in P^ε, denoted by p^*, with the property that $R^{2\varepsilon}(p^*) \subseteq \cup_i B\rho_Q(q_i, k\varepsilon)$, for some finite number of q's, $(i = 1, \ldots, n)$. The assumption that $\inf_Q \sup_P E_{r_w}(p, q) = b$, entails that for each q-strategy there is a p-strategy where $E_{r_w}(p, q) \geq b - \varepsilon$. Hence, for each q_i $(i = 1, \ldots, n)$ there is a p_i where $E_{r_w}(p_i, q_i) \geq b - \varepsilon$. But then if $q \in B\rho_Q(q_i, k\varepsilon)$, $q \notin I^\varepsilon(p_i)$. In other words, $B\rho_Q(q_i, k\varepsilon) \cap I^\varepsilon(p_i) = \varnothing$. (This follows because $\forall q \in I^\varepsilon(p_i), E_{r_w}(p_i, q) \leq a + \varepsilon$; however, $\forall q \in B\rho_Q(q_i, k\varepsilon), E_{r_w}(p_i, q) > b - (k + 1)\varepsilon$.) Thus, $R^{2\varepsilon}(p^*) \cap [\cap_i I^\varepsilon(p_i)] = \varnothing$. According to Lemma 2.4, then $I^\varepsilon(p^*) \cap [\cap_i I^\varepsilon(p_i)] = \varnothing$, which contradicts Lemma 2.3.

In light of Theorem 2.3, denote by V_w the value of the game. It is a simple corollary that, provided Condition $\mathcal{A}$ obtains, symmetric games are fair, i.e., for symmetric games $V_{0.5} = 0$.

Corollary 2.2. Under the condition for Theorem 2.3, if $S = T$ and $f(s, t) = -f(t, s)$, so the game is symmetric, then $V_{0.5} = 0$, i.e., then the $E_{r_{0.5}}$-game is fair.

Proof. By the symmetry of the game, and since the two players have identical strategy spaces ($P = Q$), observe that for each pair (p, q) $E_{r_{0.5}}(p, q) = -E_{r_{0.5}}(q, p)$. We argue indirectly. Suppose $\sup_P \inf_Q E_{r_w}(p, q) = \inf_Q \sup_P E_{r_w}(p, q) = b > 0$. Then there exists a p' strategy such that for all q strategies, $E_{r_{0.5}}(p', q) \geq b/2 > 0$. But p' is available to the q-player, denoted now as strategy q', and from the foregoing observation, for all $p E_{r_{0.5}}(p, q') \leq -b/2 < 0$. This contradicts T.2.3. □

Under the same Condition $\mathcal{A}$, next, we show that there exist solutions to the E_{r_w}-game within the metric completions of ρ_P and ρ_Q.

(Condition $\mathcal{A}$ is assumed for the complete spaces.) Not only does the game have a value, but minimax strategies exist. The central idea in this theorem is an application of a result about the nonempty intersection of closed sets in a complete metric space, due to Kuratowski (1968), using the closure of the sets $I^\varepsilon(p)$. The next lemma, attributed to Kuratowski, is the metrical analog to compactness for the familiar and elementary result that, in a compact space, if a family of closed sets has the finite intersection property, then the family has a nonempty intersection.

Following Kuratowski, denote by $\alpha(x)$ the greatest lower bound of numbers ε such that set x can be decomposed into a finite union of sets of diameter $< \varepsilon$.

Lemma 2.5 (Kuratowski, 1968, vol. 1, p. 412, the Corollary). In a complete metric space, let $\{F_\iota\}$ be a family (of arbitrary cardinality) of closed sets with the finite intersection property. If there are sets F_ι with arbitrarily small $\alpha(F_\iota)$, then the entire family has nonempty intersection. □

Consider the topology on (the equivalence classes of) P (or on Q) induced by ρ_P (or ρ_Q), respectively. The product topology on $P \times Q$ can be metrized by (among others):

$$\rho_{P\times Q}[(p_1, q_1), (p_2, q_2)] = \max\{\rho_P(p_1, p_2), \rho_Q(q_1, q_2)\}.$$

Take the metric completions ρ_{P^*}, ρ_{Q^*}, and $\rho_{P\times Q^*}$ obtained by embedding the sets P, Q, and $P \times Q$ in the space of bounded, continuous (real-valued) functions (on P, Q, and $P \times Q$) using the supremum metric. (See Dugundji, 1968, p. 304). These completions are related by the next lemma.

Lemma 2.6. The space $(P \times Q)^*$ is (identically) isometric with the product space $P^* \times Q^*$, where $\rho_{P\times Q^*}$ is the common metric. □

Proof. By corollary 5.3, p. 303, of Dugundji (1968).

Thus, we may identify limits from the space $(P \times Q)^*$ by taking limits from each player's space of strategies. We use this metric completion to produce (extended) strategies p^* and q^* that solve the E_{r_w}-game.

Theorem 2.4.

i. Provided that Condition $\mathcal{A}$ obtains, $(\exists q^* \in Q^*)$ $(\forall p \in P^*)$ $E_{r_w}(p, q^*) \leq V_w$.

ii. Likewise, provided Condition $\mathcal{A}$ obtains with the players' roles reversed, $(\exists p^* \in P^*)(\forall q \in Q^*)E_{r_w}(p^*, q) \geq V_w$.

iii. So, if Condition $\mathcal{A}$ obtains for both players, $E_{r_w}(p^*, q^*) = V_w$, and the strategy pair (p^*, q^*) solves the E_{r_w}-game.

Proof. Regarding (i), we apply Kuratowksi's Lemma 2.5 to the *closed* sets $\{\mathbf{I}^\varepsilon(p): p \in P^*, \varepsilon > 0\}$, where closure is with respect to the complete space P^*. However, in order to do this, first we state a property concerning the extension of expectations from $(P \times Q)$ to $(P \times Q)^*$. The following two lemmas are easy to prove, the second being an immediate consequence of the first. (Also, they appear in Kretkowski and Telgárski, 1983.)

Lemma 2.7. E_{r_w} has a unique, uniformly continuous extension form $(P \times Q)$ to $(P \times Q)^*$. □

Lemma 2.8.

$$\sup_P \inf_Q E_{r_w}(p, q) = \sup_{P^*} \inf_{Q^*} E_{r_w}(p, q)$$

and

$$\inf_Q \sup_P E_{r_w}(p, q) = \inf_{Q^*} \sup_{P^*} E_{r_w}(p, q). \quad \square$$

Next, we duplicate Lemmas 2.2 and 2.3 for the space of metric completions P^* and Q^*. The proofs of Lemmas 2.9 and 2.10 follow exactly those of 2.2 and 2.3, respectively, and are omitted.

Lemma 2.9. Let $\sup_{P^*}\inf_{Q^*} E_{r_w}(p, q) = a$. Then there is a finitely additive measure, μ, on the space of Q^*-strategies with the property that, $\forall p \in P^*, \int_{Q^*} E_{r_w}(p, q)\, d\mu \leq a$. □

Based on Lemma 2.9, define the function $\vartheta^*(p) = \int_{Q^*} E_{r_w}(p, q)\, d\mu$ and denote by $\mathbf{I}^\varepsilon(p)$ the closed set $\mathbf{I}^\varepsilon(p) = \{q \in Q^*: |E_{r_w}(p, q) - \vartheta(p)| \leq \varepsilon\}$.

Lemma 2.10. The family $\{\mathbf{I}^\varepsilon(p): p \in P^*, \varepsilon > 0\}$ has the finite intersection property. □

Then part (**i**) of the theorem follows from Lemma 2.5, since Condition $\mathcal{A}$ assures the existence of closed sets $\mathbf{I}^{\varepsilon}(p)$ with arbitrarily small "α" (in Kuratowski's notation).

Part (**ii**) of the theorem is demonstrated by reversing the players' roles and part (**iii**) is an immediate consequence of (**i**) and (**ii**). □

II.2. *Non-convexity of the* E_{rw}*-Games, as a Function of* w

Our purpose in this subsection is to indicate, by example, that the minimax solution to an E_{rw}-game may fail to be a convex combination of the extreme solutions (where $w = 0$ and $w = 1$).

Example 2.1. Consider a modification of Wald's game where (as in "Pick the Bigger Integer") $f(s, t) = -1$ if $t > s$, $f(s, t) = 0$ if $s = t$, and $f(s, t) = 1$ if $2 \le t$ and $s > t$, but (unlike Wald's game) $f(s, t) = -0.5$ for $1 = t < s$. The row corresponding to the pure strategy ($t = 1$) is altered.

It is straightforward to show that the game has values: $V_0 = -1$, $V_1 = -0.2$, and $V_{0.5} = -1/3$, for the parameter settings $w = 0$, $w = 1$, and $w = 0.5$, respectively. Let p_d *and* q_d denote any diffuse (p.f.a.) mixed strategies on the integers. Then the minimax strategies for the games are:

$$p_0^* \text{ is arbitrary (as } q_0^* \text{ is an "equalizer" strategy)}, \quad q_0^* = q_d$$

for $w = 0$;

$$p_1^* = 0.4p_d + 0.6\,(s = 1), \quad q_1^* = 0.8q_d + 0.2\,(t = 1)$$

for $w = 1$;

$$p_{0.5}^* = (2/3)p_d + (1/3)\,(s = 1), \quad q_{0.5}^* = (1/3)q_d + (2/3)\,(t = 1)$$

for $w = 0.5$.

Note that $V_{0.5} \ne 0.5V_0 + 0.5V_1$. That is, the value of the game for $w = 0.5$ is not the equal mixture of the values for the two extreme games. Though $V_{0.5} = (1/6)V_0 + (5/6)V_1$ and $q_{0.5}^* = (1/6)q_0^* + (5/6)q_1^*$, $p_{0.5}^*$ is not a similar mixture of p_0^* and p_1^*, for any p_0^*. The strategies for the mixed game are not a mixture of the strategies for the extreme games. In short, our proposal to use joint distributions which are the convex combination of two extreme distributions ($w = 0$ and $w = 1$), the E_{rw}-game, results in a non-convexity of minimax values and minimax strategies with respect to the parameter w.

Condition $\mathcal{A}$ applies (and in either order, provided $w > 0$) to the strategy spaces for this game. For example, with $w = 0.5$, $p_{0.5}^*$ equalizes on the two strategies used to define $q^*_{0.5}$: the diffuse q_d and the pure $(t = 1)$. Thus, each mixed strategy $xq_d + (1 - x)\ (t = 1)$ (for $0 \leq x \leq 1$) is a "good" response to $p^*_{0.5}$. Any other "good" response to $p^*_{0.5}$ is ρ_Q-close to one of these mixtures. Moreover, these mixtures are contained within n closed ρ_Q-balls (of radius $1/n$), where each closed ball is centered at the strategy $q_i = [i/n]q_d + [(n - i)/n]\ (t = 1)$ (for $1 \leq i \leq n$). To see that $\mathcal{A}$ is satisfied, choose $k = 3$ and, given $\varepsilon > 0$, let $n \geq 1/\varepsilon$. Thus, if $q \in R^{2\varepsilon}(p^*_{0.5})$ then $q \in \cup_i B\rho_Q(q_i, 3\varepsilon)$.

III. SOME COMPARISONS WITH OTHER WORK ON FINITELY ADDITIVE MIXED STRATEGIES

Investigation of finitely additive strategies in zero-sum, two-person games dates (at least) from Samuel Karlin's (1950) essay. Karlin, in turn, responds to Ville's (1938) game without a value, which uses a bounded, discontinuous payoff function on the unit square. By a separating hyperplane argument, Karlin shows that Ville's game does have a value when the players use finitely additive strategies and these are composed into a joint distribution *with the order of expectations fixed*. That is, Karlin's expectation for a (bounded) payoff function f on $[0, 1]^2$, using the finitely additive strategy pair (p, q), is based on the analysis of e.g., $E_p[E_q f(x, y)]$. Also, Karlin's proof relies on an assumption that the space of pure strategies available to a player is compact.

Heath and Sudderth (1972) extend Karlin's result to all two-person games with bounded payoff functions $f(s, t)$, again using a separating hyperplane argument. It is reported here as Theorem 2.1. [Their proof avoids Karlin's assumption that the space of pure strategies is compact.] But, like Karlin's work, their joint distributions are based on a fixed order of expectations. In terms of our proposal to use the family:

$$E_{r_w} = wE_q[E_p f(s,t)] + (1-w)E_p[E_q f(s,t)],$$

Heath and Sudderth's theorem establishes the (von Neumann) minimax result only for the two extreme cases: $w = 0$ and $w = 1$.

E. B. Yanovskaya (1970) expresses dissatisfaction with the required asymmetry of these solutions. (J. Kindler, 1983, adopts Yanovskaya's approach for this reason.) Yanovskaya introduces a method for assigning values to a strategy pair (p, q), and here that value is denoted $E^Y(p, q)$, which we paraphrase as follows: When $E_q[E_p f(s, t)] = E_p[E_q f(s, t)]$, then $E^Y(p, q)$ is defined by integration in either order. However, when $E_q[E_p f(s,t)] \neq E_p[E_q f(s,t)]$, then $E^Y(p, q)$ is stipulated to be some (real) quantity a. With this method, Yanovskaya shows that the minimax theorem (as in Theorem 2.3) obtains for a nonempty set of a values: either the set is a closed interval $[a_1, a_2]$, or it is a single value $[a]$. The special value $a^* = [a_1 + a_2]/2$ then is uniquely determined by an additional appeal to three invariance/symmetry conditions. In short, with Yanovskaya's proposal, all symmetric (bounded) zero-sum (two-person) games are fair, i.e., $a^* = 0$.

We are completely sympathetic with the objection that fixing the order of integration introduces undesirable asymmetries, as in Heath and Sudderth's solutions to Wald's game. However, we find Yanovskaya's proposal unsatisfactory, for the following reason. The $E^Y(p, q)$-numbers do not satisfy (finitely additive) expected utility theory. (The cogency of the three invariance conditions for choosing the midpoint of the $[a_1, a_2]$-interval is a different matter altogether.)

For instance, in connection with Wald's game (for which $a_1 = -1$, $a_2 = 1$ and thus $a^* = 0$), consider the three strategy pairs (p_1, q_d), (p_d, q_d), and (p_x, q_d), with $(0 < x < 1)$, where: p_d and q_d are diffuse, p_1 is the pure strategy $\{s = 1\}$, and $p_x = xp_1 + (1 - x)p_d$. Then $E^Y(p_1, q_d) = -1$, but $E^Y(p_d, q_d) = E^Y(p_x, q_d) = a^* = 0$. However, p_x is the simple mixture of two strategies p_1 and p_d; hence, according to (even finitely additive) expected utility theory, the values ought to satisfy: $E(p_x, q_d) = xE(p_1, q_d) + (1 - x)E(p_d, q_d)$. We understand this expectation feature of expected utility theory to provide the justification (in fact, von Neumann's justification) for assigning values to (countably additive) mixed strategies based on the values of pure strategies. It seems clear to us that in extending the value structure to include finitely additive mixed strategies, the simple expectation property (described above) is to be respected. Therefore we do not accept Yanovskaya's method for solving finitely additive games.

When discussing how to prove minimax theorems like Theorem 2.3 for infinite games (under the assumption that all strategies are countably additive), Karlin offers this advice.

> In the theory of infinite games, the truth of [the minimax theorem] is a deep question, requiring some kind of assumption of continuity [for joint expectations] and the restriction that at least one of the spaces [P and Q] is compact space in some suitable sense. (Karlin, 1959, p. 23)

We were unsuccessful in finding a way to duplicate the proofs given in section 2.1 (based on metric completions) using compactification of the spaces P, Q, and $P \times Q$. For one example, using Stone-compactification in Wald's game leads to a failure of the counterpart to Lemma 2.6 (see Glicksberg's theorem, 1959). For a second example, using the topology of "pointwise" convergence with Wald's game invalidates the counterpart to Lemma 2.7. Instead of compactification, we settled on Condition $\mathcal{A}$ as a useful alternative.

Last, it is straightforward to show that, provided Condition $\mathcal{A}$ obtains (or if $\mathcal{A}$ obtains with P and Q spaces reversed), the set of minimax strategies for q-player (or for p-player) is precompact with respect to ρ_Q (or ρ_P). This is in contrast with Fan's (1953) T.3(ii), minimax theorem. Fan's result requires (for our case) the assumption that E_{r_w} be an almost periodic function on the product set $P \times Q$. "Almost periodicity" is both necessary and sufficient to make *all* of $P \times Q$ into a precompact, uniform space. Also, it is a symmetric condition on the product: right almost periodicity is equivalent to left almost periodicity. Neither of these feature is a consequence of Condition $\mathcal{A}$.

IV. SOME CONNECTIONS BETWEEN "IMPROPER" PRIORS AND F.A. MINIMAX STRATEGIES

Wald (1950) uses game theory to model nonsequential, statistical decisions, roughly, as follows: Player 1 is Nature. A pure strategy is the determination of a parameter $\theta \in \Theta$. Nature's mixed strategies are ("prior") distributions $p(\theta)$. The Statistician is Player 2. The Statistician observes an r.v. X, whose distribution $F_\theta(x)$ is given as a function of the parameter θ. (That is, the statistical model fixes the "likelihood" function for the game.) The Statistician has options, terminal decisions, $d^t \in D^t$, where a pure strategy d (a nonrandomized decision function) for the Statistician is a function from X to D^t. Payoffs to Nature ("losses" to the Statistician) are indicated by non-negative real numbers, $L(\theta, d^t) \geq 0$. Also, Wald's theory assumes losses are in "regret" form, i.e., $\forall \theta \, \exists d^t L(\theta, d^t) = 0$.

In a typical statistical decision problem each side has infinitely many pure strategies. Wald's treatment of statistical games imposes (asymmetric) mathematical conditions on the strategy sets for the two players. These conditions prove sufficient to insure that the game has a value. But the asymmetry leads to existence of a (mixed) minimax solution for the Statistician only: only Player 2 is assured a solution, using countably additive mixed strategies.

Example 4.1. *Point Estimation of a normal mean parameter* (known variance) *with squared-error loss.* (Strictly, this problem is not treated by Wald's theory, since squared-error loss is not bounded. That mathematical detail is irrelevant to the point illustrated here, however.)

"Point estimation" is the variety of problem where, given a potential observation x, the Statistician must propose a value for the (real-valued) parameter. The Statistician's pure strategies are of the form $d{:}X \rightarrow \Theta$. Squared-error loss (to the Statistician) is the payoff $(d - \theta)^2$, understood as Nature's gain.

Let $F_\theta(x)$ be the normal distribution with mean θ and unit variance $X \sim N(\theta, 1)$. Under squared-error loss the game is determined, with value $V = +1$. The Statistician's minimax strategy, a pure strategy, is the intuitive rule $d^*(x) = x$ – posit the observed value. However, no σ-additive mixed strategy for Nature has this large a maximin value. [Note: $\forall\theta\ E_\theta\,(d^* - \theta)^2 = 1$; so , d^* equalizes "risk."] Of course, Nature can approximate the maximin value +1 using countably additive mixed strategies. Specifically, if Nature chooses the mixed strategy $\pi_n(\theta) \sim N(0, n)$, then

$$\inf_{Dt} E_{\pi_n}(d-q)^2 = n/(n+1) < 1 = V.$$

Thus, against π_n the Statistician can improve on d^* only by the amount $1/(n + 1)$.

Consider the sequence π_n of mixed (prior probability) strategies that approximate the value of the game for Nature:

$$\pi_n(\theta) \sim N(0, n), \quad n = 1, \ldots$$

The π_n strategies converge (weak-star) to a uniform measure on Θ: an "improper prior" on Θ, represented by (σ-finite) Lebesgue measure. The sequence π_n of $N(0, n)$ probabilities has subsequences converging (weak-star) to diffuse f.a. probabilities p^*: distributions that assign zero probability to each unit interval, $\forall k p^*(k \le \theta \le k + 1) = 0$. Moreover, $\inf_{Dt} E_p{}^*(d - q)^2 = 1$. So, (p^*, d^*) solves this estimation game. Nature needs to play a (diffuse) finitely additive, mixed strategy to achieve its maximin value for the game.

V. CONCLUSIONS AND OPEN QUESTIONS

First we address a question about what might be intended by a f.a. mixed strategy. We know of five philosophical considerations that lead some to the use of f.a., rather than σ-additive probabilities. In increasing importance for our discussion here, they are:

1. Measurability precludes non-trivial, σ-additive probabilities on uncountable sets (Ulam, 1930). As is well known, a σ-additive probability can be extended to the power set provided the extension may be only finitely additive (as follows from the Hahn-Banach theorem). DeFinetti (1972, p. 201) and Dubins and Savage (1976, p. iii) have voiced this theme in justification of f.a. probability. In rebuttal, we point to the important result of Solovay (1970) which shows that every subset of the real line may be Lebesgue measurable if the Axiom of Choice is weakened and a suitable large-cardinal is introduced. There is more than one way to solve the measurability question.
2. Limits of relative frequencies (over infinite sets of possible outcomes) are not generally σ-additive measures. In rebuttal to this familiar observation there is the position (often voiced by "Bayesian" statisticians) that probability is not a limit of relative frequencies – the limiting frequency interpretation is not a useful one.
3. Some, important decision theoretic treatments of personal probability do not require more than finite additivity. For example, deFinetti's (1974) theory of coherent previsions and Savage's (1954) normative theory of preference allow for (merely) finitely additive personal probabilities, based on principles of rational choice.
4. Familiar (classical) statistical techniques often have Bayesian "models" that use diffuse finitely additive "prior" probabilities. These take the form of "improper" priors, e.g., in Jeffreys' (1971) theory.
5. Wald's treatment of statistical games often leads to maximin strategies for Nature which are purely f.a., as illustrated by Example 4.1 (above). The same example suggests that diffuse maximin strategies may be approximated by countably additive mixed strategies, as the π_n approximate p^*. Recall, however, Wald's game (Example 1.1), "Pick the Bigger Integer," and our fair diffuse minimax (and

maximin) solutions to the fully symmetric version, $E_{r0.5}$. No countably additive strategy is a good approximation to these. Nonetheless, the infinite game may be approximated by a sequence of trivial, symmetric finite games – "Pick the Biggest Integer $\leq k$." Obviously, each of these trivial, fair games is solved with a pure, point-mass strategy, $\{k\}$. But (ignoring the difference between choosing k for certain and choosing k with probability 1), each (weak-star) limit of this sequence of pure-strategies is a diffuse probability, corresponding to our proposal for f.a. solutions to the infinite game. Thus, the f.a. solutions to the infinite game "Pick the Bigger Integer" are approximated by θ-additive solutions to finite approximating games, rather than being approximated by σ-additive strategies within the infinite game. This observation leads us to our first open question:

Issue 1. How (and when) can f.a. minimax strategies be approximated by σ-additive ones? When is the approximation by a sequence of σ-additive strategies within the infinite game (as is possible in Example 4.1, but not in Example 1.1) and when is the approximation by a sequence of bounded games (as is possible in both examples)?

We conclude with several questions abut the adequacy of our mathematical approach to solving games using f.a. probabilities.

Issue 2. How generous is Condition $\mathcal{A}$? We have not indicated (because we do not know) when an infinite game satisfies $\mathcal{A}$. It may be worthwhile to investigate this question, even if all strategies are required to be σ-additive, as in traditional game theory. We say this because the minimax theorem, Theorem 2.3, obtains for countably additive game theory. Lemmas 2.2–2.4 apply when P and Q are sets of σ-additive probabilities. (Specifically, the f.a. measure μ of Lemma 2.2. appears there as a computational device for defining the integral $\vartheta(p)$ and the set $I^{\varepsilon}(p)$.) Of course, Condition $\mathcal{A}$ does not obtain in Wald's game, Example 1.1, when strategies are σ-additive; but it does obtain when strategies are f.a. probabilities. We would like to understand the circumstances that make $\mathcal{A}$ hold.

Issue 3. What are the mathematical entities introduced in the metric completion of Theorem 2.4? We mean to ask both: (i) When are the spaces $P \times Q$ metrically complete (for $\rho_{P\times Q}$)? and (ii) Are the minimax "strategies" formed by the completion of $\rho_{P \times Q}$ f.a. probabilities over some interesting algebra of events?

We hope to address these topics in our future work.

REFERENCES

deFinetti, B. (1972). *Probability, Induction, and Statistics.* New York: Wiley.

deFinetti, B. (1974). *The Theory of Probability* (2 volumes). New York: Wiley.

Dubins, L. E. (1975). "Finitely Additive Conditional Probabilities, Conglomerability and Disintegrations," *Annals of Prob.* **3**, 89–99.

Dubins, L. E. and Savage, L. J. (1976). *Inequalities for Stochastic Processes: How to Gamble If You Must.* New York: Dover.

Dugundji, J. (1968). *Topology.* Boston: Allyn and Bacon.

Dunford, N. and Schwartz, J. T. (1958). *Linear Operators. Part I: General Theory.* New York: Interscience.

Fan, K. (1953). "Minimax Theorems," *Proc. National Acad. Sci.* **39**, 42–47.

Glicksberg, I. (1959). "Stone-Cech Compactifications of Products," *Trans. Amer. Math. Soc.* **90**, 369–382.

Heath, D. and Sudderth, W. (1972). "On a Theorem of deFinetti, Oddsmaking, and Game Theory," *Annals of Math. Stat.* **43**, 2072–2077.

Jeffreys, H. (1971). *The Theory of Probability*, 3d edition. Oxford: Oxford University Press.

Karlin, S. (1950). "Operator Treatment of the Minimax Principle," in *Contributions to the Theory of Games* (H. W. Kuhn and A. W. Tucker, eds.), pp. 133–154. Princeton: Princeton University Press.

Karlin, S. (1959). *Mathematical Methods and Theory in Games, Programming, and Economics*, vol. 2. Reading: Addison-Wesley Publishing Co.

Kindler, J. (1983). "A General Solution Concept for Two-Person Zero Sum Games," *J. Optimization Theory and Applications* **40**, 105–119.

Kretkowski, R. and Telgárski, R. (1983). "On Totally Bounded Games," *Math. Solovaca* **33**, 381–387.

Kuratowski, K. (1968). *Topology*, vol. 1. New York: Academic Press.

Savage, L. J. (1954). *The Foundations of Statistics.* New York: Wiley.

Schervish, M. J., Seidenfeld, T., and Kadane, J. B. (1984). "The Extent of Non-Conglomerability of Finitely Additive Probabilities," *Z. Wahrscheinlichkeitstheorie* **66**, 205–226.

Solovay, R. M. (1970). "A Model of Set Theory in Which Every Set of Reals Is Lebesgue Measurable," *Ann. of Math.*, 1–56.

Ulam, S. (1930). "Zur Masstheorie in der algemeinen Mengenlehre," *Fund. Math.* **16**, 140–150.

Ville, J. (1938). "Sur la théorie générale des jeux où intervient l'habilité des joueurs," in *Traité du Calcul des Probabilités et de ses Applications* (E. Borel et al.), vol. 2, Part 5, pp. 105–113. Paris.

von Neumann, J. and Morgenstern, O. (1947). *Theory of Games and Economic Behavior*, 2d edition. Princeton: Princeton University Press.

Wald, A. (1950). *Statistical Decision Functions*. New York: Wiley.

Yanovskaya, E. B. (1970). "The Solution of Infinite Zero-Sum Two-Person Games with Finitely Additive Strategies," *Theory of Prob. with Applications* **15**, 153–158.

Yoshida, K. and Hewitt, E. (1952). "Finitely Additive Measures," *Trans. Amer. Math. Soc.* **72**, 46–66.

3.4

Randomization in a Bayesian Perspective

JOSEPH B. KADANE AND TEDDY SEIDENFELD

> Applying the theory (of personal probability) naively one quickly comes to the conclusion that randomization is without value for statistics. This conclusion does not sound right; and it is not right. Closer examination of the road to this untenable conclusion does lead to new insights into the role and limitations of randomization but does by no means deprive randomization of its important function in statistics.
>
> L. J. Savage (1961)

> Though we all feel sure that randomization is an important invention, the theory of subjective probability reminds us that we have not fully understood randomization. . . . The need for randomization

This research was sponsored in part by the Office of Naval Research under Contract N00014-85-K-0539 and in part by the Ethics and Values in Science and Technology Program of the National Science Foundation and the National Endowment for the Humanities under Grant ISP-8116810. Other scholars working on the latter grant are: David Kairys, Ken Schaffner, and Nell Sedransk, advised by Thomas J. J. Blanck, Eugenie S. Casella, Jack Coulehan, Alan Meisel, Preston Covey, A. John Popp, Jerome J. DeCosse, John C. Ruckdeschel, Arvin S. Glicksman, Katheryn D. Katz and Rachelle Hollander. The test case at Johns Hopkins University is being conducted by Drs. Thomas J. J. Blanck and Eugenie S. Casella. The authors are grateful for the helpful comments of Morris DeGroot, Persi Diaconis and Dennis Lindley. While all these have contributed ideas that may appear here, none but the authors should be held responsible.

Reprinted from *Journal of Statistical Planning and Inference*, Volume 25, Joseph B. Kadane and Teddy Seidenfeld, "Randomization in a Bayesian Perspective," pp. 329–345.

presumably lies in the imperfection of actual people and, perhaps, in the fact that more than one person is ordinarily concerned with an investigation.

L. J. Savage (1962)

Randomization has thus been a puzzle for Bayesian theory for many years. In this essay, we give our current views on this subject.

There are two principal arguments for randomization that we are familiar with. The first is to support a randomization-analysis of the data. This notion goes back to Fisher, and is exposited in a series of papers by Kempthorne (1955, 1966, 1977). It asks whether what is observed is surprising given all the other designs that might have been randomly selected and data that might have been observed, but were not. By its appeal to what did not occur, such an analysis violates the likelihood principle; hence, it is not compatible with Bayesian ideas. Many have criticized randomization-analysis for failing the likelihood principle: see Basu (1981) and Bunke and Bunke (1978) for illustrations. This reply to Fisher, Kempthorne, and others who defend randomization-analysis is, we think, what Savage means by the "naive" Bayesian rejection of randomization.

We explore the argument for randomization-analysis in detail in Section I. It is our purpose there to distinguish two cases. In one (illustrated with Example A), the randomized inference depends upon what is wholly irrelevant evidence according to the likelihood principle, evidence that we define as "suppressible". In the second case (Example B), the randomized inference ignores ancillary information which is *not* suppressible. The two cases are different ways of violating the likelihood principle with a randomization-analysis. Case 2, the theme of validating randomization by ignoring ancillary but not suppressible evidence, is central to our subsequent discussion of how randomized experimental designs are justified (Example C).

The second of the two arguments we know for randomization is that, in design, it is thought to provide methodological insurance against a variety of observer "biases". In Section II we explain what bias is about, why, and *for whom* it is a problem. The latter question can be motivated with the aid of the following:

Randomization amounts to deciding by some random device, whose outcomes are out of the control of the researcher-analyst, which units to study, say which people to interview in a survey, or which treatments to assign to each patient in a clinical trial. Thought of as a decision, each such choice or assignment has some expected utility (see Lindley (1972) for a clear exposition of how that expected utility is calculated). If one choice has higher expected utility than all others, why not choose it? If many are equally good, choosing randomly among them is optimal, but so is choosing with certainty any of the optimal choices. Thus, it would appear that randomization is always unnecessary and sometimes suboptimal. Without loss of utility, why then cannot decisions be chosen without randomization?

The hallmark of this reasoning, in our opinion, is that it applies to what we call "experiments to learn". That is, the objective of the researcher is to inform himself or herself. In such a case, which is typical for example of pilot studies in many disciplines, it is indeed unnecessary and quite possibly suboptimal to randomize. *When only one decision-maker is relevant*, we accept this analysis and would not randomize.

Observer bias, however, seems to us to concern a different goal for experiments, which we call "experiments to prove". This, we think, is what Savage intends as the "closer examination". With "experiments to prove", several decision-makers are involved. In Section III we discuss some Bayesian facts about decisions involving more than one decision-maker. And, in Section IV, we consider randomization in this light. We conclude that randomization has no particular merit in creating "evidence to prove". Alternative, non-randomized designs are available for this purpose. In Section V we illustrate this for clinical trials, where ethical considerations can be given priority over unrestricted randomization without creating a "biased" design.

I. THE "NAIVE" BAYESIAN THEORY OPPOSES THE SUPPRESSION OF EVIDENCE

The "naive" application of the theory of personal probability to which Savage refers is, we suspect, the appeal to "ancillarity" as a refutation of randomization. That argument proceeds as follows.

Suppose hypotheses of interest are indexed by a parameter θ and there are new data d. If we use Bayes' rule (conditionalization) to update our "prior" opinion about θ, $p(\theta)$, to a "posterior" probability given the new data d, $p_d(\theta)$, then the "posterior" is proportional to the product of the "likelihood" and "prior":

$$p_d(\theta) \propto p(d|\theta)\,p(\theta).$$

A statistic t is called *ancillary for* θ if its likelihood is constant, i.e., if

$$p(t|\theta) = p(t),$$

independent of θ. Then, by Bayes' rule, ancillary data are irrelevant:

$$p_t(\theta) \propto p(t|\theta)p(\theta) = p(t)p(\theta) \propto p(\theta).$$

If the new data d can be written as a conjunction of the two statistics s and t, $d = (s, t)$, with t ancillary for θ, then (by Bayes' rule) the updated probability for θ depends only on the (conditional) likelihood for s, given t and θ:

$$\begin{aligned} p_d(\theta) \propto p(d|\theta)p(\theta) &= p(s|t,\theta)p(t|\theta)p(\theta) \\ &= p(s|t,\theta)p(t)p(\theta) \\ &\propto p(s|t,\theta)p(\theta). \qquad (*) \end{aligned}$$

Thus, whatever there is of relevance in data d to hypotheses θ is contained in the statistic s, given the ancillary data t.

Define t to be *suppressible* in the presence of s for θ if t is ancillary and also

$$p(s|t,\theta) = p(s|\theta), \quad \text{independent of } t.$$

Then we have that the data d may be contracted to s, without loss of relevant evidence about θ since

$$\begin{aligned} p(\theta|s,t) &\propto p(s|t,\theta)p(\theta) && \text{as } t \text{ is ancillary} \\ &\propto p(s|\theta)p(\theta) && \text{as } t \text{ is suppressible} \\ &\propto p(\theta|s) && \text{by Bayes' theorem.} \end{aligned}$$

That is, when t is suppressible in the presence of s, s is sufficient for θ:

$$\begin{aligned} p(t|s,\theta) &= p(s|t,\theta)\, p(t|\theta)/p(s|\theta) && \text{by Bayes' theorem} \\ &= p(s|t,\theta)\, p(t)/p(s|\theta) && \text{since } t \text{ is ancillary} \\ &= p(t) && \text{as } t \text{ is suppressible.} \end{aligned}$$

Thus, $p(d|s, \theta) = p(d|s)$ independent of θ, as required for sufficiency.

We may apply this directly both (i) to overturn "orthodox" procedures based on randomized data analysis, and (ii) to refute the "classic" (Fisher's) argument: that randomization in experimental design offers methodological insurance against a biased sample. Let us illustrate each of these criticisms. In Example A below, the random digit y is suppressible. However, in Example B, the random pairing is ancillary but not suppressible.

There are two familiar uses of randomization in "orthodox" (Neyman–Pearson) testing and interval estimation. (A) It can be the basis of a "most powerful" test. That is, the "best" test (at a fixed size, $\alpha = \alpha_0$) may be a mixed test. (B) Second, a randomized test may have an exact size whereas no non-randomized one does; hence, in such problems, only "mixed" confidence intervals have exact "coverage" probabilities.

IA. *Randomization May Improve the Power of a Test by Simulating Continuity for Discrete Random Variables*

Example A. Let x be a random quantity with $x \in \{0, 1, \ldots, 10\}$. Consider tests of a simple null hypothesis h_0 versus a simple rival hypothesis h_1. Under the null hypothesis the distribution of x, $p(x|h_0)$, is:

$$p_0(0) = 0.05; \quad p_0(n) = n(0.019)/1.1 \quad (n = 1, \ldots, 10).$$

Under the rival hypothesis the distribution of x, $p(x|h_1)$, is:

$$p_1(0) = 0.5; \quad p_1(n) = n(0.01)/1.1.$$

When $\alpha = 0.02$ (the probability of a "type 1" error), the only "pure" test (based on an observation of x) meeting the size restriction rejects h_0 if $x = 1$. Its "power" is a paltry 0.009, so that this pure test is "biased". (One is more likely to reject the null hypothesis when it is true than when it is false!)

Suppose, instead, you can augment the observation, x, with a randomly chosen digit $y \in \{0, \ldots, 9\}$. Then there is a mixed test ($\alpha = 0.02$), with "power" 0.2: reject h_0 provided $x = 0$ and $y \in \{0, 1, 2, 3\}$. (This is the best $\alpha = 0.02$ test, mixed or pure, based on x.) The increase in power results from using the randomizer to simulate a continuous random quantity z with the same likelihood as x. Then, by the Neyman–Pearson lemma, the best test with z is a likelihood ratio test, i.e. let the rejection region consist of z-outcomes with lowest likelihood ratio, $p(z|h_0)/p(z|h_1) < d$. In the example, this region corresponds to $x = 0$.

Of course, with data (x, y), the random digit y is ancillary to the hypotheses of interest. By the result (*) above, from a Bayesian point of view only the conditional likelihood ratio matters:

$$p(x|y, h_0)/p(x|y, h_1) = p(x|h_0)/p(x|h_1).$$

Thus, the random digit y is suppressible in the presence of x.

IB. *Randomization May Provide a Basis for Exact Confidence Levels by Bypassing "Nuisance" Factors*

Example B. Let x_i be a sequence of n (identically, independently) normally distributed random quantities, $N(\mu_x, \sigma_x^2)$. Likewise, let y_i be a sequence of n (i.i.d.) normal $N(\mu_y, \sigma_y^2)$ quantities. The parameter of interest is the difference in population means, $\delta = \mu_x - \mu_y$, and the (unknown) population variances (σ_x^2, σ_y^2) are "nuisance" factors. Use random numbers from the set $\{1, \ldots, n\}$ to randomly match the x's and y's. This forms n-pairs (x_j, y_j) $(j = 1, \ldots, n)$. Define the n-differences $z_j = x_j - y_j$. Then the random variable z is normally distributed, $N(\delta, \sigma_z^2)$, where $\sigma_z^2 = \sigma_x^2 + \sigma_y^2$. An exact confidence interval for δ is created from the familiar t-test based on the data z_j. (This analysis corresponds to a permutation-test solution of the Behrens–Fisher problem.) Thus, with the aid of randomization, the "orthodox" analysis bypasses the "nuisance" factor σ_x/σ_y.

This randomized t-test is not valid from the Bayesian point of view and for two reasons. One objection, which is not germane to our discussion of randomization, is the fact that the reduction of the data (x_j, y_j) to the paired differences z_j fails to preserve all the relevant information about δ.

For instance, assume the (x_j, y_j) have a bivariate normal distribution $(\rho = 0)$ and the familiar "improper" prior $(\propto 1/\sigma_x\sigma_y)$ is used in order to reproduce the t-test analysis. Still, we find that the posterior distribution based on all the data differs from the posterior distribution based on just the paired differences, $p(\delta|x_j, y_j) \neq p(\delta|z_j)$. The difference amounts to a reduction by $\frac{1}{2}(n-1)$ in the degrees of freedom of the estimate for $\sigma_x^2 + \sigma_y^2$ as provided by the z_j-data. Put another way, $p(\delta|x_j, y_j)$ is a function of both the paired sums $w_j = x_j + y_j$ and the paired differences, z_j. (See DeGroot (1980) for an interesting analysis of the inference problem about an unknown ρ, given "broken" pairs.)

Let us sidestep this objection and, in the fashion of Lindley's (1965, Vol. 2, p. 84) analysis, consider the reduction from data (x_j, y_j) to data z_j an acceptable approximation – as it becomes with increasing n. The objection to the randomized t-test that we focus upon is concerned with the *suppression* of the ancillary pairing that results when the data are contracted to the paired differences, z_j, alone.

The t-test using the z_j takes the form $(n-1)^{1/2}(\bar{z}-\delta)/s_z$. That is, the t-test uses the further reduction of the data to the two statistics $(\bar{z}, s_z^2)$. Alternative (random) pairings of the original data leave $\bar{z}$ unchanged but alter the variance, s_z^2. There are $n!$ pairings of the data and thus (barring ties) there are $n!$ possible variances as rank ordered by the relative magnitude of the resulting s_z^2. Randomization picks out one of these, with probability $1/n!$, yielding the ancillary data $t = \text{rank}(s_z^2)$. That is, the evidence available includes both $s = (\bar{z}, s_z^2)$ and the ancillary statistic, t, of which one among the $n!$ possible s_z^2-ranks was selected.

According to the result (*) above,

$$p(\delta \mid s, t) \propto p(s \mid \delta, t)\, p(\delta).$$

However, $p(s \mid \delta, t) \neq p(s \mid \delta)$, and t is not suppressible. Specifically, $p(s \mid \delta, t)$, but not $p(s \mid \delta)$, involves the nuisance factor σ_x/σ_y. That is, the randomized t-test analysis, based on $p(s \mid \delta)$, is invalid because it mistakenly takes the ancillary, randomized choice of t to be suppressible. Hence, using the randomization to eliminate the nuisance parameter is unwarranted from this Bayesian point of view.

Let us illustrate our analysis in the simple case of two observations from each population, $n = 2$. There are two, alternative pairings of the x's and y's. In the one case we get the smaller of the two possible s_z^2, to wit, with $\text{rank}(s_z^2) = 1$, we have the "t-statistic":

$$[(x_1 + x_2) - (y_1 + y_2)] / \big| |x_1 - x_2| - |y_1 - y_2| \big|.$$

And with $\text{rank}(s_z^2) = 2$, using the larger of the two possible s_z^2, we have:

$$[(x_1 + x_2) - (y_1 + y_2)] / [|x_1 - x_2| + |y_1 - y_2|].$$

Unfortunately, the distributions of these statistics depend upon the nuisance factor σ_x/σ_y. In particular, the second of these corresponds to the Behrens–Fisher statistic. See, for example, Fisher's discussion (1973, pp. 97–103) for its exact distribution.

Thus, given the ancillary statistic t, the posterior distribution $p(\delta \mid s, t)$ remains a function of the agent's beliefs about the unknown ratio σ_x/σ_y. From a Bayesian point of view, the randomization does not succeed in eliminating the nuisance parameter. A Bayesian can use the permutation test as a solution to the problem of Example B only by censoring the (ancillary) datum t. We find it is counterintuitive to adopt a strategy to willfully censor evidence which, though ancillary, is not suppressible.

I. J. Good (1971, #679) observes that often a Bayesian can make sense of "orthodox" statistical procedures by avoiding parts of the data. In this case Good could employ his "Statistician's Stooge" to carry out the random pairing for him, but to report (to Good) only the values z_j. (A related idea is found in Levi's (1980, Chapter 17) use of "data as input", rather than "data as evidence".) That would allow Good to respect the total evidence principle while offering him the convenience of the permutation test analysis. Next, we explore the censoring of ancillary but not suppressible data in connection with our discussion of the role of randomization for sound experimental design.

IC. *Randomized Designs as Methodological Insurance Against a "Biased" Sample*

In his pedagogically influential discussion of an experiment for testing an hypothesis that a Lady can taste the difference between tea made with milk added first (rather than second), Fisher (1971, Chapter 2) raises three methodological concerns to which randomization is the supposed remedy:

1. By presenting the Lady with 8 cups (4 prepared in each way) *in a random order*, the researcher is (supposed to be) assured that the Lady's success is reasonably attributed to her professed ability to taste differences and is not the result, instead, of her ability to anticipate the pattern of cups when that is left to the experimenter's imagination.
2. By randomly assigning each preparation to four of the eight cups, the study is (supposed to be) protected against a confounding of uncontrolled factors with what is tested for. For example, suppose the Lady reacts to small differences in the cups rather than to the preparation of the tea. Then not knowing which are the relevant cup characteristics is of no concern since, when randomized, the chance is negligible that just those cups to which she would react "milk-first" get milk-first.
3. By randomly assigning treatments to cups the experimenter is (supposed to be) warranted in adopting a particular statistical distribution corresponding to the "null" hypothesis. And for problems

where the "alternative" hypothesis is well-formed, the randomization justifies adopting a particular statistical model for the experimental outcomes. Accordingly, the probability is 1/70 that the Lady correctly identifies all four "milk-first" of the eight randomly prepared cups, given the null hypothesis that she cannot discriminate (and that she knows four of eight were so made).

Observe that, according to Fisher's randomization solution, the three problems are treated exactly the same in their first- and third-person interpretations.

Under randomization, the researcher and the reader each believes the Lady's success at discerning the two kinds of tea infusions is separated from her ability to anticipate subconsciously preferred patterns – for the simple reason that the allocation scheme used is known to be a mechanism without preferences. Of course, when there can be no game between researcher and subject, e.g., when the responses are involuntary (because the subjects are plots of land with no interest in the outcomes), this problem doesn't even arise.

Similarly, the researcher and the reader are to agree that, if the cups are randomized between the two preparations and if the Lady has no real ability to discriminate "milk-first" tea, then:

- "uncontrolled" factors are uncorrelated with her accuracy – "nuisance" factors are eliminated from consideration;
- and there is an established chance of 1/70 that she correctly identifies all four cups prepared with milk-first – there is consensus on a simple statistical model for her responses.

Unfortunately, these arguments are invalid once the researcher (or the reader) is made aware of the (ancillary) outcome of the randomization. That is, just as in Example B, these analyses are cogent (only) prior to observing the particular allocation arrived at by randomization. If the random allocation directs serving all four cups of one kind first, or directs putting the "milk-first" tea into the cups which are seen to be the brightest in color, is there any ground for continued agreement on the (conditional) odds of 1/70 that the Lady correctly identifies the tea (given this allocation and the null hypothesis)? Even if the researcher can convince himself these factors are irrelevant to the Lady's performance, what features of the randomization lend credence to this judgment? What is it concerning randomization that makes such a judgment (of irrelevance of the allocation to the test outcomes) compelling for the reader? We can find none and suspect that randomization has little to do with whatever grounds there are for the belief that the allocation is irrelevant to the test results.

This, then, is the "naive" Bayesian criticism of randomization. Of course, if neither the researcher nor the reader learns the (ancillary) details of the allocation, if they know only that allocations are made "at random" and they learn the frequency data of "favorable" test responses, then there is a Bayesian interpretation of the "orthodox" claims. Then there is a Bayesian reconstruction of how randomization supports a consensus of low probability that allocations are "biased", etc. Rubin's (1978) Bayesian account of randomization is based on this fact. By *failing* to make a record of the ancillary information in a randomized allocation, both the "assignment" and "recording" mechanisms are "ignorable" (in Rubin's terms); what we call a "transparent" allocation. Also, then there is consensus on a statistical distribution for the recorded data, given the null hypothesis, since they are simplified by the elimination of "nuisance" factors – exactly as in Example B.

This is shown, as follows. Suppose the Lady will specify one of the 70 possible sets of four cups in response to the test question, "Which four are made 'milk-first'?" Let the Lady's response that it is the i-th quadruple be denoted by Q_i and let T_j designate the allocation of "milk-first" to the j-th quadruple of cups. Under the null hypothesis (that the Lady has no ability to identify which are "milk-first" cups), these are independent:

$$p(Q_i \mid T_j) = p(Q_i),$$

where p is "subjective". Then, with a randomized allocation of the two treatments, four cups prepared each way,

$$\begin{aligned} p(\text{“the lady gets all four right”}) &= \sum_i p(T_i \,\&\, Q_i) \\ &= \sum_i p(Q_i \mid T_i)\, p(T_i) \\ &= \frac{1}{70} \sum_i p(Q_i \mid T_i) \\ &= \frac{1}{70} \sum_i p(Q_i) \\ &= 1/70, \end{aligned}$$

independent of the "subjective" probability, $p(Q_i)$. A similar analysis shows that

$$p(\text{"the lady gets just three right"}) = 16/70$$

and

$$p(\text{"the lady gets just two right"}) = 36/70,$$

all in accord with the "usual" distribution for the recorded data under the null hypothesis.

Thus, if the recorded data include only the number correctly identified, while the (ancillary) statistic of which cups were treated "milk first" is censored, then, under the null hypothesis, there is a consensus about a simple statistical distribution for the experimental outcomes. Nuisance factors, such as the effect of the thickness or color of the cups on the Lady's response, are eliminated – just as σ_x/σ_y is eliminated in Example B.

In this example, the randomization-plus-censoring does not produce a consensus model under the alternative hypothesis, given that the Lady has some ability to identify "milk-first" tea. The difficulty is that the claim "some ability" remains too vague. In other settings, however, for instance in sampling problems where the quantities of interest are population frequencies for particular characteristics, randomized-sampling-plus-censoring (of the ancillary information of which items were sampled) justifies a multinomial statistical model for the recorded data of the sample frequencies.

The "closer examination" called for by Savage is, we suppose, based on the idea that sometimes it may be efficient to seek interpersonal agreement about what the data show by using a simple randomized design (with the concomitant censoring of ancillary data) instead of an experiment which maximizes what the researcher (alone) can expect to learn. In an experiment that optimizes what the researcher expects to learn, there is an obstacle to consensus (a limitation on what he can expect to prove). Then there is no *robust* Bayesian analysis of the experimental data: different "prior" opinions about "nuisance" factors (themselves unaddressed by the experiment) can interfere with a shared interpretation of experimental results. Also, if the investigator attends to the cost of calculation and deliberation, then it may be efficient *to learn* (first-person) from a randomized experiment (in which the ancillary data are unrecorded). The non-randomized design may require costly calculations to analyze its data. Rubin (1978, p. 54) put the point this way: "A comparable non-randomized design would generally be substantially more difficult to execute and analyze because of the need to deal explicitly with all covariates being balanced."

Reliance on a method that requires and encourages censoring of evidence is a problem not just because of the lost opportunity to squeeze a bit more out of the data. Such a method actively encourages researchers to avoid reporting their data and beliefs fully, for fear that the additional evidence will destroy the inference they wish to make. Our discomfort with this incentive for data censoring should not be taken to mean that we think that all details of a study must, or should, or can be reported. The details of a study that are omitted should be those that don't matter, those that are suppressible. Here, to the contrary, details are omitted because they do or might matter.

However, there exist other strategies for designing experiments that promote a consensus of posterior probabilities and which avoid randomization and censoring data: we discuss one of these in Section v. Moreover, the alternative designs are ethically superior in the setting of clinical trials, as we argue below. Thus, even a sophisticated (rather than naive) Bayesian defense of randomization, one which emphasized the goal of experiments "to prove" rather than "to learn", fails to establish it as a sine qua non of sound experimental methods.

II. OBSERVER BIAS

Suppose that you are the reader of a paper I have written about the comparison of two treatments for a psychological ailment. Suppose also that I am the inventor and proponent of treatment A, which, in my study, did better than treatment B (I am pleased to report). Suppose also that patients who have the ailment severely are known to be very difficult to treat, but that patients who have the ailment only mildly are quite tractable. What might you make of my results? How can I make a study that will be persuasive to you? If I could decide which patient went into each treatment, the possibilities for mischief are huge. Even if I assigned patients to whatever treatment I think will be best for them (a not uncommon, nor unworthy thing to do), how can I (or, what is even more difficult, you) sort out what happened afterward, especially in the circumstance that a perfect measure of severity of illness is unavailable? How can I (or, what is even more difficult, you) be sure that I have not essentially built the apparent superiority of treatment A into the study by how the patients were chosen? This is the claimed role for randomization, and it is one that deserves to be taken seriously.

We have built into the example above the motive of being a proponent of treatment A. But this is really unnecessary to the argument. If it is, as the Bible tells us, the truth that will make us free, as researchers it is the surprising truth that will make us famous. Thus each researcher may have, or may be suspected of having by a skeptical and critical reader, a motive to exaggerate the truth. We are not writing here of deliberate and flagrant falsification, but of the much more subtle ways in which we can fool ourselves, and try to fool others.

Thus to leave the assignment of patients to treatments to the researcher seems perilous. One method around this is randomization, which would assign patients to treatments with a probability mechanism outside the control of the researcher. And this is what is meant, as we understand it, by using randomization to avoid bias. Nothing in this argument suggests that randomization is unique in this; for example designs that maximize balance might also deal with observer bias.

There are two ways that we see to view this argument. One concentrates on the researcher, and says that because he or she may not be aware of the implicit biases inherent in the experimental procedures used, randomization, by relieving the experimenter of the responsibility, can aid good design. This argument is within "experiments to learn", but relies on the inability of the researcher to know his or her own mind. A second view of observer bias involves more than one decision-maker, at least an experimenter and a reader, who may have different beliefs. Consequently this argument is in the "experimentation to prove" category. In order to analyze it in a Bayesian perspective, we first review some literature on the Bayesian analysis of problems involving more than one decision-maker.

III. BAYESIAN ANALYSIS OF DECISIONS INVOLVING MORE THAN ONE DECISION-MAKER

The Bayesian literature on problems that involve two or more decision-makers stems from the same principles that animate the analysis of one-decision-maker problems. Given that I must make a decision among various alternatives, my outcome, that is, the utility of my action, depends on what you do. I do not know what you will do, but, as a Bayesian, I do have a probability distribution over your possible acts. Then, not surprisingly, the Bayesian norm is that I should decide in such a way that I maximize my subjective expected utility, where here the expectation is over the probability distribution generated by my uncertainty about your behavior.

It is important that I do not have to model your behavior as if you were a Bayesian. As is usual in the Bayesian paradigm, I am entitled to think whatever I think about you. I might think that you are a Bayesian, with a known probability distribution and a utility function uncertain to me, but in my opinion, with probability $\frac{1}{3}$ it is absolute error and with probability $\frac{2}{3}$ it is squared error. Then I am saying that with probability $\frac{1}{3}$ you will choose the median of this known probability distribution, and with probability $\frac{2}{3}$ you will choose the mean. Of course, I may be quite wrong about how you will decide, but this is an ever-present risk to a Bayesian. Adherence to the axioms only guarantees coherence (Lindley, 1972), a kind of consistency, but not correctness. But I may, equally as coherently, believe that you do not make your decisions in a Bayesian way. I must still have a probability distribution over your actions, but that is the only constraint. In fact, for the purpose of making my decision optimally, all that matters is my probability distribution, and not the underlying theory that led me to it (and of course, my utility function).

While seemingly innocuous, this principle has interesting consequences in the social sciences. Many social scientists apparently believe that uncertainty about the actions of other people is fundamentally different from other kinds of uncertainty, and that other rules of optimal behavior ought to apply. For some flavor of this debate, see Kadane and Larkey (1982, 1983), Harsanyi (1982) and Aumann (1987). The debate has been particularly sharp in game theory. One of the games that most sharply raises the issue is the prisoner's dilemma. In this two-player game, each player faces a choice between two alternatives, a cooperative choice and a competitive choice. If both choose to cooperate, both are rewarded. If both choose to compete, both are punished. But if one chooses to compete and the other to cooperate, the former is greatly rewarded while the latter is severely punished. In a single play of this game, both traditional game theory as inherited from von Neumann and Morgenstern and Bayesian game theory prescribe the competitive response. However, in multiple-play games, only the traditional theory recommends competition, while Bayesian game theory, taking account as it must of the likely effect of my action now on your action later, can recommend either move, depending on the prior employed. See Wilson (1986) for more on the Bayesian approach to the prisoner's dilemma. A very different game analyzed in the Bayesian framework can be found in DeGroot and Kadane (1983). Important discussion has also been undertaken in the economics literature under the label of the principal–agent problem (Pratt and Zeckhauser, 1985). Actual modelling of real decision-making in a clinical trial, for example, is enormously complicated, and is still in its infancy.

IV. BAYESIAN ANALYSIS OF OBSERVER BIAS

In an "experiment to prove", there are at least two actors: the reader and the author. Applying the theory of Section III, the reader may model the author as a Bayesian whose utility reflects a desire to prove a hypothesis, and who knows more about the phenomenon at hand, having studied it more closely; the experimental design becomes a legitimate source of worry to the reader. Consequently the author must explain just how the sample was chosen, or just how patients were assigned treatment. A method that allows the author's judgment to enter the design then imposes on the reader a substantial cost of elicitation and analysis. A reader who is sufficiently interested could do such an analysis, but it is wise of an author to try to arrange things so that such an effort is not required. This would reduce the entry cost of reading the work, and consequently encourage readership. Being read is also an element in becoming famous.

All that is required to achieve this simplicity of analysis, however, is that the author not be able to control the design once it is set in motion; that is, the author should not be able to achieve the goal of being able to assign the less sick patients to the favored treatment. Randomization is one available method to do this, but it is not unique in this respect. There are other designs that can do this.

We have been preceded by Stone (1969) in finding that a Bayesian analysis of randomization requires more than one decision-maker. However, Stone's analysis and conclusions differ from ours in several important respects. Stone presupposes a Bayesian B who is doing the study, and another Bayesian A who is reading it, in a finite sampling context. Bayesian B reports his prior π_B, as well as the results of his sampling experiment. Stone then wonders how Bayesian A is to use the information contained in π_B to analyze the data. Stone argues that only random sampling performs this function:

If B selects his n units by simple random sampling and A knows this, A's problems are resolved. A then can and should use his own prior π_A in conjunction with the . . . likelihood function. Note that simple random sampling is the only sampling scheme that performs this resolution perfectly. If B used a random sampling scheme specified by probabilities $\{P(S) | S$ any sample of size $n\}$ with $P(S)$ dependent on S, A would be back in a position of uncertainty about how to deal with the information in the different $P(S)$ values, no matter how close these values were to constancy. Such is the argument for simple random sampling, requested by A and understood by B.

Within this argument, the sample size n chosen by B may well carry information about his prior and utility on the parameters of interest. Consequently even a randomized trial with the sample size chosen by B in an unknown way leaves the reader A in the very position of uncertainty that Stone claims randomization avoids.

By contrast, our argument is that any mechanism that is transparent to A in that it makes B's choice of sample depend only on measured covariates in a functional or probabilistic way will suffice to remove both π_B and B's utility function from A's inference problem. Thus, we argue that many other designs beyond simple random sampling achieve the same protection from observer bias as does randomization. This would be of only theoretical interest were it not for the existence of practical situations in which this consideration is important, one of which is described in Section v.

V. DESIGN OF A MORE ETHICAL CLINICAL TRIAL

While randomization may have the advantage of familiarity to statisticians, it can have other, counterbalancing disadvantages. One area where those disadvantages are especially acute occurs in clinical trials of new medical treatments on people. Here, the proposition is to be put to patients that their treatment is to be decided by a random device. In the United States, a patient must be informed of all information that reasonably bears on the decision to participate in a clinical trial. Many trials now make clear how the assignment of treatments is to be done, and it is quite possible that all informed consent statements will be required to reveal this information. What might a knowledgeable patient make of such information?

Many potential patients who have been trained in statistics and the making of optimal decisions under uncertainty might respond to such a proposal by saying in effect, "Doctor, you know about me and about my disease. You must have a hunch about which treatment would be better for me. Please give me that treatment and forget about flipping coins." This line of argument, which is essentially the same as the experiments-to-learn argument against randomization, indicates that such randomization is actively harmful to the patient except in that rare case in which the expected utilities of the two treatments are exactly balanced. To make this point is not to advance the wise treatment of patients, however, if the upshot is to prevent carefully designed clinical trials from being conducted. Thus the burden of proof is on the critic to propose a better system.

A system that purports to be an improvement was proposed in Kadane and Sedransk (1980). It involves the following steps:

1. The appointment of a small number of experts on the disease and treatments under study.
2. Agreement in that group on the single indicator of outcome most reasonably of concern to a patient in the trial.
3. Agreement on a few measurable diagnostic indicators possibly linked to outcome as measured in (2) above.
4. Elicitation of each expert's opinion about the outcome indicator as a (probabilistic) function of the diagnostic indicators and treatment.
5. Agreement on a likelihood function to update the expert's opinions as data become available.
6. Actual updating of those opinions in a computer.

Thus when a new patient is to be assigned, an updated opinion is available that arguably represents each expert's current opinion. We call a treatment recommended by at least one (updated) expert admissible. Different sets of treatments might be admissible at any given time for patients that differ in the values of their diagnostic variables, and of course the set of admissible treatments changes over time as data are collected. The fundamental ethical stance proposed is that patients be assigned only admissible treatments. Within this constraint, treatments could be assigned in many ways, including randomly. However, the restriction to admissible treatments, which depends on the expert's priors (as updated by the data collected to date), is different from classical randomization which allows no such dependence. An exposition of the reasoning behind these ideas, with stress on the ethical side, can be found in Kadane (1986).

A group of doctors, lawyers, philosophers and statisticians met in the late 1980s and 1990s to examine and refine these ideas. The results of these meetings were published in Kadane (1996). Due to a fortunate happenstance, it became possible to do a test case of this method at Johns Hopkins Hospital. The trial compares verapamil and nitroprusside infusions as treatment for hypertension after separation from cardiopulmonary bypass during cardiac surgery. There are five experts from a variety of relevant specialties, only one of whom is from Johns Hopkins. Four independent variables were used to predict a function of the patient's blood pressure just after open heart surgery. As a function of these four independent variables, a normal linear model was proposed. The expert's priors were elicited using the methods of Kadane et al. (1980). The design, after taking account of the restriction to admissible treatments, stresses balance among the independent variables (Sedransk, 1973).

The nature of this design is that it is a compromise between the desire to respect better the right of a patient to the best available treatment, and the need for interpretable data. Since it is a compromise, it is liable to attack from both sides, those who feel that patients should have the right to choose their own treatment, and those who feel that only an unconstrained randomization can be allowed scientifically. This paper is aimed at the second group, and bases its appeal on the fact that unlike patient or physician choice, the design used here is a known, albeit complex and dynamic, function of measured patient characteristics. Consequently, if the analysis respects the necessary conditioning on the four independent variables representing those measured patient characteristics, the treatment assigned carries no information about the parameters of interest, which measure the effectiveness of the treatments.

Stone asserts that only randomization permits the reader (Bayesian A) to ignore the experimenter's (Bayesian B's) priors and utility functions. We argue that it is a useful property of a design that the Bayesian reader need not model the author's prior and utility: but we disagree with Stone that only classically randomized designs achieve this goal. The clinical trial, discussed above, does not require such modelling on the reader's part, whether or not it uses randomization to assign admissible treatments to patients.

VI. CONCLUSION

In view of these comments, what are we to make of Savage's conjecture that the Bayesian explanation of randomization must lie in human imperfection and that more than one decision-maker is involved? With the latter, we heartily agree. However, the former seems to us too harsh. The desire that the analysis of an experiment not involve a judgment of the motivation of the experimenter seems natural for science. All too much outside of science rests on ad hominem arguments, which science tries to avoid when it can. Of course, in judging research proposals, the likely payoffs from proposed experiments are exactly the issue, and hence judgments of the people involved are inevitable. But it is certainly a comfort that the results of an experiment can be analyzed without having to impugn the experimenter. This consideration independent of the prior and utility function of the experimenter can be conducted under any design that puts the experiment on "automatic pilot" once it is started, and hence does not allow the experimenter to get a finger on the scales. Randomization is one way to accomplish this, but it is not unique in having this virtue. We join with Savage and many others, however, in his respect for randomization as a statistical tool for enhancing interpersonal communication.

REFERENCES

Aumann, R. (1987). Correlated equilibrium as an expression of Bayesian rationality. *Econometrica* **55**, 1–18.

Basu, D. (1981). Randomization analysis of experimental data: The Fisher randomization test (with discussion). *J. Amer. Statist. Assoc.* **75**, 575–595.

Bunke, H. and Bunke, O. (1978). Randomization. Pro and contra. *Math. Operationsforsch. Statist. Ser. Statist.* **9**, 607–623.

DeGroot, M. (1980). Estimation of the correlation coefficient from a broken random sample. *Ann. Statist.* **8**, 264–278.

DeGroot, M. and Kadane, J. B. (1983). Optimal sequential decisions involving more than one decision maker. In: H. Rizvi, J. S. Rustagi and D. Siegmund, Eds., *Recent Advances in Statistics – Papers Submitted in Honor of Herman Chernoff's Sixtieth Birthday*. Academic Press, New York, 197–210.

Fisher, R. A. (1971). *The Design of Experiments*, 9th ed. Hafner Press, New York.

Fisher, R. A. (1973). *Statistical Methods and Scientific Inference*, 3rd enlarged ed. Hafner Press, New York.

Good, I. J. (1971). Twenty-seven principles of rationality. In: V. P. Godambe and D. A. Sprott, Eds., *Foundations of Statistical Inference*. Holt, Rinehart, and Winston, Toronto, 124–127.

Good, I. J. (1974). Random thoughts about randomness. In: K. Schaffner and R. Cohen, Eds., *PSA 1972*. Reidel, Dordrecht, 117–135.

Harsanyi, J. C. (1982). Subjective probability and the theory of games: Comments on Kadane and Larkey's paper. *Management Sci.* **28**, 121–124.

Kadane, J. B. (1986). Toward a more ethical clinical trial. *J. Medicine and Philos.* **11**, 385–404.

Kadane, J. B. (Ed.) (1996). *Bayesian Methods and Ethics in a Clinical Trial Design.* J. Wiley and Sons, New York.

Kadane, J. B., Dickey, J., Winkler, R., Smith, W., and Peters, S. (1980). Interactive elicitation of opinion for a normal linear model. *J. Amer. Statist. Assoc.* **75**, 845–854.

Kadane, J. B. and Larkey, P. D. (1982). Subjective probability and the theory of games. *Management Sci.* **28**, 113–120.

Kadane, J. B. and Larkey, P. D. (1983). The confusion of is and ought in game theoretic contexts. *Management Sci.* **29**, 1365–1379.

Kadane, J. B. and Sedransk, N. (1980). Toward a more ethical clinical trial. In: J. Bernardo, M. DeGroot, D. Lindley and A. Smith, Eds., *Bayesian Statistics*. University Press, Valencia, 329–338.

Kempthorne, O. (1955). The randomization theory of experimental inference. *J. Amer. Statist. Assoc.* **50**, 946–967.

Kempthorne, O. (1966). Some aspects of experimental inference. *J. Amer. Statist. Assoc.* **16**. 11–34.

Kempthorne, O. (1977). Why Randomize? *J. Statist. Plann. Inference* **1**, 1–25.

Levi, I. (1983). Direct inference and randomization. In: P. Asquith and T. Nickles, Eds., *PSA 1982*, Vol. 2. Edward Brothers, Ann Arbor, MI, 447–463.

Lindley, D. V. (1969). *Introduction to Probability and Statistics from a Bayesian Viewpoint* (2 Vols.). Cambridge University Press, Cambridge.

Lindley, D. V. (1972). *Bayesian Statistics, A Review*. SIAM, Philadelphia, PA.

Lindley, D. V. (1983). The role of randomization in inference. In: P. Asquith and T. Nickles, Eds., *PSA 1982*, Vol. 2. Edward Brothers, Ann Arbor, MI, 431–446.

Lindley, D. V. and Novick, M. R. (1981). The role of exchangeability in inference. *Ann. Statist.* **9**, 45–58.

Pratt, J. W. and Zeckhauser, R. J. (Eds.) (1985). *Principals and Agents: The Structure of Business.* Harvard School of Business Press, Boston, MA.

Rubin, D. B. (1978). Bayesian inference for causal effects: The role of randomization. *Ann. Statist.* **6**, 34–58.

Savage, L. J. (1961). The foundations of statistics reconsidered. In: J. Neyman, Ed., *Proceedings of the Fourth Berkeley Symposium on Mathematical Statistics and Probability*, Vol. 1. University of California Press, Berkeley, CA, 575–586.

Savage, L. J. (1962). Subjective probability and statistical practice. In: M. S. Bartlett, Ed., *The Foundations of Statistical Inference*. Methuen, London, 33–34.

Sedransk, N. (1973). Allocation of sequentially available units to treatment groups. *Internat. Statist. Inst. Proc.* **2**, 393–400.

Seidenfeld, T. (1981). Levi on the dogma of randomization in experiments. In: R. Bogdan, Ed., *Henry E. Kyburg, Jr. and Isaac Levi*. Reidel, Dordrecht, 263–291.

Stone, M. (1969). The role of experimental randomization in Bayesian statistics: Finite sampling and two Bayesians. *Biometrika* **56**, 681–683.

Suppes, P. (1983). Arguments for randomizing. In: P. Asquith and T. Tickles, Eds., *PSA 1982*, Vol. 2. Edward Brothers, Ann Arbor, MI, 464–475.

Swijtink, Z. (1982). A Bayesian argument in favor of randomization. In: P. Asquith and T. Nickles, Eds., *PSA 1982*, Vol. 1. Edward Brothers, Ann Arbor, MI, 159–169.

Wilson, J. G. (1986). Subjective probability and the prisoner's dilemma. *Management Sci.* **32**, 45–55.

3.5

Characterization of Externally Bayesian Pooling Operators

CHRISTIAN GENEST, KEVIN J. MCCONWAY, AND MARK J. SCHERVISH

ABSTRACT

When a panel of experts is asked to provide some advice in the form of a group probability distribution, the question arises as to whether they should synthesize their opinions before or after they learn the outcome of an experiment. If the group posterior distribution is the same whatever the order in which the pooling and the updating are done, the pooling mechanism is said to be externally Bayesian by Madansky (1964). In this chapter, we characterize all externally Bayesian pooling formulas and we give conditions under which the opinion of the group will be proportional to the geometric average of the individual densities.

I. INTRODUCTION

Let (Θ, μ) be a measure space and let Δ be the class of μ-measurable functions $f: \Theta \to (0, \infty)$ such that $f > 0$ μ-a.e. and $\int f d\mu = 1$. In the language of multiagent statistical decision theory (cf. Weerahandi and Zidek, 1981), a *pooling operator* is any function $T: \Delta^n \to \Delta$ which may

Thanks are due to M. H. DeGroot (Carnegie-Mellon University), C. Sundberg (University of Tennessee), S. Dharmadhikari (Southern Illinois University, Carbondale), and J. V. Zidek (University of British Columbia) for fruitful discussions in the course of writing this article. The first author's investigation was supported in part by an N.S.E.R.C. postdoctoral fellowship which was held at Carnegie-Mellon University. Part of the second author's work was performed at the University of Washington, which generously provided research facilities.

Reprinted from *The Annals of Statistics*, 14, no. 2 (1986): 487–501, with permission of the Institute of Mathematical Statistics.

be used to extract a "consensus" $T(f_1, \ldots, f_n)$ from the different subjective opinions $f_1, \ldots, f_n \in \Delta$ of the n members of a group. The current interest for pooling operators seems to stem from a theorem due to Wald (1939) concerning the optimality of Bayesian decision rules. When formulated in the context of a group decision problem, this theorem suggests that at least in the case where all the members of the group have the same utility function, it is generally preferable for them to agree on an "average opinion" $T(f_1, \ldots, f_n)$ and to adopt that action which maximizes their common utility with respect to $T(f_1, \ldots, f_n)$, rather than to take an "average decision" based on the optimal decisions of each of the individuals (see deFinetti, 1954).

A few years ago, Madansky (1964, 1978) suggested the use of pooling formulas, T, which have the following property:

$$T\left(lf_1 \Big/ \int lf_1 \, d\mu, \ldots, lf_n \Big/ \int lf_n d\mu\right) = lT(f_1, \ldots, f_n) \Big/ \int lT(f_1, \ldots, f_n) d\mu, \qquad \mu\text{-a.e.}, \tag{1.1}$$

whenever $l: \Theta \to (0, \infty)$ is such that

$$0 < \int lf_i \, d\mu < \infty, \qquad i = 1, \ldots, n. \tag{1.2}$$

A function l which satisfies condition (1.2) above is hereafter called a *likelihood for* $(f_1, \ldots, f_n)$. Implicit in the statement of (1.1) is that the integral $\int lT(f_1, \ldots, f_n)\, d\mu$ is finite whenever l is a likelihood for $(f_1, \ldots, f_n)$.

A different but obviously equivalent formulation of Madansky's condition would be to specify that

$$f_1/g_1 \propto \ldots \propto f_n/g_n \propto l \Rightarrow T(f_1, \ldots, f_n)/T(g_1, \ldots, g_n) \propto l. \tag{1.3}$$

Pooling operators which obey (1.1) or (1.3) are called *externally Bayesian*. By adopting an externally Bayesian formula, a group assures itself that when additional information l is perceived jointly, their collective opinion can be modified (using Bayes' rule), producing the same result as if the pooling operator had been applied after each individual distribution has been revised. Raiffa (1968, pages 221–226) shows with reference to a concrete example how using a pooling formula which fails (1.1) may lead the members of the group to act strangely. In Raiffa's example, as it were, all the individuals try to increase the influence of their opinion on the consensus by insisting that it be computed before the outcome of an experiment is known. This happens because the members of the group know that whether the f_i's are updated or not, the consensus will be computed using the same weighted arithmetic average of their opinions, viz.

$$T(f_1, \ldots, f_n) = \sum_{i=1}^{n} w_i f_i, \qquad \mu\text{-a.e.,} \tag{1.4}$$

a formula which violates (1.1) unless $w_i = 1$ for some i and $w_j = 0$ for all $j \neq i$.

To ensure that the order in which the pooling and updating are done is immaterial, Bacharach (1972) suggests that the consensus should be computed using a logarithmic pooling operator, viz.

$$T(f_1, \ldots, f_n) = \prod_{i=1}^{n} f_i^{w_i} \Big/ \int \prod_{i=1}^{n} f_i^{w_i} \, d\mu, \qquad \mu\text{-a.e.} \tag{1.5}$$

where $w_1, \ldots, w_n$ are nonnegative weights such that $\Sigma_{i=1}^{n} w_i = 1$ as in (1.4). According to Bacharach, it is Peter Hammond who first observed that the operator (1.5) is externally Bayesian. In a recent article, Genest (1984b) has shown that this is also the only solution of the functional equation (1.1) when there exists a function G: $(0, \infty)^n \to (0, \infty)$ which is Lebesgue measurable and such that

$$T(f_1, \ldots, f_n)(\theta) \propto G(f_1(\theta), \ldots, f_n(\theta)), \qquad \mu\text{-a.e.,} \tag{1.6}$$

where the proportionality constant must be independent of θ. (A precise statement of this result is to be found in Section II.) The formula (1.6) means that, except for a normalizing constant, the value of T at a particular θ depends on the f_i's only through their values at θ. The import of this theorem is still limited, however, especially because the proof given in Genest (1984b) does not apply when the space (Θ, μ) is purely atomic, an assumption which excludes the important case where Θ is finite or countable.

In this essay, we will provide all the solutions of the functional equation (1.1), that is, without restriction to those operators which obey (1.6) and without imposing regularity conditions on the space (Θ, μ). The form of the solutions is given in (2.1) below and is worked out explicitly in Section III for the case where individual opinions are represented as probabilities of a single event. When the function G in (1.6) is indexed by θ and (Θ, μ) can be partitioned into at least four nonnegligible sets, we show that all the solutions must be of the form

$$T(f_1, \ldots, f_n) = g\prod_{i=1}^{n} f_i^{w_i} \Big/ \int g\prod_{i=1}^{n} f_i^{w_i}\, d\mu, \qquad \mu\text{-a.e.}, \qquad (1.7)$$

g being an arbitrary bounded function on Θ and $w_1, \ldots, w_n$ being arbitrary weights adding up to 1. Pooling operators of the form (1.7) have already been characterized by McConway (1978), but only in the case where the measure μ is purely atomic (thereby forcing Θ to be countable). Obviously, the weights in (1.7) can be negative, but only when Θ is finite.

It may not always be reasonable to require a pooling formula to satisfy the criterion of Madansky (1964, 1978), viz. (1.1). Such would be the case if, for instance, the group itself were not required to make a decision, but rather were only asked to provide a group opinion to an external decision maker. In situations where a decision maker is present, French (1981, 1985) and Lindley (1985) have argued rightly that it would seem more reasonable for this decision maker to adopt the Bayesian approach and to treat the opinions of the members of the group as data which he/she would use to update his/her own subjective probability distribution. In certain circumstances, French and Lindley have shown how the formula $T(f_1, \ldots, f_n)$ representing the decision maker's opinion after hearing the experts could well violate (1.1). In this essay, however, we address what French (1985) would call the "group decision problem," that in which there is no natural decision maker and the group is either unwilling or unable to provide one.

Whether it involves a decision maker or not, the problem of determining a sensible formula for representing the opinions of a group has received a lot of attention in recent years. Let us mention, for example, the papers of French (1980, 1981), Morris (1974, 1977), Winkler (1968, 1981), and Genest and Schervish (1985), all of which adopt the Bayesian viewpoint. References on the so-called "group problem" include Laddaga (1977), McConway (1981), Wagner (1982, 1984), and Genest (1984a, c). An extensive bibliography has recently been compiled on both versions of the problem by Genest and Zidek (1986).

II. CHARACTERIZING EXTERNALLY BAYESIAN POOLING OPERATORS

First observe that the class Δ is nonempty if and only if the measure μ is σ-finite, and that the operator (1.7) is well-defined since

$$0 < \int g \prod_{i=1}^{n} f_i^{w_i} d\mu \le \|g\|_\infty \prod_{i=1}^{n} \left[\int f_i d\mu\right]^{w_i} = \|g\|_\infty < \infty$$

by Hölder's inequality. For convenience, we will assume that every singleton subset of Θ is measurable. The following theorem is a slight modification of a theorem of Genest (1984b).

Theorem 2.1. *Let* (Θ, μ) *be a measure space, and suppose that* μ *is not purely atomic. Let also* $T: \Delta^n \to \Delta$ *be a pooling operator for which* (1.6) *holds. Then* T *is externally Bayesian if and only if it is logarithmic, i.e., if and only if there exist nonnegative weights* $w_1, \ldots, w_n$ *such that* $\Sigma_{i=1}^n w_i = 1$ *for which* (1.5) *holds.*

Proof. The proof is the same as in Genest (1984b), since the σ-field on which μ is defined *always* contains nonnegligible sets with arbitrarily small measure, except in the case where the measure μ is purely atomic. (See Halmos (1950), Exercise 1, page 174). □

In the following, our main objective is to generalize Theorem 2.1 by characterizing all the pooling operators which have property (1.1) without imposing condition (1.6), and without restricting the underlying measure space (Θ, μ). To do this, we first consider the case in which the "group" consists of a single expert, and we show that every externally Bayesian pooling operator is then of the form (1.7).

Theorem 2.2. *Let* $T: \Delta \to \Delta$ *be a pooling operator. Then* T *is externally Bayesian if and only if there exists a bounded function* $g: \Theta \to [0, \infty)$ *such that* $g > 0$ μ*-a.e. and*

$$T(f) = gf \Big/ \int gf \, d\mu, \qquad \mu\text{-a.e.},$$

for all μ*-densities* f *in* Δ.

Proof. Let f and h be arbitrary in Δ. Set $g = T(h)/h$ and consider the likelihood function $l = f/h$. Since T is externally Bayesian, we have

$$T(f) = T\left(lh \Big/ \int lh \, d\mu\right) = lT(h) \Big/ \int lT(h) d\mu = gf \Big/ \int gf \, d\mu, \qquad \mu\text{-a.e.},$$

for all $f \in \Delta$. We also have $\int gf d\mu < \infty$ for all f, which implies that g is essentially bounded. (See, for example, Theorem 20.15 in Hewitt and

Stromberg (1965).) The definition of T does not depend on the choice of h, since g is unique up to a constant multiple. □

In the case where $n > 1$, the basic idea consists of reducing the problem to the context of Theorem 2.2 by dividing the domain of T into equivalence classes in such a way that, given the value of T at one member of an equivalence class, the externally Bayesian property defines the value of T at all other members of that class.

Definition 2.3. Two vectors $(f_1, \ldots, f_n)$ and $(f_1^*, \ldots, f_n^*)$ in Δ^n are said to be equivalent and we write $(f_1, \ldots, f_n) \sim (f_1^*, \ldots, f_n^*)$ if and only if

$$\forall i, j \exists c_{i,j} > 0 \quad \text{such that} \quad f_i / f_j = c_{i,j}(f_i^* / f_j^*), \qquad \mu\text{-a.e.}$$

It is obvious that $\sim$ is an equivalence relation on Δ^n and that two vectors of densities belong to the same eqivalence class if and only if there exists a likelihood function l: $\Theta \to (0, \infty)$ such that

$$f_i^* = lf_i \Big/ \int lf_i \, d\mu, \qquad \mu\text{-a.e.},$$

for all $i = 1, \ldots, n$. In the sequel, we use the Greek letter α to refer to an arbitrary element of the quotient-space $\mathcal{A} = \Delta^n/\sim$. For each $\alpha \in \mathcal{A}$, we denote $(f_1^\alpha, \ldots, f_n^\alpha)$ an arbitrary but fixed vector of μ-densities in α. That such a representative can be chosen for each equivalence class α follows from the Axiom of Choice.

Example 2.4. Let $(f_1^\alpha, \ldots, f_n^\alpha)$ be a vector of exponential densities such that $f_i^\alpha(\theta) = \lambda_i \exp\{-\lambda_i \theta\}$ for $\theta > 0$, λ_i being a nonnegative parameter, $i = 1, \ldots, n$. It is easy to see that a vector $(f_1, \ldots, f_n)$ belongs to the same equivalence class as $(f_1^\alpha, \ldots, f_n^\alpha)$ if and only if $\int f_1 \exp\{(\lambda_1 - \lambda_k)\theta\} d\theta < \infty$ for all $k = 1, \ldots, n$, and $f_i \propto f_1 \exp\{(\lambda_1 - \lambda_i)\theta\}$, μ-a.e. for $i > 1$. The condition of f_1 means that its moment generating function is finite at each point $\lambda_1 - \lambda_k$, $k = 2, \ldots, n$.

We are now in a position to state the main result of this section.

Theorem 2.5. *Let T: $\Delta^n \to \Delta$ be an arbitrary pooling operator. Then T is externally Bayesian if and only if*

$$T(f_1, \ldots, f_n) \propto b_\alpha v_\alpha f_1 / f_1^\alpha, \qquad \mu\text{-a.e.}, \tag{2.1}$$

where (using the above notation) α represents the equivalence class of $(f_1, \ldots, f_n)$, and for each α, b_α is some essentially bounded function and v_α is some function such that $v_\alpha \geq \max\{f_1^\alpha, \ldots, f_n^\alpha\}$, μ-a.e.

Proof. It is easy to see that for any fixed b_α and v_α, pooling operators of the form (2.1) are well-defined and externally Bayesian. The crux of the proof consists in showing that these are the *only* ones.

For each $\alpha \in \mathcal{A}$, denote $h_\alpha = T(f_1^\alpha, \ldots, f_n^\alpha)$. For an arbitrary $(f_1, \ldots, f_n)$ which is equivalent to $(f_1^\alpha, \ldots, f_n^\alpha)$, consider the likelihood function $l = f_1/f_1^\alpha$. Since $f_i^\alpha/f_1^\alpha = c_{i,1} f_i/f_1$, μ-a.e., one has $\int l f_i^\alpha \, d\mu = c_{i,1} < \infty$ and $l f_i^\alpha / \int l f_i^\alpha \, d\mu = f_i$ for all $i = 1, \ldots, n$. Since T is externally Bayesian, it follows that

$$T(f_1, \ldots, f_n) \propto l T(f_1^\alpha, \ldots, f_n^\alpha) = f_1 h_\alpha / f_1^\alpha, \qquad \mu\text{-a.e.},$$

and this remains true as long as $(f_1, \ldots, f_n)$ belongs to the equivalence class α.

To complete the proof, assume that $v_\alpha \geq \max\{f_1^\alpha, \ldots, f_n^\alpha\}$, μ-a.e. To show that $h_\alpha/v_\alpha = b_\alpha$ is essentially bounded, pick an arbitrary g in Δ and define $f_1 = g f_1^\alpha / v_\alpha$, μ-a.e. Since $\int f_1 f_i^\alpha / f_1^\alpha \, d\mu < \infty$, for all $i = 1, \ldots, n$, we can define $f_2, \ldots, f_n \in \Delta$ such that $(f_1, \ldots, f_n) \sim (f_1^\alpha, \ldots, f_n^\alpha)$ by putting $f_i \propto f_1 f_i^\alpha / f_1^\alpha$ for $i \geq 2$. Then $T(f_1, \ldots, f_n) \propto g h_\alpha / v_\alpha$ and hence

$$\int g h_\alpha / v_\alpha \, d\mu < \infty$$

for all $g \in \Delta$. The conclusion now follows from Theorem 20.15 in Hewitt and Stromberg (1965). □

At first glance, it may appear as though formula (2.1) is based solely on the opinion f_1 of the first expert, which is intriguing. In reality, however, the opinions of all the individuals influence the consensus since it is necessary to know them all in order to determine the equivalence class corresponding to the vector $(f_1, \ldots, f_n)$. Note that each equivalence class is characterized by an arbitrary vector $(f_1^\alpha, \ldots, f_n^\alpha)$ in the class or, equivalently, by a component f_k^α and the ratios f_i^α/f_k^α, which ratios are invariant (up to constant multiples) within a given class. For this reason, the operator (2.1) could also be written in the form

$$T(f_1, \ldots, f_n) \propto b_\alpha v_\alpha f_k / f_k^\alpha, \qquad \mu\text{-a.e.},$$

or alternatively as

$$T(f_1, \ldots, f_n) \propto b_\alpha v_\alpha \prod_{i=1}^{n} [f_i/f_i^\alpha]^{w_i}, \qquad \mu\text{-a.e.}, \tag{2.2}$$

provided that the weights w_i add up to 1. In principle, a different set of weights could even be chosen for each vector $(f_1, \ldots, f_n)$, since the ratios f_i/f_i^α are eqùal to one another up to constant multiples. Another trivial observation is that every measurable function is bounded when Θ is finite. In this case, therefore, the requirement that b_α be bounded is vacuous.

Formula (2.1) is more general than the logarithmic opinion pool (1.7), but this operator is included as a particular case. To verify this, it suffices to choose the function b_α in (2.2) equal to $g\Pi_{i=1}^{n}(f_i^\alpha)^{w_i}/v_\alpha$, where g is essentially bounded. This choice is legitimate since $\Pi_{i=1}^{n}(f_i^\alpha)^{w_i}/v_\alpha \leq \Pi_{i=1}^{n}(f_i^\alpha)^{w_i}/\max\{f_1^\alpha, \ldots, f_n^\alpha\} \leq 1$, μ-a.e. More generally, the operator

$$T(f_1, \ldots, f_n) = g_\alpha \prod_{i=1}^{n} f_i^{w_i(\alpha)} \Big/ \int g_\alpha \prod_{i=1}^{n} f_i^{w_i(\alpha)} d\mu, \qquad \mu\text{-a.e.},$$

is well-defined and is externally Bayesian when the functions g_α are essentially bounded and $\Sigma_{i=1}^{n} w_i(\alpha) = 1$. That is, the function g and the weights w_i in (1.7) may vary with the equivalence class to which the vector $(f_1, \ldots, f_n)$ belongs without conflicting with (1.1) or (1.3).

In order to synthesize the group's opinions using a formula of the form (2.1), it will generally be necessary to first determine the equivalence class to which the observed vector of opinions belongs. In the following section, we will show what this involves in the specific case where each individual in the group is asked to provide his/her subjective probability for the realization of an event of interest.

III. THE EVENT CASE

In this section, we consider in detail the special case in which the space Θ consists of only two points, say $\Theta = \{0, 1\}$. This is the case in which each expert opinion can be thought of as the probability assigned to the event $E = \{\theta = 1\}$. In this case, the equivalence class to which a particular vector of probability assessments belongs is easy to construct. First note that each probability assessment consists of two positive numbers adding to unity, say p and $1 - p$. (The requirement that p be strictly between 0 and 1 is equivalent to the general assumption that the experts' densities are strictly positive almost everywhere.) A vector $\mathbf{p} = ([p_1, 1 - p_1], \ldots, [p_n, 1 - p_n])$ of such assessments is equivalent to another such vector $\mathbf{q} = ([q_1, 1 - q_1], \ldots, [q_n, 1 - q_n])$ in the sense of Definition 2.3 if and only if, for each $i = 2, \ldots, n$, we have a constant k_i such that

$$p_i / p_1 = k_i q_i / q_1, (1 - p_i)/(1 - p_1) = k_i (1 - q_i)/(1 - q_1). \qquad (3.1)$$

If we let $\mathcal{P}_i$ and $\mathcal{L}_i$ be the odds ratios $p_i/(1 - p_i)$ and $q_i/(1 - q_i)$, respectively, (3.1) simplifies to $\mathcal{P}_i/\mathcal{P}_1 = \mathcal{L}_i/\mathcal{L}_1$ for $i = 2, \ldots, n$. Hence two vectors of probability assessments for an event are equivalent if and only if the pairwise ratios of assessed odds within one set are equal to the corresponding ratios in the other set. For example, with $n = 2$, the two vectors $([0.5, 0.5], [0.7, 0.3])$ and $([0.75, 0.25], [0.875, 0.125])$ are equivalent since the odds ratio of the second expert is 7/3 times as big as the odds ratio of the first expert in each case. The equivalence class $\alpha(\mathbf{p})$ to which a vector $\mathbf{p}$ of probability assessments belongs can be characterized by the $n - 1$ coordinates of the vector $\mathcal{P} = (\mathcal{P}_2/\mathcal{P}_1, \ldots, \mathcal{P}_n/\mathcal{P}_1)$. That is, the equivalence classes are in one-to-one correspondence with the $n - 1$ dimensional vectors of positive numbers. The equivalence class corresponding to a vector $(\alpha_2, \ldots, \alpha_n)$ contains all vectors of probability assessments such that $\mathcal{P}_i = \alpha_i \mathcal{P}_1$. A canonical representative $\mathbf{p}^\alpha$ can be chosen from each class α with $\mathcal{P}_1 = 1$, and we can identify α with the vector $(\alpha_2, \ldots, \alpha_n)$.

Next, we will look at what the externally Bayesian formulas are. Without loss of generality, we can assume that v_α in Theorem 2.5 is identically 1 for all α, since all densities are probabilities in this problem. It follows that each externally Bayesian formula can be represented as

$$T(\mathbf{p}) = \left[\frac{b_\alpha(1) p_1}{b_\alpha(1) p_1 + b_\alpha(0)(1 - p_1)}, \frac{b_\alpha(0)(1 - p_1)}{b_\alpha(1) p_1 + b_\alpha(0)(1 - p_1)} \right], \qquad (3.2)$$

where b_α is an arbitrary (bounded) function on Θ for each α, and $\alpha = (\mathcal{P}_2/\mathcal{P}_1, \ldots, \mathcal{P}_n/\mathcal{P}_1)$. Another way to express (3.2) is to say that the odds ratio for $T(\mathbf{p})$ is $\mathcal{P}_1 b_\alpha(1)/b_\alpha(0)$. Now it is trivial to see that we can assume $b_\alpha(0) = 1$, without loss of generality, by simply altering $b_\alpha(1)$. So the odds ratio for $T(\mathbf{p})$ is just $\mathcal{P}_1 b_\alpha(1)$. For example, if $b_\alpha(1) = 1$, then T is the dictatorship that simply follows expert 1. Or if $b_\alpha(1) = \alpha_i$, then T is the dictatorship that simply follows expert i. If $b_\alpha(1) = r\Pi_{i=2}^n \alpha_i^{w_i}$ for arbitrary numbers w_i, $i = 2, \ldots, n$, and $r > 0$ then

$$T(\mathbf{p}) = \left[\frac{q\prod_{i=1}^{n} p_i^{w_i}}{q\prod_{i=1}^{n} p_i^{w_i} + (1-q)\prod_{i=1}^{n}(1-p_i)^{w_i}},\right.$$

$$\left.\frac{(1-q)\prod_{i=1}^{n}(1-p)^{w_i}}{q\prod_{i=1}^{n} p_i^{w_i} + (1-q)\prod_{i=1}^{n}(1-p_i)^{w_i}}\right], \quad (3.3)$$

where $w_1 = 1 - \Sigma_{i=2}^{n} w_i$ and $q = r/(1+r)$. Formula (3.3) is an analogue of (1.7) in the event case.

The form of the most general externally Bayesian rule (3.2) is very general. It is possible, for example, for $b_\alpha(1)$ to vary in an arbitrary fashion as a function of α. However, it would not usually be desirable for a small change in $\mathbf{p}$ to produce a large change in $T(\mathbf{p})$. It is easy to see that (3.2) is continuous as a function of $\mathbf{p}$ within each equivalence class α. In order for T to be continuous overall as a function of $\mathbf{p}$, all that is required is that $b_\alpha(1)$ be a continuous function of α (with the convention that $b_\alpha(0) = 1$). For example, we could let the weights w_i in (3.3) be continuous functions of α and obtain a generalized logarithmic pool with weights which depend on the degrees to which the experts differ.

IV. THE LOGARITHMIC OPINION POOL

Genest (1984b) conjectured that if an externally Bayesian pooling operator $T: \Delta^n \to \Delta$ were such that

$$T(f_1, \ldots, f_n)(\theta) = G(\theta, f_1(\theta), \ldots, f_n(\theta)) \Big/ \int G(\cdot, f_1, \ldots, f_n)\, d\mu, \qquad \mu\text{-a.e.}, \quad (4.1)$$

for some $\mu \times$ Lebesgue measurable function $G: \Theta \times (0, \infty)^n \to (0, \infty)$, then T must be a "modified logarithmic opinion pool," viz.

$$T(f_1, \ldots, f_n) = g\prod_{i=1}^{n} f_i^{w_i} \Big/ \int g\prod_{i=1}^{n} f_i^{w_i}\, d\mu, \qquad \mu\text{-a.e.}, \quad (4.2)$$

where g is an arbitrary bounded function on Θ and $w_1, \ldots, w_n$ are (not necessarily nonnegative) weights summing up to one. This result was actually proven by McConway (1978) in the case where Θ is countable and μ is a counting-type measure. In this section, we will extend this result by removing the restriction that the measure space should be purely atomic. Indeed, the only assumption which we will make here is that (Θ, μ) can be partitioned in at least four nonnegligible sets. We call a measure space which has this property *quaternary*, by analogy with the term *tertiary* introduced by Wagner (1982).

Our proof is a hybrid of that of Theorem 2.1 in Genest (1984b) and an argument which was developed by McConway (1978) for the countable case. In accordance with the convention adopted at the beginning of Section II, each $\theta \in \Theta$ will either be an atom or will have measure zero. We begin by addressing the case in which the measure space contains atoms.

Lemma 4.1. *Let (Θ, μ) be quaternary and let $T: \Delta^n \to \Delta$ be an externally Bayesian pooling operator. Suppose that (Θ, μ) contains at least two atoms and that there exists a $\mu \times$ Lebesgue measurable function $G: \Theta \times (0, \infty)^n \to (0, \infty)$ such that (4.1) holds for all $f_1, \ldots, f_n \in \Delta$. Then for every pair of atoms (θ, η) in Θ^2, the identity*

$$\frac{T(f_1, \ldots, f_n)(\theta)}{T(f_1, \ldots, f_n)(\eta)} = \frac{T(h_1, \ldots, h_n)(\theta)}{T(h_1, \ldots, h_n)(\eta)} \tag{4.3}$$

holds for all densities $f_1, \ldots, f_n$ and $h_1, \ldots, h_n \in \Delta$ for which

$$f_i(\theta)/f_i(\eta) = h_i(\theta)/h_i(\eta), \qquad i = 1, \ldots, n. \tag{4.4}$$

Before we present the proof of this lemma, let us mention parenthetically that the implication (4.4) $\Rightarrow$ (4.3) is weaker but similar in spirit to the so-called axiom of relative propensity consistency (RPC) of Genest, Weerahandi, and Zidek (1984). In view of Lemma 4.1, it is not surprising, therefore, that their RPC condition should imply a logarithmic opinion pool, at least when the space (Θ, μ) is purely atomic.

Proof. First observe that if two vectors of μ-densities $(f_1, \ldots, f_n)$ and $(h_1, \ldots, h_n)$ satisfy (4.4) and belong to the same equivalence class, the conclusion is immediate from (2.1) in Theorem 2.5 Hence, the proof will be complete if we can show that whenever there exist $f_1, \ldots, f_n$, $h_1, \ldots, h_n$ in Δ such that $f_i(\theta)/f_i(\eta) = h_i(\theta)/h_i(\eta)$, $i = 1, \ldots, n$, such densities exist which belong to the same equivalence class. To do this, first set $x_i = f_i(\theta)$, $y_i = f_i(\eta)$, $x_i^* = h_i(\theta)$, and $y_i^* = h_i(\eta)$ so that $x_i/x_i^* = y_i/y_i^* = t_i$, say, for $i = 1, \ldots, n$. Since (Θ, μ) is σ-finite and quaternary, there must exist a partition of Θ into four sets A_j with $0 < \mu(A_j) < \infty$, for $j = 1, 2, 3$, $\mu(A_4) > 0$, and such that $x_i\mu(A_1) + y_i\mu(A_2) < 1$, and $x_i^*\mu(A_1) + y_i^*\mu(A_2) < 1$, for all $i = 1, \ldots, n$. In particular, we can take $A_1 = \{\theta\}$ and $A_2 = \{\eta\}$. We will construct densities $\bar{f}_1, \ldots, \bar{f}_n$ and a likelihood l for $(\bar{f}_1, \ldots, \bar{f}_n)$ such that $\bar{f}_i(\theta) = x_i$, $\bar{f}_i(\eta) = y_i$ and $\bar{h}_i(\theta) = x_i^*$, $\bar{h}_i(\eta) = y_i^*$, where $\bar{h}_i = l\bar{f}_i/\int l\bar{f}_i d\mu$, $i = 1, \ldots, n$.

Denote $\gamma_i = x_i\mu(A_1) + y_i\mu(A_2)$, and note that $0 < \gamma_i < \min\{t_i, 1\}$ for each i. This is true because

$$x_i\mu(A_1) + y_i\mu(A_2) = t_i[x_i^*\mu(A_1) + y_i^*\mu(A_2)] < t_i,$$

for all $i = 1, \ldots, n$. Choose $0 < \lambda, \xi < \infty$ such that

$$\lambda < \min_{i=1,\ldots,n}\{[1-\gamma_i]^{-1}(t_i - \gamma_i)\} \leq \max_{i=1,\ldots,n}\{[1-\gamma_i]^{-1}(t_i - \gamma_i)\} < \xi.$$

Fixing $h \in \Delta$ an arbitrary density, now define

$$\bar{f}_i = x_i I(A_1) + y_i I(A_2) + \frac{t_i - \gamma_i - \lambda(1-\gamma_i)}{\mu(A_3)(\xi - \lambda)} I(A_3) + \frac{\xi(1-\gamma_i) - (t_i - \gamma_i)}{R(\xi - \lambda)} hI(A_4), \tag{4.5}$$

where $R = \int I(A_4)hd\mu > 0$ and, in general, $I(A)$ denotes the indicator of the set A. Clearly, $\int \bar{f}_i d\mu = 1$ and $\bar{f}_i(\theta) = x_i$, $\bar{f}_i(\eta) = y_i$, $i = 1, \ldots, n$. To generate the $\bar{h}_i$'s, consider the likelihood

$$l = I(A_1) + I(A_2) + \xi I(A_3) + \lambda I(A_4). \tag{4.6}$$

It is easy to see that $\int l\bar{f}_i d\mu = t_i$, so that $\bar{h}_i(\theta) = x_i^*$ and $\bar{h}_i(\eta) = y_i^*$, $i = 1, \ldots, n$. □

Lemma 4.1 shows that if $T: \Delta^n \to \Delta$ is externally Bayesian and satisfies (4.1) for some function G on $\Theta \times (0, \infty)^n$, then for all pairs of atoms (θ, η) in Θ^2, there must exist a Lebesgue measurable function $Q(\theta, \eta)$: $(0, \infty)^n \to (0, \infty)$ such that for all $f_1, \ldots, f_n \in \Delta$,

$$\frac{T(f_1, \ldots, f_n)(\theta)}{T(f_1, \ldots, f_n)(\eta)} = Q(\theta, \eta)\left[\frac{f_1(\theta)}{f_1(\eta)}, \ldots, \frac{f_n(\theta)}{f_n(\eta)}\right]. \tag{4.7}$$

We will now derive a more specific form for the right-hand side of (4.7).

Lemma 4.2. *In adition to the conditions of Lemma* 4.1, *suppose that* (Θ, μ) *contains at least three atoms. Then, there exist constants* $v_1, \ldots, v_n$, *such that for all* $x_1, \ldots, x_n > 0$ *and every pair of atoms* (θ, η) *in* Θ^2, *we have*

$$Q(\theta,\eta)(x_1,\ldots,x_n) = Q(\theta,\eta)(\mathbf{1})\prod_{i=1}^{n} x_i^{\nu_i}, \tag{4.8}$$

where **1** *denotes the n-dimensional vector* $(1, \ldots, 1)$.

Proof. In the same manner as Genest, Weerahandi, and Zidek (1984), define new functions $NQ(\theta, \eta)\colon (0, \infty)^n \to (0, \infty)$ by

$$NQ(\theta,\eta)(x_1,\ldots,x_n) = \frac{Q(\theta,\eta)(x_1,\ldots,x_n)}{Q(\theta,\eta)(1,\ldots,1)}$$

for all atoms $\theta \neq \eta$. Let θ, η, and ζ be three distinct atoms in Θ and pick $\varepsilon > 0$ small enough that there exist densities in Δ which assume any of the values ε, εx_i, or ε/y_i at any of these three atoms. Writing $\boldsymbol{\varepsilon} = (\varepsilon, \ldots, \varepsilon)$, $\mathbf{x} = (x_1, \ldots, x_n)$, and $\mathbf{y} = (y_1, \ldots, y_n)$, and assuming that all operations on vectors are performed componentwise, we have

$$\begin{aligned} NQ(\theta,\eta)(\mathbf{xy}) &= \frac{G(\theta,\varepsilon\mathbf{x})/G(\eta,\varepsilon/\mathbf{y})}{G(\theta,\boldsymbol{\varepsilon})/G(\eta,\boldsymbol{\varepsilon})} \\ &= \frac{G(\zeta,\boldsymbol{\varepsilon})/G(\eta,\varepsilon/\mathbf{y})}{G(\zeta,\boldsymbol{\varepsilon})/G(\eta,\boldsymbol{\varepsilon})}\,\frac{G(\theta,\varepsilon\mathbf{x})/G(\zeta,\boldsymbol{\varepsilon})}{G(\theta,\boldsymbol{\varepsilon})/G(\zeta,\boldsymbol{\varepsilon})} \\ &= NQ(\theta,\zeta)(\mathbf{x})NQ(\zeta,\eta)(\mathbf{y}) \end{aligned}$$

for all $\mathbf{x}$ and $\mathbf{y}$ in $(0, \infty)^n$. The argument now proceeds exactly as that beginning at (2.1) of Genest, Weerahandi, and Zidek (1984), except that in our case, the nonmeasurable solutions are automatically eliminated because G, Q, and hence NQ were assumed to be Lebesgue measurable. It follows that (4.8) holds for all $x_1, \ldots, x_n > 0$ and all pairs of atoms θ and η. □

To complete the proof of (4.2) for atoms, fix ζ an atom in Θ and choose ε strictly between 0 and $1/\mu(\zeta)$. For all atoms $\theta \in \Theta$, now define $g(\theta) = Q(\theta, \zeta)(\mathbf{1})G[\zeta, \boldsymbol{\varepsilon}]\varepsilon$, where $\nu = \Sigma_{i=1}^{n}\nu_i$. Then for all atoms θ, we find

$$G(\theta, x_1, \ldots, x_n) = g(\theta)\prod_{i=1}^{n} x_i^{\nu_i},$$

for all $0 < x_1, \ldots, x_n < 1/\mu(\theta)$, which implies that

$$T(f_1, \ldots, f_n)(\theta) = g(\theta)\prod_{i=1}^{n} f_i(\theta)^{v_i} \Big/ \int g\prod_{i=1}^{n} f_i^{v_i}\,d\mu \qquad (4.9)$$

for all atoms θ.

Next, we derive a formula similar to (4.9) for those θ which are not atoms.

Lemma 4.3. *Let $T: \Delta^n \to \Delta$ be externally Bayesian and assume that* (4.1) *holds for some $\mu \times$ Lebesgue measurable function $G: \Theta \times (0, \infty)^n \to (0, \infty)$. Assume also that (Θ, μ) is not purely atomic. Let N be the complement of the set of atoms. Then*

$$T(f_1, \ldots, f_n)(\theta)$$
$$= g(\theta)\prod_{i=1}^{n} f_i(\theta)^{w_i} \Big/ \int G(\cdot, f_1, \ldots, f_n)\,d\mu, \qquad \mu\text{-a.e. on } N, \qquad (4.10)$$

for some nonnegative weights $w_1, \ldots, w_n \in \mathbb{R}$ adding up to unity.

Proof. Define a new function $NG: \Theta \times (0, \infty)^n \to (0, \infty)$ by

$$NG(\theta, z_1, \ldots, z_n) = G(\theta, z_1, \ldots, z_n)/G(\theta, 1, \ldots, 1)$$

for all $z_1, \ldots, z_n \in (0, \infty)$. It will be enough to show that $NG(\theta)(z_1, \ldots, z_n)$ is a function of the z_i's only, say $NG(z_1, \ldots, z_n)$. For, once this is done, we can define a new pooling operator $T^*: \Delta^n \to \Delta$ by

$$T^*(f_1, \ldots, f_n)(\theta) = NG(f_1(\theta), \ldots, f_n(\theta)) \Big/ \int NG(f_1, \ldots, f_n)\,d\mu.$$

It is easy to see that T^* is externally Bayesian and of the form (1.6) with G replaced by NG. We can then apply Theorem 2.1 to conclude that

$$NG(z_1, \ldots, z_n) = \prod_{i=1}^{n} z_i^{w_i}$$

for some nonnegative constants $w_1, \ldots, w_n$ adding up to one. Letting $g(\theta) = G(\theta, 1, \ldots, 1)$ for all θ in N, we arrive at (4.10).

To see that $NG(\theta, z_1, \ldots, z_n)$ is a function of the z_i's alone, we proceed along the same lines as in the proof of Lemma 4.1. Given $z_1, \ldots, z_n > 0$, choose $\varepsilon > 0$ such that $\varepsilon < \min_{i=1,\ldots,n}\{1/2, 1/2z_i\}$ and let A_j, $1 \le j \le 4$, be a partition of N such that $0 < \mu(A_j) < \varepsilon$ for $j = 1, 2$;

$0 < \mu(A_3) < \infty$ and $\mu(A_4) > 0$. That such a partition of N exists follows from the fact that (Θ, μ) is σ-finite and nonatomic on N.

Next, define $\bar{f}_i$ as in (4.5) using $x_i = y_i = 1$ and $t_i = 1/z_i$, $i = 1, \ldots, n$. If the likelihood for $(\bar{f}_i, \ldots, \bar{f}_n)$ is defined by (4.6), we have $\int l\bar{f}_i \, d\mu = 1/z_i$, $i = 1, \ldots, n$. Letting $\bar{h}_i = l\bar{f}_i / \int l\bar{f}_i \, d\mu$, we have that $\bar{f}_i(\theta) = 1$ and $\bar{h}_i = z_i$ for $\theta \in A_1 \cup A_2$ and $i = 1, \ldots, n$. Now, since T is externally Bayesian, we know that

$$\frac{T(\bar{h}_1, \ldots, \bar{h}_n)(\theta)}{l(\theta)T(\bar{f}_1, \ldots, \bar{f}_n)(\theta)}$$

is constant μ-almost everywhere. Also, since $l(\theta) = 1$ on $A_1 \cup A_2$ and since (4.1) holds μ-almost everywhere on N, it follows that

$$NG(\theta, z_1, \ldots, z_n) = \frac{T(\bar{h}_1, \ldots, \bar{h}_n)(\theta)}{l(\theta)T(\bar{f}_1, \ldots, \bar{f}_n)(\theta)}$$

on $A_1 \cup A_2$. Hence $NG(\theta, z_1, \ldots, z_n)$ is essentially constant as a function of θ on $A_1 \cup A_2$. Denote the constant value $NG(z_1, \ldots, z_n)$. To see that NG is essentially constant on N as a function of θ, assume to the contrary that there exists a subset B of N with $\mu(B) > 0$ and such that $NG(\theta, z_1, \ldots, z_n) > (<) NG(z_1, \ldots, z_n)$ for almost all θ in B. Since μ is nonatomic on N, choose a subset of B with positive measure at most ε. Let this new set be A_2, and repeat the above construction, keeping A_1 the same as before. The conclusion still holds that $NG(\theta, z_1, \ldots, z_n)$ is essentially constant on $A_1 \cup A_2$ in contradiction to the assumption that $NG(\theta, z_1, \ldots, z_n) > (<) NG(z_1, \ldots, z_n)$ on A_2. □

We are now in a position to prove the main result of this section.

Theorem 4.4. *Let (Θ, μ) be a quaternary measure space and let $T: \Delta^n \to \Delta$ be an externally Bayesian pooling operator If there exists a $\mu \times$ Lebesgue measurable function $G: \Theta \times (0, \infty)^n \to (0, \infty)$ such that (4.1) holds for all vectors of opinions $(f_1, \ldots, f_n)$ in Δ^n, then T is of the form (1.7), i.e.,*

$$T(f_1, \ldots, f_n) = g\prod_{i=1}^{n} f_i^{w_i} \bigg/ \int g\prod_{i=1}^{n} f_i^{w_i} d\mu, \qquad \mu\text{-a.e.}, \qquad (4.11)$$

for some essentially bounded function $g: \Theta \to (0, \infty)$ and some constants $w_1, \ldots, w_n \in \mathbb{R}$ such that $\Sigma_{i=1}^{n} w_i = 1$. Moreover, the weights w_i are nonnegative unless Θ is finite or there does not exist a countably infinite partition of (Θ, μ) into nonnegligible sets.

Proof. If (Θ, μ) does not contain any atoms, (4.11) is immediate from Lemma 4.3. If (Θ, μ) is purely atomic, then (4.11) derives easily from (4.9) with $w_i = v_i$, $i = 1, \ldots, n$. It is straightforward to see that $\Sigma_{i=1}^{n} w_i = 1$ from the fact that T is externally Bayesian.

More difficult is the case in which μ has atoms but is not purely atomic. In this case, we can use Lemma 4.3 to obtain the result on the set N, the complement of the set of atoms of μ. Label the atoms θ_1, $\theta_2, \ldots$ and let $G_i(x_1, \ldots, x_n)$ denote $G(\theta_i, x_1, \ldots, x_n)$ for all $x_1, \ldots, x_n$ strictly between 0 and $1/\mu(\theta_i)$. From the definition of externally Bayesian, we have

$$\frac{l(\theta)G(\theta, f_1(\theta), \ldots, f_n(\theta))}{G(\theta, h_1(\theta), \ldots, h_n(\theta))} = \text{constant a.e.}\ \mu, \tag{4.12}$$

whenever h_i is proportional to lf_i for all i. From (4.10), we have that on N, the left-hand side of (4.12) equals $\Pi_{i=1}^{n} t_i^{w_i}$, where t_i is the integral of lf_i for each i.

Now fix $t_1, \ldots, t_n$ and pick a single atom θ_j. Let ε be small enough so that ε/t_i is strictly between 0 and $1/\mu(\theta_j)$ for each i. Let $A_1 = \{\theta_j\}$ and construct the same densities as in the proof of Lemma 4.1, setting $x_i = \varepsilon$ and $x_i^* = \varepsilon/t_i$ for each i. (Here A_2, A_3, A_4, and the y_i's and y_i^* are arbitrary sets and values satisfying the restrictions described in the proof of Lemma 4.1.) Also construct the same likelihood l as in that proof. Using the fact that (4.12) holds on all of Θ, we have

$$\frac{G_j(\varepsilon, \ldots, \varepsilon)}{G_j(\varepsilon/t_1, \ldots, \varepsilon/t_n)} = \prod_{i=1}^{n} t_i^{w_i}. \tag{4.13}$$

It follows directly from (4.13) that for all $z_1, \ldots, z_n$ between 0 and $1/\mu(\theta_j)$, $G_j(z_1, \ldots, z_n) = G_j(\varepsilon, \ldots, \varepsilon)\varepsilon^{\ 1}\Pi_{i=1}^{n} z_i^{w_i}$. By letting $g(\theta_j) = G_j(\varepsilon, \ldots, \varepsilon)\varepsilon^{\ 1}$, we have proven (4.11).

Finally, note that the weights are automatically nonnegative provided that μ is not purely atomic. If (Θ, μ) is purely atomic but includes a countably infinite number of atoms, it is fairly easy to construct densities $f_1, \ldots, f_n$ which will make the integral

$$\int g \prod_{i=1}^{n} f_i^{w_i} \, d\mu \tag{4.14}$$

infinite and lead to a contradiction unless all the weights are nonnegative. Of course, (4.14) is always finite when Θ is finite and μ is some sort of counting measure, so negative weights cannot be ruled out in this case. Similarly, g must be essentially bounded, or else there exists

f such that (4.14) is infinite when all of the f_i are equal to f (c.f. Theorem 20.15 of Hewitt and Stromberg, 1965). □

Unless the group of experts is reporting its opinions to an outside decision maker who chooses to aggregate them using (4.9), it is difficult to interpret g, let alone offer advice as to how it could be selected. To some extent, the same applies to the interpretation and determination of the weights, although some heuristics are available (for example, see Winkler, 1968, or Genest and Schervish, 1985). In fact, even if the pooling operator (4.9) is adopted by a decision maker, it is not so clear what g and the weights stand for. In particular, we should guard from concluding too hastily that the function g represents the decision maker's "prior." After all, there is no reason why a prior density should necessarily be bounded, nor is every bounded function a possible prior density.

One way around the choice of g in (4.9) would be to insist that the pooling operator T preserves unanimity. In general, an aggregation procedure $T: \Delta^n \to \Delta$ preserves unanimity if and only if $T(f, \ldots, f) = f$ for all $f \in \Delta$. As the following corollary indicates, this is enough to reduce (4.9) to an ordinary logarithmic opinion pool.

Corollary 4.5. *Let* (Θ, μ) *be a quaternary measure space and let* $T: \Delta^n \to \Delta$ *be an externally Bayesian pooling operator which preserves unanimity. If there exists a* $\mu \times$ *Lebesgue measurable function* $G: \Theta \times (0, \infty)^n \to (0, \infty)$ *such that* (4.1) *holds for all vectors of opinions* $(f_1, \ldots, f_n)$ *in* Δ^n, *then* T *is a logarithmic opinion pool, i.e.,*

$$T(f_1, \ldots, f_n) = \prod_{i=1}^{n} f_i^{w_i} \Big/ \int \prod_{i=1}^{n} f_i^{w_i}\, d\mu, \qquad \mu\text{-a.e.},$$

for some arbitrary reals $w_1, \ldots, w_n$ *adding up to* 1. *Moreover, the weights* w_i *are nonnegative unless* Θ *is finite or there does not exist a countable partition of* (Θ, μ) *into nonnegligible sets.*

V. DISCUSSION

The purpose of this essay has been twofold. On the one hand we have provided a characterization of all externally Bayesian pooling operators in Theorem 2.5. The form of these operators is quite general, as it was derived under the sole assumption that all densities are strictly positive almost everywhere. On the other hand, when the space Θ is assumed to be quaternary and the operator is required to satisfy a "locality" condition (4.1), we show that the operator must be logarithmic in the sense of (1.7). This latter result does not apply to the cases in which Θ consists of merely two or three atoms.

If the space Θ contains only two points, then it is trivial to see that every pooling operator satisfies (4.1). In this case, one can easily construct externally Bayesian operators which are not logarithmic. One such example is constructed from (3.2) as follows. Let $b_\alpha(0) = 1$ and let $b_\alpha(1) = \max\{1, \alpha_2, \ldots, \alpha_n\}$. It is easy to see that $T(\mathbf{p})$ equals $[\max_i\{p_i\}, 1 - \max_i\{p_i\}]$, which is externally Bayesian, satisfies (4.1), and is clearly not logarithmic. If (Θ, μ) consists of only three atoms, it is not known whether logarithmic opinion pools are the only externally Bayesian operators which satisfy (4.1). Theorem 2.5 still holds in this case, however. The case of one atom is left to the reader.

The theorems of this essay shed some light on the mechanics of externally Bayesian behavior. If, however, a decision maker wishes to treat the opinions of a group of experts as data, little is known about the implications of the externally Bayesian criterion for the modeling process. In the Bayesian model proposed by Lindley (1985), conditions are given under which the decision maker's posterior distribution would be externally Bayesian, but his conditions are not easily interpretable.

REFERENCES

Bacharach, Michael (1972). Scientific disagreement. Unpublished manuscript.

deFinetti, Bruno (1954). Media di decisioni e media di opinioni. *Bull. Inst. Internat. Statist.* **34** 144–157.

French, Simon (1980). Updating of belief in the light of someone else's opinion. *J. Roy. Statist Soc. Ser. A* **143** 43–48.

French, Simon (1981). Consensus of opinion. *European J. Oper. Res.* **7** 332–340.

French, Simon (1985). Group consensus probability distributions: A critical survey. In *Bayesian Statistics* **2** (J. M. Bernardo, et al., eds.) 183–201. North-Holland, Amsterdam.

Genest, Christian (1984a). A conflict between two axioms for combining subjective distributions. *J. Roy. Statist. Soc. Ser. B* **46** 403–405.

Genest, Christian (1984b). A characterization theorem for externally Bayesian groups. *Ann. Statist.* **12** 1100–1105.

Genest, Christian (1984c). Pooling operators with the marginalization property. *Canad. J. Statist.* **12** 153–163.

Genest, Christian and Schervish, Mark J. (1985). Modeling expert judgments for Bayesian updating. *Ann. Statist.* **13** 1198–1212.

Genest, Christian and Zidek, James V. (1986). Combining probability distribu-

tions: A critique and an annotated bibliography (with discussion). *Statist. Sci.* **1** 114–148.
Genest, C., Weerahandi, S. and Zidek, J. V. (1984). Aggregating opinions through logarithmic pooling. *Theory and Decision* **17** 61–70.
Halmos, Paul R. (1950). *Measure Theory*. Van Nostrand, New York.
Hewitt, Edwin and Stromberg, Karl (1965). *Real and Abstract Analysis*. Springer, New York.
Laddaga, Robert (1977). Lehrer and the consensus proposal. *Synthese* **36** 473–477.
Lindley, Dennis V. (1985). Reconciliation of discrete probability distributions. In *Bayesian Statistics* **2** (J. M. Bernardo, et al., eds.) 375–390. North-Holland, Amsterdam.
Madansky, Albert (1964). Externally Bayesian groups. Technical Report RM-4141-PR, RAND Corporation.
Madansky, Albert (1978). Externally Bayesian groups. Unpublished manuscript, University of Chicago.
McConway, Kevin J. (1978). The combination of experts' opinions in probability assessment: Some theoretical considerations. Ph.D. thesis, University College London.
McConway, Kevin J. (1981). Marginalization and linear opinion pools. *J. Amer. Statist. Assoc.* **76** 410–414.
Morris, Peter A. (1974). Decision analysis expert use. *Management Sci.* **20** 1233–1241.
Morris, Peter A. (1977). Combining expert judgments: a Bayesian approach. *Management Sci.* **23** 679–693.
Raiffa, Howard (1968). *Decision Analysis: Introductory Lectures on Choices under Uncertainty*. Addison-Wesley, Reading, Mass.
Wagner, Carl G. (1982). Allocation, Lehrer models, and the consensus of probabilities. *Theory and Decision* **14** 207–220.
Wagner, Carl G. (1984). Aggregating subjective probabilities: Some limitative theorems. *Notre Dame J. Formal Logic* **25** 233–240.
Wald, Abraham (1939). Contributions to the theory of statistical estimation and testing hypotheses. *Ann. Math. Statist.* **10** 299–326.
Weerahandi, Samaradasa and Zidek, James V. (1981). Multi-Bayesian statistical decision theory. *J. Roy. Statist. Soc. Ser. A* **144** 85–93.
Winkler, Robert L. (1968). The consensus of subjective probability distributions. *Management Sci.* **15** B61–B75.
Winkler, Robert L. (1981). Combining probability distributions from dependent information sources. *Management Sci.* **27** 479–488.

3.6

An Approach to Consensus and Certainty with Increasing Evidence

MARK J. SCHERVISH AND
TEDDY SEIDENFELD

ABSTRACT

We investigate conditions under which conditional probability distributions approach each other and approach certainty as available data increase. Our purpose is to enhance Savage's (1954) results, in defense of "personalism", about the degree to which consensus and certainty follow from shared evidence. For problems of consensus, we apply a theorem of Blackwell and Dubins (1962), regarding pairs of distributions, to compact sets of distributions and to cases of static coherence without dynamic coherence. We indicate how the topology under which the set of distributions is compact plays an important part in determining the extent to which consensus can be achieved. In our discussion of the approach to certainty, we give an elementary proof of the Lebesgue density theorem using a result of Halmos (1950).

I. INTRODUCTION

In his classic discussion of Bayesian inference, L. J. Savage (1954, Sections 3.6 and 4.6) illustrates how (finitely many) different investigators come to agree on the truth of one hypothesis, given an increasing sequence of shared observations. More precisely, Savage's result is this. Assume the following two conditions.

Reprinted from *Journal of Statistical Planning and Inference*, Volume 25, M. Schervish and T. Seidenfeld, "An Approach to Consensus and Certainty with Increasing Evidence," pp. 401–414, Copyright 1990, with kind permission from Elsevier Science – NL, Sara Burgerhartstraat 25, 1055 KV Amsterdam, The Netherlands.

1. The agents' initial opinions over a (finite) set of rival hypotheses are not too discrepant – there is agreement on which hypotheses have probability 0.
2. There is agreement also on the (distinct) likelihoods for these hypotheses over an infinite sequence of observations which are identically, independently distributed given an hypothesis.

Then, almost surely, the sequence of conditional probabilities for the hypotheses *converge* (given more of these data) *to a common*, 0–1 distribution focused on the true hypothesis.

Savage offers this finding as a partial rebuttal to the charge that his Bayesian theory of *personal* probability is overly "subjective" – that it cannot explain how scientific methods may be a source of "objective" knowledge. In short, he uses this result to explain how interpersonal agreements about what is practically certain can arise within the Bayesian paradigm. The object of our discussion here is to indicate how the two parts to Savage's conclusion – *consensus* and *certainty* of opinions – obtain (asymptotically) under more general conditions than are permitted by (1) and (2). That is, we indicate how Savage's reply may be enhanced.

The first part of Savage's conclusion, (almost certain) consensus for a pair of agents, does not depend upon the assumptions (2) so long as the hypotheses of interest are expressible in terms of (perhaps infinite) sets of observables, as was shown by Blackwell and Dubins (1962). (Also, like Blackwell and Dubins, we avoid Savage's restriction to conditional probability given non-null events but, instead, we require that probabilities are countably additive, unlike in Savage's argument.) We discuss how compactness of a set C of (mutually, absolutely continuous) probabilities affects the kind of consensus that may be achieved with increasing shared evidence. When the extreme points of C are compact in the discrete topology, consensus ("almost everywhere") follows from the theorem of Blackwell and Dubins: Corollary 1. When the extreme points of C are compact in the uniform-distance topology, convergence ("in-probability", but not "almost everywhere") of sequences of pairs of probabilities is proven: Corollaries 2, 3, and Example 3. And when the extreme points of C are weak-star compact, no consensus is assured at all: Example 2.

The second part of Savage's conclusion, that the conditional probabilities of a measurable event E converge (almost surely) to 1 or 0 as E occurs or fails to occur, follows from coherence alone. We offer two arguments for the approach to certainty. One proof (which we suspect is known to many) is as a consequence of Doob's martingale convergence theorem. The other argument uses just a basic result governing the extension of σ-finite measures from an algebra to its smallest σ-algebra. (A similar result is needed, also, in the application of Doob's martingale theorem.) We show how the approach to certainty for conditional probabilities provides an elementary proof of the Lebesgue density theorem: Corollary 6. Example 4 illustrates the importance of the measurability assumption for certainty, even with exchangeable probability distributions.

In addition, we show that consensus and certainty, viewed as claims about the asymptotic behavior of the agents' unconditional probabilities with increasing data, do not require the full force of temporal conditionalization – they do not require the use of Bayes' theorem to revise degrees of belief.[1] That is, though we do require (static) coherence, we do not assume that, over time, an agent updates his personal probability by conditionalization. We do not impose a constraint of (full) dynamic coherence. For consensus, it suffices that the agents use conditional probabilities arbitrarily chosen from a class C enveloped by finitely many (mutually absolutely continuous) distributions. Under the conditions of Corollary 1, asymptotic certainty follows from static coherence: Corollary 4.

II. THE STRUCTURAL ASSUMPTIONS FOR THE SPACE $(X, \mathcal{B}, P)$

II.1. *The Measurable Space $(X, \mathcal{B})$*

Consider a denumerable sequence of sets X_i $(i = 1, \ldots)$ with associated σ-fields $\mathcal{B}_i$. Form the infinite Cartesian product $X = X_1 \times \ldots$ of sequences $(x_1, x_2, \ldots) = x \in X$, where $x_i \in X_i$, that is, where each x_i is an atom of its algebra $\mathcal{B}_i$. (This is mild as the $\mathcal{B}_i$ may be unrelated.) In the usual fashion, let the measurable sets in X (the events) be the elements of the σ-algebra $\mathcal{B}$ generated by the set of measurable rectangles. (A measurable rectangle $A = A_1 \times \ldots$ is one where $A_i \in \mathcal{B}_i$ and $A_i = X_i$ for all but finitely many i.) Thus, $(X, \mathcal{B})$ is a measurable space.

Define the spaces of histories $(H_n, \mathcal{H}_n)$ and futures $(F_n, \mathcal{F}_n)$ where $H_n = X_1 \times \ldots \times X_n$, $\mathcal{H}_n = \mathcal{B}_1 \times \ldots \times \mathcal{B}_n$, and where $F_n \times X_{n+1} \times \ldots$ and $\mathcal{F}_n = \mathcal{B}_{n+1} \times \ldots$. Identify these as sub-$\sigma$-fields of $(X, \mathcal{B})$ by writing

$G_n \in \mathcal{H}_n$ as $G_n \times X_{n+1} \times \ldots$ and $E_n \in \mathcal{F}_n$ as $X_1 \times \ldots \times X_n \times E_n$. We shall be concerned, in particular, with the (increasing) sequence of histories $h_n \in H_n$ and the (decreasing) sequence of future events $E_n \in \mathcal{F}_n$, as these are judged given each history.

II.2. *The Probability* P

Let P be a (countably additive) probability over the measurable space $(X, \mathcal{B})$. Assume P is *predictive* (Blackwell and Dubins, 1962), so that there exist conditional probability distributions of events given past events, $P^n(\cdot|\mathcal{H}_n)$.[2] In particular, given a history h_n, there is a conditional probability distribution for the future, $P^n(\cdot|h_n)$ on $\mathcal{F}_n$.

Next, we show that conditional probability given some history, $P^n(\cdot|h_n)$ on $\mathcal{B}$, is characterized by the conditional probability for the future, $P^n(\cdot|h_n)$ on $\mathcal{F}_n$. We observe that when $P(B|\cdot)$ is defined as the Radon–Nikodym derivative of $P(\cdot \cap B)$ with respect to $P(\cdot)$, then if $D \cap E = \emptyset$: (i) $P(D \cup E|C) = P(D|C) + P(E|C)$ [a.e. P] and (ii) $P(D|E) = 0$ [a.e. P]. Thus, for all $A \in \mathcal{B}$ and for almost all x,

$$P^n(A|h_n) = P^n(A \cap h_n|h_n) + P^n(A \cap h_n^c|h_n) \quad \text{by (i)}$$

$$\text{and} \qquad = P^n(A \cap h_n|h_n) \quad \text{by (ii)}.$$

However, $A \cap h_n$ can be written as $(x_1, \ldots, x_n) \times E_n$, where $E_n \in \mathcal{F}_n$, as desired. This elementary result will be helpful in our discussion (in Section IV) of convergence to certainty.

III. CONSENSUS THROUGH MARTINGALES

Consider any probability Q which is in agreement with P about events of measure 0 in $\mathcal{B}$, i.e., $\forall E \in \mathcal{B}$, $P(E) = 0$ iff $Q(E) = 0$, so that P and Q are mutually absolutely continuous. Then Q, too, is predictive with conditional probability distributions $Q^n(\mathcal{F}_n|h_n)$. In their important paper of 1962, Blackwell and Dubins establish (almost sure) asymptotic consensus between the conditional probabilities P^n and Q^n. In particular, they show:

Theorem 1. *For each P^n there is a Q^n so that, almost surely, the distance between them vanishes with increasing histories*:

$$\lim_{n\to\infty} \rho(P^n, Q^n) \to 0 \quad [\text{a.e. } P \text{ or } Q],$$

where ρ is the uniform distance metric between distributions. That is, with μ and ν defined on the same measure space $(M, \mathcal{M})$, $\rho(\mu, \nu)$ is the l.u.b., *over events $E \in \mathcal{M}$, of $|\mu(E) - \nu(E)|$.*

(Blackwell and Dubins prove this result about consensus by a "slightly generalized" martingale convergence theorem – their Theorem 2 (1962, p. 883).)

What can be said about consensus when considering a set C of mutually absolutely continuous probabilities? Quite obviously, unless C is closed, there may be no consensus among conditional probabilities. In this vein, we note a simple corollary to Theorem 1.

Corollary 1. *Let C be a closed, convex set of probabilities all mutually absolutely continuous, and generated by finitely many of its extreme points. Denote this finite set by $C = \{P_1, \ldots, P_k\}$. Then, asymptotically, the conditional probabilities in C achieve consensus uniformly. That is, for almost all $x \in X$, $\forall \varepsilon > 0, \exists m, \forall n > m, \forall P, Q \in C, \rho(P^n, Q^n) < \varepsilon$.*

Proof. Because C is finite, by Theorem 1, for almost all $x \in X$,

$$\forall \varepsilon > 0, \exists m, \forall n > m, \max_{P_i, P_j \in C} \rho(P_i^n, P_j^n) < \varepsilon.$$

Recall that if $\rho(T, U) < \varepsilon$, then $\rho(T, S) < \varepsilon$ and $\rho(S, U) < \varepsilon$ for each convex combination $S = \alpha T + (1 - \alpha)U$, $0 \le \alpha \le 1$. Recall also that

$$\forall P \in C, \forall h_n, \exists \alpha_1, \ldots, \alpha_k \; (\alpha_i \ge 0, \textstyle\sum_i \alpha_i = 1)$$

$$P^n(\cdot | h_n) = \textstyle\sum_i \alpha_i P_i^n(\cdot | h_n), \quad i = 1, \ldots, k.$$

The corollary is immediate from these two observations. □

Note also that the corollary ensures consensus when agents are merely statistically coherent but take their updated probabilities from those in C.

Example 1. Let x_i be the i.i.d. Normal (μ, σ^2), where the conjugate priors for these parameters have hyperparameters which lie in a compact set. Given the observed history, h_n, let two agents choose probabilities P^n, Q^n (given h_n) in any manner (even as a function of h_n) from this set. Then, consensus obtains, almost surely, with observation of the x_i's.

The next result, which has a weaker conclusion, is true and helps to identify the role played by the topology in fixing closure of C.

Corollary 2. *Let C be a compact set (under the topology induced by ρ) of mutually absolutely continuous, predictive probabilities on the space $(X, \mathcal{B})$. Let $\{P_n, Q_n\}$ be any sequence of pairs from C. Then, $\forall R \in C$,*

$$\rho(P_n^n, Q_n^n) \xrightarrow{R} 0 \quad \text{as } n \to \infty.$$

That is,

$$\forall \varepsilon > 0, \lim_{n \to \infty} R(\{h_n : \rho(P_n^n, Q_n^n) > \varepsilon\}) = 0.$$

Proof. The result follows from the claim: $\rho(P_n^n, R^n) \xrightarrow{R} 0$ as $n \to \infty$. That is,

$$\forall \varepsilon > 0, \lim_{n \to \infty} R(\{h_n : \rho(P_n^n, R^n) > \varepsilon\}) = 0.$$

[The claim suffices for the corollary, since $\forall h_n$, $\rho(P_n^n, Q_n^n) \leq \rho(P_n^n, R^n) + \rho(R^n, Q_n^n)$.] We demonstrate the claim using Blackwell–Dubins' theorem, Theorem 1, and a simple lemma.

Lemma. *For any predictive probabilities S and T,*

$$\text{if } \rho(S, T) < \alpha\beta, \text{ then } \forall n, S(\{h_n : \rho(S^n, T^n) > \alpha\}) < \beta.$$

The proof of the lemma is straightforward and is omitted.

Proof (of the claim). We argue indirectly. Suppose there is a subsequence, denoted by $\{n_i\}$, where $\exists \delta > 0$, $\forall n_i$, $R(\{h_{n_i} : \rho(P_{n_i}^{n_i}, R^{n_i}) > \varepsilon\}) > \delta$. C is compact. So it is sequentially compact and every sequence contains a convergent subsequence. Hence, $\{P_{n_i}\}$ contains a ρ-convergent subsequence, which we denote by $\{P_{m_j}\}$ where $\forall j, \exists i\ P_{m_j} = P_{n_i}$, and $\boldsymbol{P}$ is its uniform limit, which we denote by $P_{m_j} - \rho \to \boldsymbol{P} \in C$. Thus, $\forall k > 0$, $\exists i, \forall j > i$, $\rho(P_{m_j}, \boldsymbol{P}) < \varepsilon\delta/4k$. Then, by the lemma,

$$\exists i, \forall n, \forall j > i, \boldsymbol{P}\left(\left\{h_n : \rho(P_{m_j}^n, P^n) > \frac{1}{2}\right\}\right) < \delta/2k.$$

Since both $\boldsymbol{P}$ and R belong to the set C, by the Blackwell–Dubins theorem (going from "a. e." to "in probability" convergence),

$$\exists m, \forall n > m, \boldsymbol{P}\left(\left\{h_n : \rho(P^n, R^n) > \frac{1}{2}\varepsilon\right\}\right) < \delta/2k.$$

Let $m' = \max(m, m_i)$. Then, $\forall n > m'$ and $\forall m_j > m'$, $\boldsymbol{P}(\{h_n: \rho(P^n_{m_j}, R^n) > \varepsilon\}) < \delta/k$. Hence,

$$\forall \varepsilon > 0, \lim_{m' \to \infty} (\forall n > m' \, \forall m_j > m')\boldsymbol{P}(\{h_n : \rho(P^n_{m_j}, R^n) > \varepsilon\}) = 0.$$

That is, $\rho(P^n_{m_j}, R^n) \xrightarrow{P} 0$ as the pair $(m_j, n) \to \infty$. Because R is (finite and) absolutely continuous with respect to $\boldsymbol{P}$, *then* $\rho(P^n_{m_j}, R^n) \xrightarrow{R} 0$ as the pair $(m_j, n) \to \infty$. This contradicts the supposition, $\forall n_i$ $R(\{h_{n_i}: \rho(P^{n_i}_{n_j}, R^{n_i}) > \varepsilon\}) > \delta$, and proves the claim. □

However, Corollary 2 is not true when compactness of C is under the weak-star topology.[3] This is shown by a counterexample.

Example 2. Let $X_i = \{0, 1\}$ so that $(X, \mathcal{B})$ is the measurable space of (Borel sets) of infinite flips of a coin. Let R be the exchangeable probability on X given by the beta mixing prior $\alpha = \beta = 1$, the uniform distribution over the binomial parameter θ, for the deFinetti representation of R as a mixture of i.i.d. Binomial distributions.

Consider the set G_n of histories of length n for which the observed relative frequency of 1's is less than or equal to $\frac{1}{2}$. Let S_n be the exchangeable probability on X with beta mixing prior $\alpha = 6n$, $\beta = n$. Define probability P_n as follows:

$$P_n(\cdot) = \int \phi(\cdot \mid h_n) \mathrm{d}R(h_n),$$

where

$$\phi(\cdot \mid h_n) = \begin{cases} S^n_n(\cdot \mid h_n) & \text{if } h_n \in G_n, \\ R^n(\cdot \mid h_n) & \text{if } h_n \in G^c_n. \end{cases}$$

Then the sequence P_n converges weak-star to R, since each P_n agrees with R on all rectangles in $\mathcal{H}_n$ $(\supset \mathcal{H}_m, m \leq n)$. (This follows by Halmos's Theorem 13.A, which we discuss in connection with Theorem 2, below.) Therefore, the set $W = \{R, P_n \ (n = 1, \ldots)\}$ is (weak-star) closed and every sequence in W contains a (weak-star) convergent subsequence.

So W is compact in the weak-star topology.[4] The elements of W are mutually absolutely continuous. (Use the Radon–Nikodym theory with the definition of P_n and the mutual absolute continuity of the S_n with R.) However, $\forall h_n \in G_n$, $\rho(P^n_n, R^n) > \frac{1}{4}$. This is because $\forall h_n \in G_n$, $P^n_n(x_{n+1} = 1 \mid h_n)$ $(= S^n_n(x_{n+1} = 1 \mid h_n)) \geq \frac{3}{4}$ while $R^n(x_{n+1} \mid h_n) \leq \frac{1}{2}$. Moreover, for each n, $R(G_n) \geq \frac{1}{2}$.

Note also that, given G_n, the difference $|P^n_n(\cdot) - R^n(\cdot)|$ fails to converge to 0, even pointwise, over events in $(X, \mathcal{B})$. Let E be the event that the limiting relative frequency of "1" exceeds $\frac{3}{4}$. Then,

$$\lim_{n\to\infty} \inf_{h_n \in G_n} |P^n_n(E) - R^n(E)| = \frac{1}{2}.$$

Last, observe that no superset of the P_n can be compact in the topology induced by ρ as $\rho(P_n, R) \geq \frac{1}{8}$.

The corollary can be strengthed to say:

Corollary 3. *Let the set C be as in Corollary 2. Then for all R in C,*

$$\forall \varepsilon > 0, \quad \lim_{n\to\infty} \sup_{P,Q \in C} R(\{h_n: \rho(P^n, Q^n) > \varepsilon\}) = 0.$$

Proof. We argue indirectly. Suppose $\exists \varepsilon > 0$ with

$$\liminf_{n\to\infty} \sup_{P,Q \in C} R(\{h_n: \rho(P^n, Q^n) > \varepsilon\}) = \delta > 0.$$

Equivalently, $\forall m$, $\exists n_m > m$, $\exists (P_m, Q_m) \in C$ such that

$$R(\{h_n: \rho(P^{n_m}_m, Q^{n_m}_m) > \varepsilon\}) = \delta.$$

Without loss of generality, assume $n_m > n_{m-1}$. Form two new sequences of pairs of elements of C, as follows. Let $A_n = P_m$ when $n = n_m$ and $A_n = R$ otherwise. Likewise, let $B_n = Q_m$ when $n = n_m$ and $B_n = R$ otherwise. Apply Corollary 2 to the sequence of pairs (A_n, B_n). We have, $\forall \varepsilon > 0$,

$$\lim_{n\to\infty} R(\{h_n: \rho(A^n_n, B^n_n) > \varepsilon\}) = 0,$$

which contradicts the supposition,

$$\forall m R(\{h_n: \rho(A^{n_m}_{n_m}, Q^{n_m}_{n_m}) > \varepsilon\}) = \delta. \quad \square$$

Corollaries 2 and 3 guarantee convergence (in probability) for each sequence of pairs of probabilities chosen from the set C. These two results do *not* ensure the "in probability" convergence of conditional probabilities taken from C, analogous to the "almost everywhere" convergence of Corollary 1. The next example shows that compactness of C (under the uniform topology) is insufficient for "in probability" convergence of the conditional probabilities in C. Hence, trivially, ρ-compactness of C does not suffice for the "almost everywhere" convergence, which obtains according to the first corollary when C is generated by finitely many extreme points.

Example 3. Let $(X, \mathcal{B})$ and R be as in Example 2. Define the (integer-valued) function $n^*(m)$ $(m \geq 2)$, recursively, as follows:

$$n^*(2) = 1, n^*(m) = n^*(m-1) + \left[\!\left[\frac{1}{2}(m+1)\right]\!\right] \quad \text{for } m = 3, \ldots,$$

where $[\![r]\!]$ is the integer part of r. For $k = 0, \ldots, [\![\frac{1}{2}m]\!]$, define

$$A_{k,m,n} = \left\{ h_n = (x_1, \ldots, x_n): \left| (1/n)\left(\sum_{i=1}^{n} x_i\right) - k/m \right| < 1/2m + 1.001[\log\log n^*(m)/2n^*(m)]^{1/2};\ n = n^*(m), \ldots, n^*(m+1) - 1 \right\}$$

and define $m^*(n)$ to be that integer m such that $n^*(m) \leq n < n^*(m+1)$. Next, define the probability

$$Q_{k,m,n} = \int \phi(\cdot|h_n) \mathrm{d}R(h_n),$$

where

$$\phi(\cdot|h_n) = \begin{cases} S_n^n(\cdot|h_n) & \text{if } h_n \in A_{k,m,n}, \\ R^n(\cdot|h_n) & \text{if } h_n \notin A_{k,m,n}. \end{cases}$$

(This is analogous to the definition of P_n in Example 2). Since $R(A_{k,m,n}) = Q_{k,m,n}(A_{k,m,n}) \to 0$ uniformly in k and n as $m \to \infty$, any sequence of probabilities of the form $P_n = Q_{k,m,n}$, for $m = m^*(n)$ and $k \in \{0, \ldots, [\![\frac{1}{2}m]\!]\}$, will converge to R uniformly as $n \to \infty$.

By the law of the iterated logarithm for the binomial probability P_θ $(0 < \theta < 1)$,

$$P_\theta\left(\left\{h_n:\left|(1/n)\left(\sum_{i=1}^{n} x_i\right)-\theta\right|>\theta(1-\theta)(1.001)[2\log\log n/n]^{1/2}\right.\right.$$
$$\left.\left.\text{infinitely often in } n\right\}\right)=0.$$

For all n and all $\theta < \frac{1}{2}$,

$$\left\{h_n:\left|(1/n)\left(\sum_{i=1}^{n} x_i\right)-\theta\right|\leq\theta(1-\theta)(1.001)[2\log\log n/n]^{1/2}\right\}$$
$$\subset\left\{h_n:\left|(1/n)\left(\sum_{i=1}^{n} x_i\right)-\theta\right|\leq 1.001[\log\log n^*(m)/2n^*(m)]^{1/2},\right.$$
$$\left.\text{for } m=m^*(n)\right\}$$
$$\subset\left\{h_n:\left|(1/n)\left(\sum_{i=1}^{n} x_i\right)-k/m\right|\leq 1/2m+1.001[\log\log n^*(m)/2n^*(m)]^{1/2},\right.$$
$$\left.\text{for } |k/m-\theta|\leq 1/2m\right\}.$$

Now choose P_n as follows. Let $P_n = Q_{k,m,n}$ for $m = m^*(n)$ and $k = n - n^*(m) = k^*(n)$. From this it follows (by the law of the iterated logarithm) that, for all $\theta < \frac{1}{2}$, $P_\theta(\{A_{k^*(n),m^*(n),n}$ infinitely often in $n\}) = 1$. Then, we have that $R(\{A_{k^*(n),m^*(n),n}$ infinitely often in $n\}) \geq \frac{1}{2}$. Let E be the event that the limiting frequency of "1" is at most 0.65. For $n \geq 600$, and for all histories $h_n \in A_{k^*(n),m^*(n),n}$, $|P^n_n(E) - R^n(E)| \geq 0.9$. Hence,

$$\{h_n: A_{k^*(n)m^*(n),n} \text{ infinitely often}\} \subset \{h_n: \rho(P^n_n, R^n) \not\to 0\},$$

and it follows that $R(\{h_n: \rho(P^n_n, R^n) \to 0\}) \leq \frac{1}{2}$.

Next, select a sequence of probabilities, $\{P_{n,h_n}\}$, where the n-th term in the sequence depends on the history h_n, as follows. If $(1/n)(\Sigma^n_{i=1} x_i) \leq \frac{1}{2}$, let $P_{n,h_n} = Q_{k,m^*(n),n}$, where k is chosen so that $|(1/n)\Sigma^n_{i=1} x_i) - k/m^*(n)| \leq 1/2m^*(n)$. Otherwise, if $(1/n)(\Sigma^n_{i=1} x_i) > \frac{1}{2}$, let $P_{n,h_n} = R$. The set C of R together with all the P_{n,h_n} (the $Q_{k,m^*(n),n}$ for $k = 0, \ldots, [\![\frac{1}{2}m^*(n)]\!]$) is compact (in the uniform topology), since every subsequence converges uniformly to R, and the elements of C are mutually absolutely continuous. But, for $n \geq 600$, $R(\{h_n: \rho(P^n_{n,h_n}, R^n) > 0.9\}) \geq \frac{1}{2}$. Thus,

$$\liminf_{n\to\infty} R\left(\left\{h_n: \sup_{P\in C}\rho(P^n, R^n) > 0.9\right\}\right) \neq 0.$$

Therefore,

$$\sup_{P,Q \in C} \rho(P^n, Q^n) \nrightarrow 0 \text{ (in probability)}, \qquad (*)$$

in marked contrast with Corollary 3.

Last, note that the set C' consisting just of R and the P_n is ρ-compact and contains mutually absolutely continuous elements. It is easy to see (in effect by Corollary 3) that

$$\sup_{P,Q \in C'} \rho(P^n, Q^n) \xrightarrow{R} 0. \qquad (**)$$

Thus, "in probability" consensus of the form (**) does not entail "almost everywhere" consensus of the form $\rho(P_n^n, R^n) \to 0$ [a.e. R].

IV. THE ALMOST CERTAIN APPROACH TO CERTAINTY FOR EVENTS MEASURABLE WITH RESPECT TO $(X, \mathcal{B})$

Though Blackwell and Dubins do not make mention of it, another consequence of Doob's martingale convergence theorem (Doob, 1953, Theorem 7.4.1, p. 319) is the desired result about the approach to certainty in these conditional probabilities. Denote the characteristic function of a set E by $\chi_E(x) = 1$ if $x \in E$ and $\chi_E(x) = 0$ if $x \notin E$, for $x \in X$ a complete history.

Theorem 2. $\forall E \in \mathcal{B}, \lim_{n\to\infty} P^n(E|h_n) = \chi_E(x)$ [a.e. P].

Theorem 2 asserts that, for each event E and for all but a set of complete histories of P-measure 0 (depending upon E), the sequence of conditional probabilities, $P^n(E|h_n)$, converges to 1 or to 0 as E occurs or not.

Proof. Theorem 2 is a substitution instance of Doob's Theorem 7.4.3 (p. 331) which reads (in relevant parts) as follows: Let z be a random variable with finite absolute expectation, $E\{|z|\} < \infty$ and let $\mathcal{G}_1 \subset \mathcal{G}_2 \subset \ldots$ be Borel (σ-)fields of measurable sets. Let $\mathcal{G}_\infty$ be the smallest Borel (σ-)field of sets with $\cup_n \mathcal{G}_n \subset \mathcal{G}_\infty$. Then, $\lim_{n\to\infty} E\{z|\mathcal{G}_n\} = E|\{Z|\mathcal{G}_\infty\}$ [a.e.]. Theorem 2 obtains by taking $z = \chi_E$, $\mathcal{G}_n = \mathcal{H}_n$, and $\mathcal{G}_\infty = \mathcal{B}$. □

Doob's Theorem 7.4.3 is proven with Theorem 7.4.1 (his martingale convergence theorem) and a familiar measure-theoretic result (in effect, Theorem 13.A of Halmos, 1950, p. 54), which asserts that a σ-finite measure μ on an algebra $\mathcal{A}$ induces a unique σ-finite extension of μ on the smallest σ-algebra containing $\mathcal{A}$, $\sigma[\mathcal{A}]$.[5] It is our purpose, next, to show that Theorem 2 can be derived from a measure-theoretic result (Halmos, Theorem 13.D, p. 56), related to the extension Theorem 13.A, without appeal to martingale theory. But first we state without proof a simple consequence of Theorem 2 and Corollary 1.

Corollary 4. *Let the set C be as in Corollary 1. If* $f(h_n) \in \{P^n(E|h_n): P \in C\}$, *then the sequence of conditional probabilities,* $f(h_n)$, *converges to* $\chi_E(x)$ [a.e. $P \in C$}.

Thus, when the set C is as in Corollary 1, static coherence suffices for asymptotic certainty.

Besides the Theorem 13.A, which ensures the unique extension of the measure μ from $\mathcal{A}$ to $\sigma[\mathcal{A}]$, also there is an important result about approximation (in measure) of elements of $\sigma[\mathcal{A}]$ by elements of $\mathcal{A}$. This is reported by Halmos (1950, p. 56).

Theorem 13.D. *If* μ *is a* σ*-finite measure on a ring R, then for every set E of finite measure in* $\sigma[R]$ *and for every* $\varepsilon > 0$, *there is a set* $E_0 \in R$ *with* $\mu(E \triangle E_0) \leq \varepsilon$ ($\triangle$ *is symmetric difference*).

Theorem 2 can be derived from this directly (as noted by Halmos, Theorem 49.B, p. 213), without supposing P is predictive.[6]

The technique used in the alternative proof establishes the approach to certainty for the extension of $(X, \mathcal{B}, P)$ to its measure completion.

Corollary 5. *Let* $(X, \overline{\mathcal{B}}, \overline{P})$ *be the measure completion of* $(X, \mathcal{B}, P)$. *Then*

$$\forall E \in \overline{\mathcal{B}}, \lim_{n\to\infty} \overline{P}^n(E|h_n) = \chi_E(x) \quad [\text{a.e. } \overline{P}].$$

Also, Theorem 2 provides an elementary proof of the Lebesgue density theorem (see Lebesgue, 1904, and Oxtoby, 1971, pp. 16–18), which we give as a corollary. Let μ be Lebesgue measure. For each measurable set E on the real line $\mathcal{R}$, define the density at the point x by:

$$\lim_{h\to\infty} \mu(E \cap [x-h, x+h])/2h,$$

and denote by $\phi(E)$ the set of points at which E has density 1.

Corollary 6. The one-dimensional Lebesgue density theorem: *For each measurable set* $E \subset \mathcal{R}$, $\mu(E \,\Delta\, \phi(E)) = 0$.

Proof. Use Corollary 5 and the fact that the unit interval [0, 1] is Borel equivalent to 2^{ω}. (See Royden, 1968, p. 268.)

V. CONCLUSIONS

Theorem 2 asserts that, with increasing evidence, conditional probabilities for an event E approach certainty, almost surely. We alert the reader to the requirement that $E \in \mathcal{B}$, so that asymptotic certainty about a parameter μ depends upon the measurability of μ.

Example 4a. Let x_1 be binary, $X_1 = \{0, 1\}$, with probability $P(x_1 = 1) = \mu$. Consider the infinite sequence $X = (x_1, x_1, x_1, \ldots)$ generated by repeating the outcome x_1. This is (trivially) an exchangeable sequence. By deFinetti's representation theorem, the probability P on X is given as a mixture of i.i.d. Binomial distributions with some "prior" (mixing) probability distribution $\pi(\theta)$ over the binomial parameter θ. However, $\mu \neq \theta$. On the contrary, P is represented by the "prior" mixture $\pi(\theta = 0) = (1 - \mu)$, $\pi(\theta = 1) = \mu$ and, obviously, observations tells us about θ only, not about the parameter $\mu(= \pi)$ which is not measurable in the σ-field $(X, \mathcal{B})$.

Example 4b. A less obvious version of this problem is as follows. Let x_i form an exchangeable sequence where, for each integer k, $x_{i_1}, \ldots, x_{i_k}$ have the Multivariate Normal distribution $N_k = (\mu, \Sigma)$, with $\mu' = [\mu, \ldots, \mu]$ and with Σ equal to the $k \times k$ matrix having main diagonal elements all 2's and off-diagonal entries all 1's. Then μ fails to be measurable with respect to the σ-field $(X, \mathcal{B})$. Repeated observations of the x_i do not make the posterior probability concentrate about μ. Rather, the asymptotic posterior distribution of μ from these data, with limiting sample average $= \hat{x}$, is identical to the posterior one would obtain (using the same prior over μ) from a sample of two, independent Normal$(\mu, 2)$ observations with sample average $\hat{x}$.

Against the background of Theorem 2, we report conditions which guarantee asymptotic consensus (under the uniform distance metric) for conditional probabilities taken from a set C of unconditional distributions. Not surprisingly, depending upon how C is closed, different conclusions obtain.

When C is (contained within or) generated by finitely many (mutually absolutely continuous) elements, i.e., when C is convex and its extreme points form a compact set in the discrete topology, consensus of conditional probabilities occurs, almost surely. When the extreme points of C form a compact set in the uniform topology, there is "in probability" (but not "almost sure") consensus for sequences of pairs of conditional probabilities. When C is compact in the weak-star topology, there may be no limiting consensus. In this case, there can even be an event E and a sequence of pairs (P_n, Q_n) from C where the (paired) conditional probabilities of E differ by a fixed amount, $|P_n^n(E) - Q_n^n(E)| > \delta > 0$.

Of course, these "large sample" results fail to provide bounds on the rates with which consensus and certainty occur. What they do show, however, is the surprising fact that these asymptotic properties of conditional probabilities do *not* depend upon exchangeability or other kinds of symmetries of the (unconditional) probabilities in C. Rather, agreement on events of zero probability and a suitable closure suffice for consensus, while "the approach to certainty" is automatic.

NOTES

1 A careful discussion of static and dynamic coherence is given by Levi (1980, Chapter 4). Simply stated, static coherence requires that an agent have (conditional) degrees of beliefs captured by a (conditional) probability which respects the total evidence principle. Let $P_k(\cdot|\cdot)$ be one such (static) representation, with background knowledge K. If the agent learns some new evidence E, so that K' (the closure under implication of K with E) is the new knowledge, then temporal conditionalization requires identifying the updated (static) representation $P_{K'}(\cdot|\cdot)$ with the conditional probability $P_k(\cdot|\cdot, E)$. Hence, temporal conditionalization provides a dynamic constraint.

Savage's analysis (§3.6 of 1954) is ambiguous between the static and dynamic reading of the convergence in conditional probabilities. The ambiguity persists even in his subsequent essay, "Implications of Personal Probability for Induction" (1967), though we think he appears inclined there to endorse temporal conditionalization. M. Goldstein concentrates on this distinction in his "Exchangeable Belief Structures" (1986) and other essays of his cited therein. Besides Levi's clear presentation, different philosophic perspectives on the question of static vs. dynamic coherence include: Kyburg's "Conditionalization" (1980), van Fraassen's "Belief and the Will" (1984), and additional references given therein.

2 Predictive probabilities are those which admit *regular conditional distributions* in the sense of Breiman (1968, p. 77). Without regularity of conditional distributions, all that can be shown about the existence of conditional probabilities follows from the Radon–Nikodym theorem. In Section IV we point out that,

for event E, the convergence to certainty of the conditional probabilities, $P^n(E|h_n)$, does not require that P be predictive. We alert the reader to minor variations in the definitions of "predictive", as found in Breiman (1968, p. 77), Doob (1953, p. 26), and Halmos (1950, pp. 209–210). See, in particular, Doob's discussion (1953, p. 624).

Last, we note that the sufficient condition for P being predictive, as given by Breiman (1968, Theorem 4.34, p. 79), is also sufficient for extending a set function μ given "marginally" as a probability on n-dimensional sets of a ring R (that is, for each n, μ is given as a probability on n-dimensional rectangles), to a coherent probability on $\sigma[R]$, its infinite dimensional σ-ring, as shown by Halmos (1950, T.A, p. 212).

3 Diaconis and Freedman (1986, appendix) discuss "weak-star merging" of posterior probabilities in a setting with i.i.d. data. Their interesting Theorem A.1 (1986, p. 18) equates such "merging" of opinions with their rather strict notion of "consistency" of posterior probabilities.

4 Theorem I.6.13, p. 21 of Dunford and Schwartz (1958) asserts that, in a metric space, a closed and sequentially compact set is compact. For discussion of the metrizability of the weak-star topology see Dudley (1968, §1). Those results on the metrizability of the weak-star topology apply to our probabilities on $(X, \mathcal{B})$ since the unit interval $[0, 1]$, a separable metric space, is Borel equivalent to 2^ω (Royden, 1968, p. 268).

5 This note offers some details about how Doob's Theorem 7.4.3 is demonstrated in order to emphasize the role of the extension Theorem 13.A. Define y_n by $y_n = E\{\chi_E(x)|h_n\}$. Then the random variables $y_1, y_2, \ldots$ constitute a martingale. That is, as required for a martingale: (a) $E\{|y_n|\}$ $(= P(E)) < \infty$, and (b) $E\{y_{n+1}|y_1, \ldots, y_n\} = E[y_{n+1}|y_n\} = y_n$.

The latter is a special case of Doob's Example 1 (1953, p. 92), with $\eta = \chi_E$, and $\xi_n = x_n$, since $h_n = (x_1, \ldots, x_n)$.

By Doob's Theorem 7.4.1(i), $\lim_{n\to\infty} y_n = w$ exists (almost surely). But by 7.4.1(ii), w behaves exactly like the conditional probability $P(E|x)$ (which is the Radon–Nikodym integrand representation of $P(E \cap \cdot)$ over $\mathcal{B}$) for all elements of the field $\cup_n \mathcal{H}_n$. That is, $\forall n, \forall A \in \mathcal{H}_n$,

$$\int_A w\mathrm{d}P = \int_A y_n \mathrm{d}P(h_n) = \int_A P(E|x)\mathrm{d}P = P(E \cap A).$$

By the extension Theorem 13.A, a measure on $\mathcal{B}$ is uniquely determined by its values on the field $\cup_n \mathcal{H}_n$. Therefore, $w = P(E|x) = \chi_E(x)$, almost surely, as was to be shown.

6 Halmos proves Theorem 49.B by showing "almost uniform convergence". Actually, his argument stops short of that. The proof is completed by a familiar construction relating to Egoroff's theorem, e.g., Exercise 30 of Royden (1968, p. 72).

REFERENCES

Blackwell, D. and L. Dubins (1962). Merging of opinions with increasing information. *Ann. Math. Statist.* **33,** 882–887.

Breiman, L. (1968). *Probability*. Addison-Wesley, Reading, MA.
Diaconis, P. and D. Freedman (1986). On the consistency of Bayes estimates. *Ann. Statist.* **14**, 1–26.
Doob, J.L. (1953). *Stochastic Processes.* Wiley, New York.
Dudley, R.M. (1968). Distances of probability measures and random variables. *Ann. Math. Statist.* **39**, 1563–1572.
Dunford, N. and J.T. Schwartz (1958). *Linear Operators*, Part I. Interscience, New York.
van Fraassen, B.C. (1984). Belief and the will, *J. of Philos*. **81**, 235–256.
Goldstein, M. (1986). Exchangeable belief structures. *J. Amer. Statist. Assoc.* **81** (396), 971–976.
Halmos, P.R. (1950). *Measure Theory*. Van Nostrand, New York.
Kyburg, H.E.K. (1980). Conditionalization. *J. of Philos*. **77**, 98–114.
Lebesgue, H. (1904). *Leçons sur l'Intégration et la Recherche des Fonctions Primitives.* Paris.
Levi, I. (1980). *The Enterprise of Knowledge.* MIT Press, Cambridge, MA.
Oxtoby, J.C. (1971). *Measure and Category.* Springer-Verlag, New York.
Royden, H.L. (1968). *Real Analysis.* Macmillan, New York.
Savage, L.J. (1954). *The Foundations of Statistics.* Wiley, New York.
Savage, L.J. (1967). Implications of personal probability for induction. *J. of Philos.* **64**, 593–607.

3.7

Reasoning to a Foregone Conclusion

JOSEPH B. KADANE, MARK J. SCHERVISH,
AND TEDDY SEIDENFELD

ABSTRACT

When can a Bayesian select an hypothesis H and design an experiment (or a sequence of experiments) to make certain that, given the experimental outcome(s), the posterior probability of H will be greater than its prior probability? In this chapter we discuss an elementary result that establishes sufficient conditions under which this reasoning to a foregone conclusion cannot occur. We illustrate how when the sufficient conditions fail, because probability is finitely but not countably additive, it may be that a Bayesian can design an experiment to lead his/her posterior probability into a foregone conclusion. The problem has a decision theoretic version in which a Bayesian might rationally pay not to see the outcome of certain cost-free experiments, which we discuss from several perspectives. Also, we relate this issue in Bayesian hypothesis testing to various concerns about "optional stopping."

I. INTRODUCTION

In a lively (1962) discussion of some foundational issues, several noted statisticians, especially L. J. Savage, focused on the controversy of whether an experimenter's stopping rule is relevant to the analysis of his or her experimental data. Savage wrote (1962, p. 18):

The [likelihood] principle has important implications in connection with optional stopping. Suppose the experimenter admitted that he had seen 6 red-

This research was supported by National Science Foundation Grants SES-9123370, DMS-9005858, DMS-9302557, and SES-9208942 and by Office of Naval Research Contracts N00014-89-J-1851 and N00014-91-J-1024.

Reprinted with permission from the *Journal of the American Statistical Association*, 91, no. 435 (September 1996): 1228–1236.

eyed flies in 100 and had then stopped because he felt that he had thereby accumulated enough data to overthrow some popular theory that there should be about 1 per cent red-eyed flies. Does this affect the interpretation of 6 out of 100? Statistical tradition emphasizes, in connection with this question, that if the sequential properties of his experimental programme are ignored, the persistent experimenter can arrive at data that nominally reject any null hypothesis at any significance level, when the null hypothesis is in fact true. These truths are usually misinterpreted to suggest that the data of such a persistent experimenter are worthless or at least need special interpretation; see, for example, Anscombe (1954), Feller (1940), Robbins (1952). The likelihood principle, however, affirms that the experimenter's intention to persist does not change the import of his experience.

The tradition Savage refers to is, in a paraphrase of Anscombe (1954), Feller (1940), Robbins (1952), and Cornfield (1970), captured by the following imaginary circumstance. Suppose that a statistician has his designs on rejecting the null hypothesis H_0: $\theta = 0$, that the mean of iid normal data is zero. The data have known unit variance. Fix k_α so that $k_\alpha/\sqrt{n}$ corresponds to the nominal rejection point in an α-level uniformly most powerful unbiased (UMPU) test of H_0 versus the composite alternative hypothesis H_0^c: $\theta \neq 0$, based on a sample of size n. Let H_t denote the simple hypothesis that $\theta = t$. Continue observing data until the sample average, $\bar{x}_n = (x_1 + \ldots + x_n)/n$, satisfies the inequality (1), then halt:

$$|\bar{x}_n| > k_\alpha/\sqrt{n}. \tag{1}$$

The likelihood principle entails that the statistician's intent to stop only when (1) obtains is irrelevant to the "evidential import" of the data for hypotheses about θ. (See Berger 1985, sec. 7.7 for a recent view.) If, contrary to traditional theory, significance is calculated independent of the stopping rule for the experiment – a mistake by traditional theory – then when the inquiry halts, H_0 has achieved an observed significance of α, or less. Moreover, given the truth of H_0, by the law of the iterated logarithm, with probability 1 the experiment terminates; that is, almost surely the inequality (1) is eventually satisfied.

In response to this tradition, Savage asserted (1962, p. 18):

The true moral of the facts about optional stopping is that significance level is not really a good guide to "level of significance" in the sense of "degree of import," for the degree of import does depend on the likelihood alone, a theme to which I must return later in the lecture.

And later in the discussion (1962, p. 72):

It is impossible to be sure of sampling until the data justifies an unjustifiable conclusion, just as surely as it is impossible to build a perpetual motion machine. After all, whatever we may disagree about, we are surely agreed that Bayes's theorem is true where it applies.

We begin our inquiry by reviewing some of the details for what we guess Savage meant as the evident import of Bayes's theorem for solving the problem of sampling to a foregone conclusion. Because we find that the Bayesian position on foregone conclusions is complicated by mathematical conditions that are important for statistics, we rehearse the following arguments, even though their conclusions have been in the literature before (see, e.g., Kerridge 1963).

Let $(S, \mathcal{A}, P)$ be a (countably additive) probability space, which we think of as the underlying joint space for all quantities of interest. Expectations are with respect to the probability P. Unconditional expectation is denoted by $E(\cdot)$, and conditional expectation given a random variable X is denoted by $E(\cdot|X)$. Let $(\mathcal{X}, \mathcal{B})$ and (Ω, τ) be measurable spaces where $X: S \rightarrow \mathcal{X}$ is a random quantity to be learned, Θ: $S \rightarrow \Omega$ is any random quantity, and $h: \Omega \rightarrow \Re^*$ is an (extended) real-valued function whose expectation $E(h)$ exists. Then the familiar law of total probability implies that

$$E[E(h(\Theta)|X)] = E(h(\Theta)) \tag{2}$$

(see, e.g., Ash 1972, T.6.5.4, p. 257).

This says, in short, that there can be no experiment with outcome X designed (almost surely with respect to P) to drive up or drive down the conditional expectation of h, given X. Of course, Equation (2) has no special logical dependence on Bayes's theorem, except that non-Bayesian statistical methods often begin with the claim that neither the "prior" expectation $E(h(\Theta))$ nor the "posterior" expectation $E(h(\Theta)|X)$ has objective status.

As an example, suppose that h is the indicator for an hypothesis H (an unobserved "event") in Ω; that is, $h(\Theta) = 1$ if $\Theta \in H$, $h(\Theta) = 0$ otherwise. Thus $E(h(\Theta))$ is the "prior" probability of H, denoted by $P(H) = p$. Let $X_1, X_2, \ldots$ be observations that become available sequentially. To consider experimental designs that mandate a minimum sample size, $k \geq 0$, define

$$N = \inf\{n \geq k: P(H|X_1, \ldots, X_n) \geq q\},$$

where $N = \infty$ if the set is empty. That is, N identifies the first point after the kth in the sequence of X_i observations when the "posterior" probability of H reaches q, at least. The event $N = \infty$ obtains when, after k-many observations, the sequence of conditional probabilities, $P(H \mid X_1, \ldots, X_n)$ $(n \geq k)$, all remain below q. We assume that $q > p$.

Let $F_{X_1,\ldots,X_n|N}(\cdot \mid N = n)$ denote the conditional cdf of $(X_1, \ldots, X_n)$, given that $N = n$. Then,

$$\begin{aligned}
P(H) &\geq P(H, N < \infty) \\
&= \sum_{n=k}^{\infty} P(N = n) P(H|N = n) \\
&= \sum_{n=k}^{\infty} P(N = n) \int P(H|X_1 = x_1, \ldots, X_n = x_n) \\
&\quad \times dF_{X_1,\ldots,X_n|N}(x_1, \ldots, x_n|n) \\
&\geq \sum_{n=k}^{\infty} P(N = n) \int q dF_{X_1,\ldots,X_n|N}(x_1, \ldots, x_n|n) \\
&= qP(N < \infty).
\end{aligned}$$

Hence

$$P(N < \infty) \leq p/q < 1. \tag{3}$$

Thus, when $p < q$, the prior probability is less than 1 that a Bayesian will halt the sequence of experiments and conclude that the posterior probability of H has risen to q (at least). Moreover, by Doob's martingale convergence theorem (thm. 7.4.3, 1953), $\lim_{n\to\infty}\{P(H|X_1, \ldots, X_n)\}$ converges (P almost surely). Thus for (P almost all) infinite sequences $(x_1, \ldots) \in (N = \infty)$, we have that $P(H|(x_1, \ldots)) \leq q$. Hence the prior probability is less than p/q that a Bayesian will conclude that the posterior probability of H is more than q. This result obtains no matter which experimental design is adopted and allows for "infinite" samples, but it relies on Bayes's rule for updating a prior to a posterior.

The same argument provides a bound for the conditional probability of terminating the experiment in finite time, given that H is false. Assume that $P(H) < 1$ and write

$$P(N < \infty | \neg H) = \frac{P(\neg H|N < \infty)P(N < \infty)}{P(\neg H)}.$$

Note three facts: that $P(\neg H | N < \infty) \le 1 - q$ by the design of the experiment, that $P(N < \infty) \le p/q$ as just shown, and that $P(\neg H) = 1 - p$ by assumption. Then,

$$P(N < \infty | \neg H) \le \frac{p(1-q)}{q(1-p)}. \tag{4}$$

For example, assume that $0 < p \le 0.1$ and let $q/p = 10$. That is, by the choice of the stopping rule, if the experiment terminates, then the posterior probability for H increases tenfold (at least). Then the inequality (4) asserts the following: Given that H is false, the conditional probability of terminating the experiment is no more than $(0.1 - p)/(1 - p) < 0.1$.

Savage (1962, pp. 72–73) offered a similar conclusion and illustrated it with a simple case of two binomial hypotheses. Kerridge (1963) derived the same bounds as in (4) for the case of a uniform prior ($p = 0.5$). Cornfield (1970, pp. 20–21) used Kerridge's inequality to argue that as a Bayesian, you cannot be sure to defeat a true "null" hypothesis.

However, with extended Bayesian methods based on so-called "improper" priors, the situation is complicated. The following example reveals how one type of improper prior leads to a violation of (3) through the formal application of Bayes's theorem. Examples of this type exist with other sorts of data distributions as well.

Example 1.1. Let X and Y be independent Poisson random variables with X having mean θ and Y having mean $\lambda\theta$. Let the prior be the product of the "flat" improper prior, the Lebesgue measure for λ, and the proper prior gamma (2, 1) distribution for θ. Suppose that we observe $Y = y$ first. The product of the prior and likelihood densities is

$$f(y, \theta, \lambda) = \exp[-\theta(\lambda + 1)](\theta\lambda)^y \theta / y!.$$

The integral of $f(y, \theta, \lambda)$ with respect to λ is $g(y, \theta) = \exp[-\theta]$, and the integral of $g(y, \theta)$ with respect to θ is 1 for all y. Thus a formal application of Bayes's theorem with this improper prior for λ yields the gamma (1, 1) posterior (marginal) density of θ, given $Y = y$. That is, the posterior density $h(\theta | Y = y) = g(y, \theta) = \exp[-\theta]$, for all possible values y. Hence we know before observing Y that the posterior on θ will be different from the prior on θ and will not depend on the value of Y in fact observed. This is an illustration of a foregone conclusion.

Now consider the distribution of X. Prior to observing $Y = y$, we can calculate

$$P(X=2)=\int_0^{\infty}\left(\frac{\theta^2}{2}\exp[-\theta]\right)(\theta\exp[-\theta])d\theta=\frac{3}{16}.$$

After observing $Y = y$, because X and Y are independent given the parameters, we have

$$P(X=2|Y=y)=\int_0^{\infty}\left(\frac{\theta^2}{2}\exp[-\theta]\right)(\exp[-\theta])d\theta=\frac{1}{8},$$

no matter what value y is. That is, we know now that after we observe $Y = y$ (and assuming that we use the formal Bayes's theorem to update our beliefs), regardless of what y is, we will assign $\{X = 2\}$ a probability that is 2/3 the size of its prior probability. Thus we have reasoned to a foregone conclusion about the observable event $\{X = 2\}$ in addition to a foregone conclusion about the unobserved parameter θ.

Next, we begin an explanation of how improper priors can admit such reasoning to a foregone conclusion.

II. THE ROLE OF FINITE ADDITIVITY IN FOREGONE CONCLUSIONS

II.1. *Finite Additivity Allows Experimentation to Foregone Conclusions*

If the probability P is merely finitely additive and not countably additive, then (3) and (4) are not valid. That is, with a finitely additive P, it is possible for a Bayesian agent to design an experiment that surely terminates in foregone conclusions. The following illustrates the key phenomenon, what deFinetti (1974) called "non-conglomerability."

Let P be a (finitely additive) probability defined on the algebra $\mathcal{A}$. Denote expectation with respect to P by $E_P[\cdot]$. Subject to the usual account of finitely additive conditional probability given a σ algebra, denote the P expectation given X by $E_P[\cdot|X]$. (See Dubins 1975, sec. 3, and Schervish, Seidenfeld, and Kadane 1984, p. 213 for some discussion of existence of coherent conditional finitely additive probabilities.)

Definition (deFinetti). P is conglomerable in the (denumerable) partition $X = \{x_1, \ldots\}$ if for every bounded random variable Y and constants k_1 and k_2,

$$k_1 \leq \mathbf{E}_P[Y] \leq k_2 \quad \text{whenever} \quad k_1 \leq \mathbf{E}_P[Y|X = x_i] \leq k_2 \quad (i = 1, \ldots).$$

Example 2.1 (Dubins 1975). Let the set S be the product of a binary event $\{E, E^c\}$ and the countably infinite space of natural numbers, $X = \{1, 2, \ldots\}$. (We identify the events with their indicator functions.) Let the algebra $\mathcal{A}$ be the σ field of all subsets of S. Define a finitely additive P by $P(E) = P(E^c) = 0.5$; $P(X = i|E) = 2^{-i}$; and $P(X = i|E^c) = 0$. That is, given E, X is the random variable corresponding to the selection of an integer by flipping a fair coin until the first "head" results, assuming that eventually the coin lands heads up. Given E^c, X is a "uniformly" chosen positive integer. Then, in violation of conglomerability, $P(E) = 0.5$ yet $P(E|X = i) = 1$, regardless of the observed value ($i = 1, \ldots$). The experimenter who has the personal probability P recognizes that, given each finite sequence of coin flips, E is practically certain.

When probability is updated by Bayes's rule of conditional probability, the experiment X makes E a foregone conclusion, contrary to the findings in Section I. In explanation of what goes wrong with the reasoning from Section I applied to merely finitely additive probabilities, nonconglomerability creates a failure of the equality (2) with respect to the conditional probability $P(E|X = i)$. Specifically, $\mathbf{E}_P[E] = 0.5$, but $\mathbf{E}_P[E|X = x_i] = 1 (i = 1, \ldots)$. Evidently, the door to foregone conclusions is opened whenever P is not conglomerable. Moreover, in this example the conditioning events all have positive probability, $P(X = i) = 2^{-(i+1)} > 0$. There is no issue of conditional probability given events of zero measure to bother with. Bayes's theorem applies without any extra conditions; nor is a special sequential argument needed. The "foregone conclusion" E is arrived at after a single trial of the experiment X.

In earlier work (Schervish et al. 1984) we investigated when, and by how much, finitely additive probabilities are nonconglomerable for events. In short, unless P is countably additive, there is a denumerable partition X and event E for which the conditional probability $P(E|X)$ is to one side of $P(E)$ and bounded away from it. If the data to be observed specify uniquely a member of that partition, then the anomalous behavior of Dubins' example is recreated. (Of course, when P is countably additive, conglomerability is satisfied in every denumerable partition.) Then, as merely finitely additive probabilities display non-conglomerability in predictable ways, are agents whose personal probabilities are not countably additive open to the criticism that they accept sampling to foregone conclusions? Contrary to the views of deFinetti (1974) and Savage (1954), is countable additivity a requirement for a coherent personal probability? Is countable additivity justified to avoid sampling to a foregone conclusion? In Sections II.2, II.3, and II.4 we consider alternative views about finitely additive probability to determine whether they endorse reasoning to foregone conclusions.

II.2. *Finite Additivity and the Value of Experimentation*

Can it be that a Bayesian would rationally pay not to see the results of a cost-free experiment before making a decision? A classic result of Bayesian decision theory (Good 1967; Raiffa and Schlaifer, 1961, sec. 4.5.2; Ramsey 1990) is that cost-free evidence is worth waiting for in advance of making a terminal decision. If it does not cost anything to postpone a decision to conduct a (free) experiment, then, ex ante, delaying the decision to learn something new carries higher expected utility than choosing to act at once. The following is a formalization of this claim.

Suppose that a Bayesian agent has the opportunity now to postpone a terminal decision $D \in \mathcal{D}$, without penalty. Suppose also that the Bayesian agent has the chance to acquire, cost-free, new evidence X from a (finite) experiment before choosing among the same terminal decisions in $\mathcal{D}$. Assume that the agent uses Bayes's rule to update and knows this of himself or herself. (As is shown in Sec. II.3, this turns out to be an important premise for the conclusion that follows.) Then, the current value of deferring the terminal decision until after the experiment is observed is not less than the current value of making a terminal decision at once.

Example 2.2. Consider a binary decision problem: two terminal decisions, D_1 and D_2, on a binary state space E and its complement E^c. The terminal acts are functions from states to outcomes, defined in the usual way: $D_i(E) = o_{i1}$ and $D_i(E^c) = o_{i2}$ $(i = 1, 2)$. The Bayesian agent carries one personal probability P over the events and one utility $U(o_{ij}) = u_{ij}$ over outcomes and seeks to maximize (conditional) expected utility.

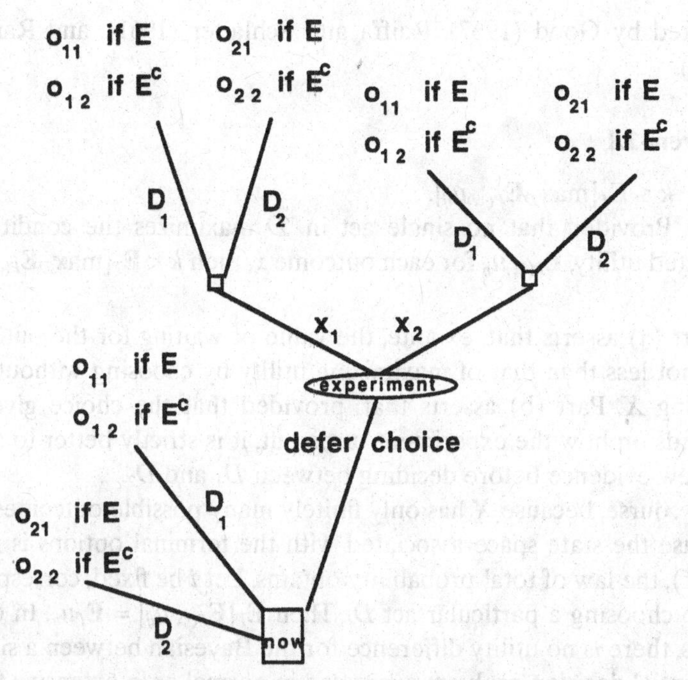

Figure 1. Extensive form for the sequential decision whether or not to experiment and observe, cost-free, the binary random variable X prior to choosing between the terminal decisions D_1 and D_2. Following standard notation, boxes denote choice nodes and ovals denote chance nodes. The payoff structure of each terminal option is provided at the end of the corresponding branch of the decision tree. The first line gives outcome when event E occurs; the second line gives outcome when event E^c occurs.

Assume, for convenience, that the agent judges the acts and states are probabilistically independent, $P(s_j|D_i) = P(s_j)$. Thus denote by $\mathbf{E}_P(u_{ij})$ the agent's expected utility for choosing D_i, $\Sigma_j P(s_j)u_{ij}$. In a terminal decision, now, the agent maximizes expected utility. Say that $\max_{\mathcal{D}}\mathbf{E}_P(u_{ij}) = k$.

Allow a binary experiment, $X = \{x_1, x_2\}$, followed by a choice between D_1 and D_2. (Because the outcome space is finite, there is no issue of countable additivity of P.) Assume that there is no cost for postponing the terminal decision – the utility of outcomes is unaffected by the experiment. The extensive form decision is depicted in Figure 1.

Denote by $\max_{\mathcal{D}}\mathbf{E}_{P(\cdot|x)}u_{ij}$ the maximum conditional expected utility of choosing between D_1 and D_2, given $X = x$. The following result was reported by Good (1967), Raiffa and Schlaifer (1961), and Ramsey (1990).

Theorem 2.1

(a) $k \leq \mathbf{E}_P[\max_{\mathcal{D}} \mathbf{E}_{P(\cdot|x)} u_{ij}]$.

(b) Provided that no single act in $\mathcal{D}$ maximizes the conditional expected utility, $\mathbf{E}_{P(\cdot|x)} u_{ij}$ for each outcome x, then $k < \mathbf{E}_P[\max_{\mathcal{D}} \mathbf{E}_{P(\cdot|x)} u_{ij}]$.

Part (a) asserts that, ex ante, the value of waiting for the outcome X is not less than that of maximizing utility by choosing without first learning X. Part (b) asserts that, provided that the choice given X depends on how the experiment turns out, it is strictly better to await the new evidence before deciding between D_1 and D_2.

Of course, because X has only finitely many possible outcomes and because the state space associated with the terminal options is finite (E, E^c), the law of total probability obtains. Let i be fixed, corresponding to choosing a particular act D_i. Then $\mathbf{E}_P[\mathbf{E}_{P(\cdot|x)} u_{ij}] = \mathbf{E}_P u_{ij}$. In other words, there is no utility difference for the Bayesian between a simple sequential decision problem expressed in normal or in extensive form. However, when the experiment X is not simple and P is merely finitely additive, the law of total probability may fail and there is the possibility that the agent will prefer strictly to decide without first learning X. In fact, the experiment X may carry negative value.

Example 2.2 (continued, itself a continuation of Example 2.1). Suppose that D_1 is a 2:1 bet on event E with stake 3 utiles, $u_{11} = 1$, $u_{12} = -2$. Suppose that D_2 is an even-odds bet against E with stake 2 utiles, $u_{21} = -1$, $u_{22} = 1$. What if the agent has the opportunity of postponing a terminal choice between these two options to learn the outcome of the experiment X? Then, ex ante, D_1 carries negative expected utility ($\mathbf{E}_P(D_1) = -0.5$), whereas D_2 has expected utility zero. That is, D_2 maximizes expected utility, given what the agent knows now. However, given $X = x_i$ (because $P(E|x_i) = 1$, $i = 1, 2$), the conditional expected utility of D_1 is 1, whereas the conditional expected utility of D_2 is -1.

Thus the agent knows that if the choice to maximize expected utility is deferred until X is learned, then D_1 will be chosen for certain. But the current value (the value now) of deferring this sure choice of D_1 is -0.5. Hence, ex ante, the value of waiting to see X and then choosing is -0.5, whereas the value of choosing now (the value of choosing D_2)

is zero. When finitely additive probability is updated by Bayes's rule, it may be better to avoid learning X than to postpone and face a foregone conclusion. The foregone conclusion leads to a choice now judged to be inferior to a current option.

There have been two important earlier attempts to mitigate the disruption to Bayesian theory caused by finite additivity. The next two sections take up each of these attempts in turn.

II.3. *Some Consequences of Goldstein's Theory for Finitely Additive Probability*

Here, we consider what Goldstein's (1983) theory has to say about the problem of reasoning to a foregone conclusion with finitely additive probability. Specifically, we apply his theory to Example 2.1. We think that it is important to review Goldstein's work because his result, (5), appears to deny what we assert about the possibility of reasoning to a foregone conclusion when probability is only finitely additive. We show that Goldstein's theory requires giving up Bayes's conditioning as the rule for updating opinion and, in fact, any rule for updating that is a function of the data.

Let the current prevision for a quantity Z be denoted by $P_N(Z)$, where the subscript N indexes the time now. Also, let $P_F(Z)$ denote the future prevision for Z with respect to some well-defined future time F, later than N. Goldstein used sequential decision reasoning to argue that so-called "G coherence" of the previsions entails the equality

$$P_N(Z) = P_N(P_F(Z)). \tag{5}$$

That is, without regard for what one might learn between now and later, and without assuming how one updates future opinions, if each of the current and future previsions is G-coherent, then the current prevision of Z must equal the current prevision of future prevision of Z.

In short, Goldstein's result, (5), establishes that one cannot (now) expect that the future previsions will suffer a foregone conclusion, regardless of how one plans to stop experimenting. But Goldstein maintained that his theory is consistent with an agent holding (merely) finitely additive previsions, just as deFinetti allowed. How can this be when, as in Example 2.1, finitely additive probabilities open the door to foregone conclusions? The answer in short, we think, is that

Goldstein's theory *prohibits* updating by Bayes's rule (in Eq. 2.1). Here is our analysis.

Suppose that we know now that we will observe X before we need to express $P_F(E)$, but that we are unsure what the total future evidence might be. We might also learn other information, but we are certain by the future time F to learn at least X and, we know this now, at N. In this case we can extend (5) to include some conditional previsions. Specifically,

$$P_N(E|X=k)=P_N(P_F(E)|X=k) \quad (k=1,\ldots) \tag{6}$$

follows within his theory, by the same reasoning that he uses to derive (5). Recall that in Example 2.1, the current probability for the event E is nonconglomerable in the X partition is $P_N(E)=0.5$, yet $P_N(E|X=k) = 1$ $(k=1,\ldots)$. Then, by (6),

$$P_N(P_F(E)|X=k)=1 \quad (k=1,2,\ldots). \tag{7}$$

Because $0 \le P_F(E) \le 1$, (7) asserts that

$$(\forall k)(\forall \varepsilon>0) \quad P_N(1\ge P_F(E)>1-\varepsilon|X=k)=1. \tag{8}$$

From (5) and (8), it follows that one now believes with probability 0.5 that $P_F(E)$ will be near zero.

Claim.

$$(\forall \varepsilon>0) \quad P_N(P_F(E)<\varepsilon)=0.5. \tag{9}$$

Proof. Note that

$$\begin{aligned}(\forall \varepsilon>0) \quad P_N(P_F(E)>1-\varepsilon) &\\ \ge \textstyle\sum_k (P_N(P_F(E)>1-\varepsilon|x=k)P_N(X=k)) &\\ =\textstyle\sum_k P_N(X=k)=0.5. &\end{aligned}$$

The inequality follows by subadditivity of finitely additive probability. The first equality follows from (8). But, by (5), $P_N(P_F(E)) = 0.5$; hence (for all $\varepsilon > 0$) $P_N(0 \le P_F(E) < \varepsilon) = 0.5$.

To discuss the variety of rules permitted in Goldstein's theory for updating probabilities, suppose that between now, N, and the future, F, one learns *only* the new information X. This simplification avoids unnecessary complications about what later is the totality of the new evidence learned.

It is evident that, because of (5), in Goldstein's theory Bayes's rule is *not* (now) an authorized rule for updating your previsions at the later time F. Specifically, Bayes' rule mandates that when X is all the new evidence acquired between N and F,

$$P_F(E) = P_N(E|X). \tag{10}$$

But as $P_N(E|X = k) = 1$ $(k = 1, \ldots)$, under Bayes's rule for updating, $P_F(E) = 1$. If one knows now that (10) applies, then $0.5 = P_N(E) \neq P_N(P_F(E)) = 1$, contradicting (5). Hence, in Example 2.1, Bayes's rule is G incoherent for updating in light of the new information X. In fact, for this example, we show shortly that every rule that is a function of X is also G incoherent for updating.

Recall that we are examining the special case where one currently believes that X is *all* the relevant information that one will learn between now and the future time F when one must assess the prevision for event E. In that case, one currently believes (9); namely, that with probability 0.5, later one will depart in an extreme way from using Bayes's rule to update the prevision for E. Later, one will assign to E a prevision that is maximally far from what Bayes's rule prescribes.

If we suppose that there is a rule for determining $P_F(E)$ that is a function of X alone, then the event $\{P_F(E) < \varepsilon\}$ is also an event of the form $\{X \in B\}$, where B is a set of integers. (This is just what is ordinarily meant by saying that $P_F(E)$ is a function of X alone.) Because $P_N(X = k) > 0$ for all k, then for each $k \in B$, $P_N(P_F(E) < \varepsilon \& X = k) = P_N(X \in B \& X = k) = P_N(X = k) > 0$. But this last inequality implies that $P_N(P_F(E) < \varepsilon | X = k) = 1$, which contradicts (8).

For example, let D_m be the event $\{X > m\}$. It is easy to calculate that $P_N(E|D_m) = (1 + 2^m)^{-1}$. One might contemplate choosing some large integer m and setting $P_F(E) = P_N(E|D_m)$ if D_m occurs and setting $P_F(E) = P_N(E|D_m^c) = 1$ if the complement of D_m, $(D_m^c = \{X \leq m\})$ occurs. If $(1 + 2^m)^{-1} < \varepsilon$, then this rule corresponds to $\{P_F(E) < \varepsilon\} = \{X > m\}$, and (8) is violated for each value $k > m$. One might wish to take the (pointwise) limit of these rules as m goes to infinity. It is easy to see that amounts to using Bayes's rule, which is also G incoherent. How strange it is that G coherence compels one to anticipate changing one's previsions in a manner that is not expressible as a function of the new evidence X that one will acquire between N and F. If it comes to pass that one does choose $P_F(E)$ close to zero, then we would be interested to hear an explanation of why one made such a choice.

The preceding example is typical of what happens in Goldstein's theory when a finitely additive probability is nonconglomerable in the margin of the data X. It is straightforward to show that if G coherence is to accommodate the simple situation where the data have positive probability, then either G coherence requires that finitely additive probabilities be conglomerable in the margin of the data to be observed, or it requires that future previsions not be a function of the data that we know now that we will observe. Next we turn to a theory of finitely additive probability that incorporates the first of these two alternatives.

II.4. *Heath–Sudderth Coherence and Reasoning with Finitely Additive Probability*

Heath and Sudderth (1978) proposed a theory of finitely additive probability based on their own account of (what we call) H–S coherence. Theirs is a *local* (rather than *global*) decision criterion involving wagers on select parameters of interest Θ and quantities to be observed X. An agent is required to indicate conditional fair odds over the sample space $\mathcal{X}$, given each $\theta \in \Theta$, and conditional fair odds over Θ, given each possible observation $x \in \mathcal{X}$. Usually, a statistical "model" fixes conditional probability over the sample space, $P(X|\theta)$, which determines the fair odds for X given Θ. Thus only the agent's "posterior" odds, $P(\Theta|x)$, are open for specification.

Let the function $sI_{X=x}[I_A - P(A|x)]$ be a (called-off) bet on event A, given $X = x$, at the agent's conditional odds $P(A|x)$, with total stake s. Thus the agent wins $s[1 - P(A|x)]$ if both $X = x$ and event A obtains, the agent loses $sP(A|x)$ if both $X = x$ and event A^c obtains, and the bet is annulled if $X \neq x$. For each possible outcome $x \in \mathcal{X}$, let $\mathcal{B}_x$ be a finite selection of bets on events involving Θ (i.e., finitely many bets on subsets of Θ) using the agent's posterior odds $P(\Theta|x)$, and let $\mathcal{B}$ be the system of these bets taken together. H–S coherence requires that there be no such system of bets $\mathcal{B}$ that have a negative *risk*; that is, no $\mathcal{B}$ with $\sup_{\theta \in \Theta} \mathbf{E}_P(\mathcal{B}|\theta) < 0$.

Heath and Sudderth showed (1978, Theorem 1) that H–S coherence is equivalent to the existence of a finitely additive probability P over the joint space, (X, Θ), such that for each bounded (measurable) function $h\ (X, \Theta)$,

$$\mathbf{E}_P[h] = \iint hP(dx|\theta)P(d\theta) = \iint hP(d\theta|x)P(dx). \qquad (11)$$

That is, for the agent to be H–S coherent, the finitely additive odds must be conglomerable in the X partition and in the Θ partition. More precisely, (11) requires that the finitely additive probability p be disintegrable in both partitions. Dubins (1975 Theorem 1) established equivalence of disintegrability in a partition and conglomerability in that partition.

Heath and Sudderth's sense of "coherence" is to be contrasted with, for instance, deFinetti's criterion of avoiding a *sure loss*. DeFinetti's criterion requires only that there be no finite set of fair unconditional, or called-off bets that lead to a negative outcome (rather than negative expectation), bounded below zero, no matter which state (say, no matter which θ) obtains. To see the difference, Heath and Sudderth's system of bets $\mathcal{B}$ includes betting on Θ given X, whereas deFinetti's indexes these as infinitely many called-off bets if X takes on more than finitely many different values. A probability P may support "coherent" fair odds in deFinetti's sense but fail to be H–S coherent – the finitely additive probability of Example 2.1 is one such P. It is evident that if a system of "fair" odds is incoherent in deFinetti's sense, then it is H–S incoherent too. In Example 2.1, event E is the subject of wagering and X is the quantity to be observed. The finitely additive probability P has the property that $0.5 = \mathbf{E}_P[E] \neq \mathbf{E}_P[\mathbf{E}_P[E|X]] = 1$; hence that P is H–S incoherent though coherent in deFinetti's sense.

When a finitely additive probability is H–S coherent, there can be no experiment the agent designs to lead to a foregone conclusion. This obtains provided, of course, that the *combined* set of quantities to be observed through experimentation falls within the scope of the Heath–Sudderth criterion. It may be that P is H–S coherent with bets on event E for observations X; likewise, P may be H–S coherent with bets on E for observations Z, though P is H–S incoherent with bets on E for the combined observations (X, Z). This is illustrated next.

Example 2.3. This example is based on example 4.2 in earlier work (Schervish et al. 1984), that a mixture of nonconglomerable distributions may be conglomerable. Let Q be a finitely additive probability on the (power set) algebra $\mathcal{A} = \{E, E^c\} \times \mathcal{X} \times \{F, F^c\}$ where E and F are (binary) events and $\mathcal{X} = \{1, 2, \ldots\}$. Let $Q(E) = Q(F) = 0.5$, and assume that $Q(E\&F) = Q(E)Q(F) = 0.25$, so that E and F are independent under Q. Let $Q(E, X|F)$ be as $P(E, X)$ in Example 2.1; that is, $Q(E\&X = k|F) = 2^{-k+1}$ and $Q(E^c\&X = k|F) = 0$. Finally, let $Q(E\&X = k|F^c) = 0$ and $Q(E^c\&X = k|F^c) = 2^{-k+1}$.

Observe that because E and F are events, there is no issue of additivity associated with Q on the four atom subalgebra $\{E, E^c\} \times \{F, F^c\}$. That is, trivially, Q is H–S coherent over this subalgebra. Also, note that the probability Q is H–S coherent over the subalgebra $\{E, E^c\} \times \mathcal{X}$. Specifically, $Q(E\&X = k) = Q(E^c\&X = k) = 2^{-k+2}$, and thus $Q(E|X = k) = Q(E) = 0.5$ $(k = 1, \ldots)$. That is, Q is conglomerable in the X partition. Similarly, $Q(F) = Q(F|X = k) = Q(F^c|X = k) = Q(F^c) = 0.5$ $(k = 1, \ldots)$, and Q is H–S coherent over the subalgebra $\mathcal{X} \times \{F, F^c\}$.

However, Q is not conglomerable in the (X, F) partition. Note that $Q(E|X = k\&F) = 1 = Q(E^c|X = k\&F^c)$ (for $k = 1, \ldots$), yet $Q(E|F) = Q(E^c|F^c) = 0.5$. Thus a bettor might display the H–S incoherence in Q by wagering with the agent using the agent's conditional "posterior" odds, once X is revealed. After X is revealed, the bettor makes a "called-off" gamble b_1 on E^c, given F, with the amount \$1 contributed to the stake by the agent against \$$\varepsilon$ wagered by the bettor. The agent wins \$$\varepsilon$ in the event that $E\&F$ occurs. The bettor wins \$1 in the event that $E^c\&F$ occurs. The gamble is called-off in the event that F^c occurs – in which case there is no exchange of funds. The bet b_1 is a favorable (conditional) wager from the standpoint of the agent's posterior Q odds after X is revealed, because, given X, $Q(E|F\&X = k) = 1$ $(k = 1, \ldots)$. For a second bet, b_2, after X is revealed, the bettor wins the agent's \$1 if $E\&F^c$ and the agent wins the bettor's \$$\varepsilon$ if $E^c\&F^c$. The bet b_2 is favorable for the agent because $Q(E^c|F^c\&X = k) = 1$ $(k = 1, \ldots)$. Let $\mathcal{B} = \{b_1 + b_2\}$ be the conjunction of these two bets. B has negative risk regardless whether E or E^c obtains:

$$\mathbf{E}_Q[\mathcal{B}|E] = \$\varepsilon Q(F|E) + \$(-1)Q(F^c|E) = \$(\varepsilon - 1)/2 < 0$$

and

$$\begin{aligned}\mathbf{E}_Q[\mathcal{B}|E^c] &= \$(-1)Q(F|E^c) + \$\varepsilon Q(F^c|E^c) \\ &= \$(\varepsilon - 1)/2 < 0\end{aligned}$$

Thus Q is H–S incoherent over $\mathcal{A}$ when both X and F are learned, though Q is H–S coherent when either one of X of F alone is observed.

The problem that we see with Heath and Sudderth's approach is that, though the agent may use the finitely additive probability Q when betting on E given X or when betting on E given F, what shall that agent do when the offer is made to learn about both X and F when only one is known? Knowing X, how does the agent assess the value of conducting an experiment to learn which of $\{F, F^c\}$ obtains? Knowing F, how does the agent assess the value of conducting an experiment to learn X? It seems difficult to insulate finitely additive probabilities against the threat of nonconglomerability by trying to determine which quantities may be H–S-coherently observed. As the example shows, nonconglomerability may surface when several quantities are observed, though no proper subset of these quantities causes H–S incoherence on its own. An H–S-coherent agent may be confronted with evidence for which his or her beliefs no longer satisfy H–S coherence.

III. CLASSICAL TESTING AND BAYESIAN INFERENCE TO FOREGONE CONCLUSIONS

In Section I we recalled the familiar concern that with classical hypothesis testing, optional stopping opens the door to foregone conclusions when classical (i.e., fixed sample size) significance levels are used to report evidential import. We have seen how the fear of sampling to a foregone conclusion can be alleviated within the (countably additive) Bayesian paradigm. In short, there are bounds on how high the probability can be of sampling until the posterior probability reaches a specified level.

At first, these two results may seem to conflict with the observation that many fixed sample-size classical procedures are Bayes or nearly Bayes procedures. For example, with normal data $X \sim N(\theta, 1)$ and the usual improper prior, the posterior probability that $\Theta \leq \theta_0$ given X is the same as the significance level for testing the hypothesis that $\Theta \leq \theta_0$ against the alternative $\Theta > \theta_0$. If sequential significance testing with a fixed sample-size tests leads to foregone conclusions, is not the same true for the Bayes procedures?

A naive response might be that the improper prior corresponds to a finitely additive prior; hence foregone conclusions are not ruled out in that case, as we discussed in Section II. However, in this particular case, the posterior inferences are coherent in the sense of Heath and Sudderth (1978), so no foregone conclusions are possible. That is, the apparent conflict is not attributable to the mere finite additivity of the probability as represented by the improper prior.

To sort out what is happening, note that we have stated the problem as one in which sampling continues until the posterior probability of the null hypothesis H: $\Theta \leq \theta_0$ rises above a specified level. It can be shown that given $\Theta = \theta \leq \theta_0$, the probability is 1 that such a sampling scheme will stop and assign a high posterior probability to the true null hypothesis. This is true whether one uses the usual improper (finitely additive) prior or a conjugate proper prior. Also, it can be shown that for these same priors, given $\Theta = \theta > \theta_0$, the probability is less than 1 that the sampling scheme will stop and assign a high posterior probability to the false null hypothesis. But if we modify the hypotheses and move the endpoint θ_0 into the alternative hypothesis, then, conditional on $\Theta = \theta_0$, there is probability 1 that the sampling scheme will stop and declare that the probability is high for the false null hypothesis $\Theta < \theta_0$. This case allows a foregone conclusion. Of course, each of the prior distributions used in this analysis puts zero probability on the event $\Theta = \theta_0$.

If one is particularly concerned about the possibility that $\Theta = \theta_0$ (as opposed to Θ being very close to θ_0), then one can avoid the foregone conclusion just mentioned by using a countably additive prior probability that puts sufficient mass at $\Theta = \theta_0$; for example, let $P(\Theta = \theta_0) > p(1 - q)/q$. This will ensure that the conditional probability given $\Theta = \theta_0$ is strictly less than 1 for stopping and declaring that the posterior probability of the false hypothesis $\Theta < \theta_0$ is high. Of course, with this changed prior, the stopping rule (based on the posterior) no longer is to sample until $\Theta < \theta_0$ achieves a preassigned classical significance level. That is our resolution to the earlier puzzle.

Note also that a similar problem does not arise in the two-sided case, because the usual fixed sample-size significance levels are not posterior probabilities, as Jeffreys discussed many years ago (1939, sec. 5.1). (See Lindley 1957 for additional commentary.) And for the special case of a point-null hypothesis with a two-sided alternative, when improper priors are used for the adjustable parameter of the alternative, it can happen that there is a foregone conclusion for the null hypothesis with only one observation. (See Jeffreys 1939, p. 251, for an analysis.)

IV. SUMMARY

In Section I we displayed some elementary reasoning, based on the law of total probability, about the protection simple Bayesian theory affords against sampling to a foregone conclusion. However, Example 1.1 alerts us to the possibility of foregone conclusions when inference is based on an improper prior. In Section II we explored limitations on the results from Section I that arise from using a finitely, but not countably additive, probability corresponding to an improper prior. We considered three perspectives on finitely additive probabilities:

- where foregone conclusions are avoided by strictly preferring; that is, paying to choose "now" rather than to collect "cost-free" evidence
- where foregone conclusions are avoided by mandating against the use of Bayes's rule for updating probabilities, as in Goldstein's theory
- where foregone conclusions are avoided by trying to circumscribe the partitions in which nonconglomerability appears, as in Heath–Sudderth coherence.

Each of these viewpoints carries a price for relaxing the principle of countable additivity.

Of course, there is a fourth perspective which also avoids foregone conclusions. This is to proscribe the use of merely finitely additive probabilities altogether. The cost here would be an inability to use improper priors. These have been found to be useful for various purposes, including reconstructing some basic "classical" inferences, affording "minimax" solutions in statistical decisions when the parameter space is infinite, approximating "ignorance" when the improper distribution is a limit of natural conjugate priors, and modeling what appear to be natural states of belief (Kadane and O'Hagan 1995).

Whether in each case the price is too high is a question for further investigation and serious debate. We need to discuss the extent to which Bayesian methodology affords or ought to afford protection against reasoning to a foregone conlusion.

REFERENCES

Anscombe, F. J. (1954), "Fixed Sample Size Analysis of Sequential Observations," *Biometrics*, 10, 89–100.

Ash, R. B. (1972), *Real Analysis and Probability*, New York: Academic Press.

Berger, J. O. (1985), *Statistical Decision Theory and Bayesian Analysis* (2nd ed.); New York: Springer-Verlag.

Cornfield, J. (1970), "The Frequency Theory of Probability, Bayes's Theorem, and Sequential Clinical Trials," in *Bayesian Statistics*, eds. D. L. Meyer and R. O. Collier, Jr., Itasca, IL: Peacock Publishing, pp. 1–28.

deFinetti, B. (1974), *The Theory of Probability*, New York: John Wiley.

Doob, J. L. (1953), *Stochastic Processes*, New York: John Wiley.

Dubins, L. E. (1975), "Finitely Additive Conditional Probabilities, Conglomerability, and Disintegrations," *Annals of Probability*, 3, 89–99.

Feller, W. (1940), "Statistical Aspects of E.S.P.," *Journal of Parapsychology*, 4, 271–298.

Goldstein, M. (1983), "The Prevision of a Prevision," *Journal of the American Statistical Association*, 78, 817–819.

Good, I. J. (1967), "On the Principle of Total Evidence," *The British Journal for the Philosophy of Science*, 17, 319–321.

Heath, D., and Sudderth, W. (1978), "On Finitely Additive Priors, Coherence, and Extended Admissibility," *The Annals of Statistics*, 6, 333–345.

Jeffreys, H. (1939), *Theory of Probability*, Oxford, U.K.: Oxford University Press.

Kadane, J. B., and O'Hagan, A. (1995), "Using Finitely Additive Probability: Uniform Distributions on the Natural Numbers," *Journal of the American Statistical Association*, 90, 626–631.

Kadane, J. B., Schervish, M. J., and Seidenfeld, T. (1986), "Statistical Implications of Finitely Additive Probability," *Bayesian Inference and Decision Techniques With Applications*, eds. P. K. Goel and A. Zellner, New York: Elsevier Science Publishers, pp. 59–76. [Chapter 2.5, this volume]

Kerridge, D. (1963), "Bounds for the Frequency of Misleading Bayes Inferences," *Annals of Mathematical Statistics*, 34, 1109–1110.

Lindley, D. V. (1957), "A Statistical Paradox," *Biometrika*, 44, 187–192.

Raiffa, H., and Shlaifer, R. (1961), *Applied Statistical Decision Theory*, Cambridge, MA: Harvard University Press.

Ramsey, F. P. (1990), "Weight or the Value of Knowledge," *The British Journal for the Philosophy of Science*, 41, 1–4.

Robbins, H. (1952), "Some Aspects of the Sequential Design of Experiments," *Bulletin of the American Mathematical Society*, 58, 527–535.

Savage, L. J. (1954), *The Foundations of Statistics*, New York: John Wiley.

Savage, L. J. with prepared contributions from Bartlett, M. S., Barnard, G. A., Cox, D. R., Pearson, E. S., and Smith, C. A. B. (1962), *The Foundations of Statistical Inference*, London: Methuen.

Schervish, M. J., Seidenfeld, T., and Kadane, J. B. (1984), "The Extent of Non-Conglomerability of Finitely Additive Probabilities," *Zeitschrift fur Wahrscheinlichkeitstheorie und verwandte Gebiete*, 66, 205–226.

3.8

When Several Bayesians Agree That There Will Be No Reasoning to a Foregone Conclusion

JOSEPH B. KADANE, MARK J. SCHERVISH, AND TEDDY SEIDENFELD

ABSTRACT

When can a Bayesian investigator select an hypothesis H and design an experiment (or a sequence of experiments) to make certain that, given the experimental outcome(s), the posterior probability of H will be lower than its prior probability? We report an elementary result which establishes sufficient conditions under which this reasoning to a foregone conclusion cannot occur. Through an example, we discuss how this result extends to the perspective of an onlooker who agrees with the investigator about the statistical model for the data but who holds a different prior probability for the statistical parameters of that model. We consider, specifically, one-sided and two-sided statistical hypotheses involving *i.i.d.* Normal data with conjugate priors. In a concluding section, using an "improper" prior, we illustrate how the preceding results depend upon the assumption that probability is countably additive.

I. EXPECTED CONDITIONAL PROBABILITIES AND REASONING TO FOREGONE CONCLUSIONS

Suppose that an investigator has his or her designs on rejecting, or at least making doubtful, a particular statistical hypothesis H. To what

This research was supported by ONR contract N00014–89-J-1851, and NSF grants DMS-9303557 and SES-9123370.

Reprinted with permission from *Philosophy of Science*, 63 (1996): S281–S289, published by the University of Chicago Press.

extent does basic inductive methodology insure that, without violating the total evidence requirement, this intent cannot be sure to succeed? We distinguish two forms of the question:

1. Can the investigator plan an experiment so that *he* or *she* is certain it will halt with evidence that disconfirms H?
2. Can the investigator plan an experiment so that *others* are certain that the investigator will halt with evidence that disconfirms H?

That is, are foregone conclusions, viewed either (1) in the first-person or (2) in the third-person perspective, precluded by fundamental principles in the design of experiments? (In both versions of the question, we understand the judgment of disconfirmation to be the investigator's.)

Related to these questions is the familiar controversy whether an experimenter's stopping rule is relevant to the analysis of his or her experimental data. Savage writes (1962, 18),

> The [*likelihood*] principle has important implications in connection with optional stopping. Suppose the experimenter admitted that he had seen 6 red-eyed flies in 100 and had then stopped because he felt that he had thereby accumulated enough data to overthrow some popular theory that there should be about 1 per cent red-eyed flies. Does this affect the interpretation of 6 out of 100? Statistical tradition emphasizes, in connection with this question, that if the sequential properties of his experimental programme are ignored, the persistent experimenter can arrive at data that nominally reject any null hypothesis at any significance level, when the null hypothesis is in fact true. These truths are usually misinterpreted to suggest that the data of such a persistent experimenter are worthless or at least need special interpretation; see, for example, Anscombe (1954), Feller (1940), Robbins (1952). The likelihood principle, however, affirms that the experimenter's intention to persist does not change the import of his experience.

Here is a simple example of what Savage refers to as "statistical tradition." Consider the null hypothesis H_0: $\theta = 0$, that the mean of *i.i.d.* Normal data is 0. The data have known unit variance. Fix k_α so that $k_\alpha/\sqrt{n}$ corresponds to the nominal rejection point in an α-level UMPU (Uniformly Most Powerful, Unbiased) test of H_0 versus the composite alternative hypothesis H_0^c: $\theta \neq 0$, based on a sample of size n. Let H_t denote the simple hypothesis that $\theta = t$. Continue observing data until the sample average $\bar{x}_n = (x_1 + \ldots + x_n)/n$, satisfies the inequality (1.1), then halt:

$$|\bar{x}_n| > k_\alpha/\sqrt{n} \tag{1.1}$$

The likelihood principle entails that the statistician's intent to stop only when (1.1) obtains is irrelevant to the "evidential import" of the data for hypotheses about θ (for a recent view, see Berger 1985, §7.7). However, the statistician has here designed an experiment that, provided it stops, yields data with a very low likelihood for H_0 versus a rival hypothesis, H_t (for $t = \bar{x}_n$). If, contrary to traditional (Neyman-Pearson) theory, the significance level is calculated independent of the stopping rule for the experiment – a mistake by that traditional theory – then when the inquiry halts H_0 has achieved an observed significance of α, or less. Moreover, given the truth of H_0, by the law of the iterated logarithm, with probability 1 the experiment terminates, i.e., almost surely the inequality (1.1) is eventually satisfied.

Traditional hypothesis testing sidesteps this foregone conclusion (of a low significance level) only by incorporating the experimenter's intention, of when to terminate sampling, as part of the relevant evidence. This is contrary to the likelihood principle. By contrast, the Bayesian answer to the first-person version of our question (1) is straightforward and elementary. In short, with a countably additive probability, the law of total probability ensures that conditional probabilities cannot lead to a foregone conclusion. Thus, when the investigator uses Bayes' rule for updating, he or she cannot plan an experiment that, by his or her own lights, leads to a posterior opinion surely below (or surely above) his or her prior opinion. This argument has been reported before. (See, e.g., D. Kerridge's 1963 note.) Kadane et al. have discussed it elsewhere (1996) and we include the bare details here for completeness of our presentation.

Let $(S, \mathcal{A}, P)$ be a (countably additive) probability space, which we think of as the underlying joint space for all quantities of interest. Expectations are with respect to the probability P. Unconditional expectation is denoted by $E(\cdot)$ and conditional expectation given a random variable X is denoted by $E(\cdot \mid X)$. Let $(\mathcal{X}, \mathcal{B})$ and (Ω, τ) be measurable spaces where,

$X: S \to \mathcal{X}$ is a random quantity to be learned,
$\Theta: S \to \Omega$ is any random quantity, and
$h: \Omega \to \Re^*$ is an (extended) real-valued function whose expectation $\mathbf{E}(h)$ exists.

Then the familiar law of total probability implies that:

$$\mathbf{E}[\mathbf{E}(h(\Theta)|X)] = \mathbf{E}(h(\Theta)). \tag{1.2}$$

(See, e.g., Ash 1972, T. 6.5.4, p. 257).

This result says that there can be no experiment with outcome X designed (almost surely with respect to P) to drive up or designed to drive down the conditional expectation of h, given X. Equation (1.2) has no special logical dependence on Bayes' theorem, except that non-Bayesian statistical methods often begin with the claim that neither the "prior" expectation $E(h(\Theta))$ nor the "posterior" expectation $E(h(\Theta)|X)$ has objective status.

For example, suppose that h is the indicator for an hypothesis H (an unobserved "event") in Ω, i.e., $h(\Theta) = 1$ if $\Theta \in H$, $h(\Theta) = 0$ otherwise. Thus, $E(h(\Theta))$ is the agent's "prior" probability of H, denoted by $P(H) = p$. Let $X_1, X_2, \ldots$ be observations which become available sequentially. In order to consider experimental designs which mandate a minimum sample size, $k \geq 0$, define $N = \inf\{n \geq k: P(H|X_1, \ldots, X_n) \geq q\}$ where $N = \infty$ if the set is empty. That is, N identifies the first point after the kth in the sequence of X_i-observations when the agent's "posterior" probability of H reaches q, at least. The event $N = \infty$ obtains when, after k-many observations, the sequence of conditional probabilities, $P(H|X_1, \ldots, X_n)$ $(n \geq k)$, all remain below q. Then,

$$P(N < \infty) \leq p/q < 1. \tag{1.3}$$

Thus, when $p < q$, the agent's prior probability is less than 1 that, using Bayes' rule for updating, he or she will halt the sequence of experiments and conclude that the posterior probability of H has risen to q, at least.

The same argument provides a bound for the conditional probability of terminating the experiment in finite time, given that H is false:

$$P(N < \infty | \neg H) \leq p(1-q)/q(1-p). \tag{1.4}$$

For example, assume $0 < p \leq 0.1$ and let $q/p = 10$. That is, by the choice of the stopping rule, whenever the experiment terminates the posterior probability for H increases tenfold (at least). Then, the inequality (1.4) asserts that, given that H is false, the conditional probability of terminating the experiment is no more than $(0.1 - p)/(1 - p) < 0.1$.

Savage (1962, 72–73) offers a similar conclusion and illustrates it with a simple case of two Binomial hypotheses. D. Kerridge (1963) derives the same bounds as in (1.4) for the case of a uniform prior

($p = 0.5$). And Cornfield (1970, 20–21) uses Kerridge's inequality to argue that as a Bayesian, you cannot be sure to defeat a true "null" hypothesis. However, these results offer answers only to the first version of our opening question. In the following section, we extend the analysis to cover the second version of our question, involving the perspective of an onlooker.

II. HYPOTHESIS TESTING AND SAMPLING TO A FOREGONE POSTERIOR PROBABILITY: A SECOND PERSPECTIVE

The results of section I are "internal" to the Bayesian agent who is designing the sequential experiment. For example, the probability bound (1.4) is for the investigator's personal probability who designed the sequential experiment. It tells the experimenter who is prepared to stop at the first observation when there is a rise of at least $q - p$ in his (her) probability of H that, given $\neg H$, it is not certain that the experiment ever halts. Apart from this "internal" check on reasoning to a foregone conclusion, what can be said from the standpoint of an onlooker who ponders whether the experimenter will come to assign a high posterior probability to a false hypothesis?

II.1. *The Probability of Halting When H Is False*

One way to answer this question is to consider a second point of view that agrees with the first about the likelihood (or sampling model) of the data, but which differs with the experimenter's prior probability of the hypotheses H and $\neg H$. This is our approach in what follows. We consider two varieties of hypotheses, where $\neg H$ is a one-sided or a two-sided alternative. Also, when $\neg H$ is two-sided, there is the distinction between an interval-valued and a (traditional) point-valued null hypothesis H. The experimenter designing the sequential study stops gathering data once his or her posterior probability for H rises to q, at least. His (her) prior probability for H is p (which is $<q$), so that the experiment terminates on the first occasion when H has gained at least $q - p > 0$ over its prior probability.

For our illustration, the possible observations, $X_1, X_2, \ldots$ are *i.i.d.* Normal $\mathbf{N}(\theta, 1)$ data. Let the experimenter have a conjugate Normal prior over Θ, $\mathbf{N}(\mu, 1/\lambda)$, with specified mean μ and precision λ. (We make the obvious adjustment and follow Jeffreys' 1939 analysis for the case where H is a point-valued hypothesis.) We do not restrict the onlooker's prior over Θ, except that we assume it is a countably additive probability. Given $\neg H$, does the onlooker believe, along with the experimenter, that with positive probability the latter will assign H probability less than q at each stage of an unending inquiry? Will both parties agree that, given H is false, it is *not* a certainty the experimenter will arrive at the foregone conclusion he (she) aims for?

ONE-SIDED TESTING. For a specified quantity θ_0, H is the hypothesis that $-\infty < \theta \leq \theta_0$ and $\neg H$ is the complementary hypothesis that $\theta_0 < \theta < \infty$. We show that no matter what the onlooker's prior probability for $\neg H$, in parallel with (1.4), he (she) too believes that, given $\neg H$, it is *less than certain* that the experimenter will ever conclude the study. We attack this problem by demonstrating a result that makes sense from a Classical point of view:

Theorem 2.1. For each $\theta > \theta_0$, P_θ(experimenter stops inquiry) < 1. □

(The proof of this theorem is given in the Appendix.) This theorem establishes that, *no matter how H fails*, still it is not a certainty that the experimenter comes to assign H (posterior) probability q, at least. Then, we have as an easy consequence:

Corollary. From the onlooker's perspective, $P(\text{experimenter halts} \mid \neg H) < 1$. □

TWO-SIDED TESTING. Next, we examine two cases: where H is an interval with non-empty interior, and the traditional point-null hypothesis (as presented in Jeffreys' work). Let H be the interval hypothesis that $-\infty < \theta_a \leq \theta \leq \theta_b < \infty$ $(\theta_a < \theta_b)$ and $\neg H$ is the complementary hypothesis that $\theta < \theta_a$ or $\theta_b < \theta$. By reasoning similar to the case of one-sided testing, we show that the onlooker agrees with the experimenter that, given $\neg H$, there is positive probability the study never terminates.

Theorem 2.2. For each $\theta > \theta_b$, and for each $\theta < \theta_a$, P_θ(experimenter stops inquiry) < 1. □

Therefore, the onlooker agrees, given $\neg H$, there is positive probability the experimenter does not arrive at the foregone conclusion that H has increased its (posterior) probability to q.

Third, we adapt the experiment's "prior" to address the traditional case of a point-valued null hypothesis using Jeffreys' (1939, Chapter 5) model of Bayesian hypothesis testing. Let H assert that $\theta = \theta_0$. The experimenter's "prior" for θ satisfies:

$$P(H) = p \quad \text{and} \quad \text{given} \neg H, \theta \sim \mathbf{N}(\theta_0, 1/\lambda).$$

Theorem 2.3. For each $\theta \neq \theta_0$ P_θ(experimenter stops inquiry) < 1. □

We have considered three cases for the "null" hypothesis H: where H is either (1) a one-sided interval, $\theta \leq \theta_0$; or (2) H is a two-sided interval, $\theta_a \leq \theta \leq \theta_b$; or (3) H is a point $\theta = \theta_0$. In each case, an onlooker who shares the same statistical model as the experimenter but who has a different prior over θ, also believes that if H fails then it is not certain that the experimenter will ever conclude sampling.

II.2. *The Probability of Halting When H Obtains*

Consider, now, the dual question, when H is true. Will the experimenter him/herself be sure of halting, given that H obtains? Will the onlooker, with a different prior for H, agree that the experimenter is sure to halt if H is true? The answer is yes to both questions, as straightforward reasoning establishes:

One-Sided Testing Lemma 2.1. Given θ, $\theta \leq \theta_0$, P_θ(experiment halts) = 1. □

Two-Sided Testing Lemma 2.2. Given θ, $\theta_a \leq \theta \leq \theta_b$, then P_θ(experiment halts) = 1. □

Thus, both the experimenter and the onlooker are certain (and they know the other is certain) that, given H, the study stops. They share a common expectation that, given H, the former comes to assign H a posterior probability of q, at least: P(experimenter halts the study $| H) = 1$.

Remark. These lemmas imply that, by continuity, unfortunately,

$$\limsup_{\theta \in \neg H} P_\theta(\text{experimenter halts}) = 1.$$

Thus, the conditional probability of stopping when H fails cannot be bounded away from 1.

II.3. *Summary of the Discussion about Hypothesis Testing*

We have examined the case of a second Bayesian, an onlooker, including as a special instance a so-called "frequentist" (whose "prior" may be concentrated on a single value of the parameter), who considers the same problem as the Bayesian experimenter. We have given details for the case of testing the mean of a normal distribution. In Section I, we saw that an experimenter who assigns an hypothesis probability p believes that the conditional probability, given that the hypothesis is false, is less than 1 that he or she will stop sampling and declare that the posterior probability is at least $q > p$. In this section, for the special case of testing a normal mean, we saw that an onlooker also believes that, given the hypothesis is false, the probability is less than 1 that the experimenter will stop and declare the probability of the hypothesis to be at least q.

We also saw that, given the truth of the hypothesis, the onlooker (and the experimenter) believe that the experimenter will stop and declare the probability of the hypothesis to be at least q. These results applied to the cases of hypotheses of the form $\theta \leq \theta_0$, $\theta_a \leq \theta \leq \theta_b$, and $\theta = \theta_0$, using natural conjugate or related priors. Since the lemma (Lemma A of the Appendix) on which these results rest applies in greater generality, it stands to reason that these results can be extended. Our purpose with these examples is to illustrate where the "internal" conclusions (of Section I) extend to the perspective of an onlooker.

III. SUMMARY AND A CAVEAT ABOUT "IMPROPER" PRIORS

In Section I, we displayed some elementary reasoning, based on the law of total probability, about the protection simple Bayesian theory affords against sampling to a foregone conclusion. In Section II, using a problem of inference about a normal mean, we extended these results to the perspective of a second investigator and to some results that apply from a traditional (Neyman-Pearson) point of view as well, i.e., results that apply regardless of the unknown value of the parameter.

We conclude, however, with a warning that the foregoing analysis depends upon the assumption that personal probability is countably additive, rather than being merely finitely additive. (Kadane et al. (1996) examine the connection between Bayesian reasoning to a foregone conclusion and the use of merely finitely additive probability.) This is particularly important when the Bayesian investigator uses a so-called "improper" prior to capture a merely finitely additive personal probability. Here, then, is an illustration of how improper priors can produce Bayesian inference to a foregone conclusion, even from the first-person perspective of the investigator.

Let $X \sim N(\theta, \tau)$ and, for a first case without reasoning to a foregone conclusion, assume that the precision τ (= variance^{-1}) is known. Suppose the prior for θ is conjugate, $\theta \sim N(\mu_0, \kappa_0)$. Then the posterior distribution of θ given $X = x$ is normal with mean

$$\mu_0\left(\frac{\kappa_0}{\kappa_0+\tau}\right)+x\left(\frac{\tau}{\kappa_0+\tau}\right) \text{and precision } \kappa_0+\tau.$$

If the prior is, on a sequence, getting more diffuse, i.e., when $\kappa_0 \to 0$, this posterior approaches a normal distribution with mean x and precision τ. This posterior can be recovered by a formal Bayes calculation with an improper, uniform distribution for θ, uniform on $(-\infty, \infty)$.

Now, for a second version where reasoning to a foregone conclusion occurs, suppose that τ is unknown as well. The conjugate prior for θ, given τ, is normal with mean μ_0 and precision $\kappa_0\tau$, and the prior on τ is the Gamma (α_0, β_0) distribution, a proper distribution! The posterior distribution is of the same form, with posterior hyperparameters:

$$\mu_1=\left(\frac{\kappa_0\mu_0+x}{\kappa_0+1}\right), \text{precision}(\kappa_0+1)\tau, \alpha_1=\alpha_0+1/2$$

$$\text{and } \beta_1=\beta_0+\frac{\kappa_0(x-\mu_0)^2}{2(\kappa_0+1)}. \tag{3.1}$$

Again, the prior on θ is taken to be diffuse, so we allow that $\kappa_0 \to 0$ in (3.1). Then, the posterior remains in the conjugate family, with:

$$\mu_1 = x, \text{precision } \tau, \alpha_1=\alpha_0+1/2, \text{and } \beta_1=\beta_0. \tag{3.2}$$

This posterior has the following peculiar implication: The prior on τ is Gamma (α_0, β_0) while the posterior (as calculated) is Gamma $(\alpha_0 + 1/2, \beta_0)$. The posterior on τ does not depend on the datum $X = x$ and is different from its prior. Thus, prior to observing $X = x$ the investigator believes τ is Gamma (α_0, β_0). However, regardless what value of X is observed, he or she knows (even in advance) that afterwards the posterior will give τ a Gamma $(\alpha_0 + 1/2, \beta_0)$ distribution. Thus, the investigator will have reasoned to a foregone conclusion about τ.

One can show, using the results of Regazzini (1987) or of Berti, Regazzini, and Rigo (1991), that the limit of priors, limit of posteriors, and likelihood in this example are jointly coherent, in deFinetti's (1974) sense. That is, there is no finite collection of pairs of events $(E_1, A_1), \ldots, (E_n, A_n)$ and constants $c_1, \ldots, c_n$ such that

$$\sum_{i=1}^{n} c_i I_{A_i}[I_{E_i} - \Pr(E_i|A_i)]$$

is uniformly negative. Thus, deFinetti's standard of *coherence* is not sufficient to ward off reasoning to a foregone conclusion.

Moreover, there exists a joint improper prior where the posterior (3.2) can be obtained by a naïve application of Bayes' theorem, namely the product of the (improper) conditional density for θ given τ, $\tau^{0.5}d\theta$, and the (proper) marginal Gamma (α_0, β_0) density for τ.

$$\tau^{\alpha 0-0.5}e^{-}\beta^{0_\tau}d\theta d\tau \tag{3.3}$$

Remark. Inspection of (3.3) shows that the improper joint density for (θ, τ) is also the product of the uniform, improper density for θ, $d\theta$, and the proper Gamma $(\alpha_0 + 0.5, \beta_0)$ distribution for τ – for which there is no reasoning to a foregone conclusion about τ based on X. A direct elicitation of the agent's prior (marginal) opinion about the parameter τ will distinguish between these two different priors. That information cannot be recovered, however, by considering the predictive distribution of the data.

In our opening section we recalled the familiar concern that, with traditional (Neyman-Pearson) hypothesis testing, optional stopping opens the door to foregone conclusions when (fixed sample-size) significance levels are used to report evidential import. We have seen how the fear of sampling to a foregone conclusion can be alleviated within the Bayesian paradigm, under the assumption that all probability is countably additive. In short, then there are bounds on how high the probability can be of sampling until the posterior probability reaches a specified level. Moreover, these safeguards are recognized also from the standpoint of an onlooker whose prior may differ from the investigator's. However, the safeguards are no longer in place when the Bayesian analysis appeals to "improper" priors (since they correspond to finitely additive probabilities that are not countably additive). Therefore, it is important, we think, to examine more carefully when the use of "improper" priors lead to foregone conclusions, even though they satisfy deFinetti's or Savage's (1954) standards of coherent opinion.

For the proof of Theorem 2.1 (Section II.1), we use a lemma, below, which is a substitution instance of Theorem 2 of Chow and Teicher (1988, p. 146):

Lemma A. Suppose that $Z_1, Z_2, \ldots$ are *i.i.d.* with $E(Z_i) = \mu > 0$ and assume that the moment generating function for Z_i exists. Let $Y_n = \sum_{i=1}^{n} Z_i$, for $n = 1, 2, \ldots$. Fix $c < 0$ and let $M = \min\{n: Y_n \leq c\}$. Then $P(M < \infty) < 1$. □.

Theorem 2.1. For each $\theta > \theta_0$, P_θ(experimenter stops inquiry) < 1.

Proof. Let $S_n = \sum_{i=1}^{n} X_i$. Then the experimenter's posterior probability distribution of Θ is

$$\Theta \sim \mathbf{N}\left(\frac{S_n + \lambda\mu}{n+\lambda}, \frac{1}{n+\lambda}\right).$$

Consider the prospect of sampling until the experimenter's posterior probability of H is at least q. The experimenter's posterior probability of H is:

$$\mathbf{\Phi}\left(\sqrt{n+\lambda}\left[\theta_0 - \frac{S_n + \lambda\mu}{n+\lambda}\right]\right),$$

where $\mathbf{\Phi}$ is the standard normal cdf. This expression is at least q if and only if

$$S_n \leq n\theta_0 - [\surd(n+\lambda)]\mathbf{\Phi}^{-1}(q) + \lambda(\theta_0 - \mu). \tag{A.1}$$

Note that as p is the experimenter's prior probability of H, $\mathbf{\Phi}^{-1}(p) = [\surd\lambda](\theta_0 - \mu)$.

To apply the lemma, choose $\theta > \theta_0$ and define $Z_i = X_i - (\theta + \theta_0)/2$. Then, $\mathbf{E}_\theta(Z_i) > 0$. Write

$$Y_n = \sum_{i=1}^{n} Z_i = S_n - n\frac{\theta_0 + \theta}{2}.$$

Then condition (A.1) becomes:

$$Y_n \leq -[\sqrt{(n+\lambda)}]\boldsymbol{\Phi}^{-1}(q) + \lambda(\theta_0 - \mu) - n(\theta - \theta_0)/2 = c_n. \qquad \text{(A.2)}$$

The experimenter's stopping rule is to halt at the first $n = N$ such that $Y_n \leq c_n$. Observe that c_n is *decreasing* at rate $O(n)$. Thus,

$$\forall(\theta > \theta_0)\exists m_\theta \text{ (both: } c_{m\theta} < 0 \text{ and } c_j < c_{m\theta} < c_i \text{ whenever } j > m_\theta > i).$$

Lemma A applies to the Y_i and Z_i with $c = c_{m_\theta}$. That is, let $M = \min\{n: Y_n \leq c_{m_\theta}\}$. Then, by the Lemma A, $P_\theta(M < \infty) < 1$.

Remark. If $q \geq 0.5$ then $\boldsymbol{\Phi}^{-1}(q) \geq 0$. Since $q > p$, $[\sqrt{\lambda}]\boldsymbol{\Phi}^{-1}(p) - [\sqrt{(n + \lambda)}]\boldsymbol{\Phi}^{-1}(q) < 0$. Thus, $\lambda(\theta_0 - \mu) - [\sqrt{(n + \lambda)}]\boldsymbol{\Phi}^{-1}(q) < 0$ and $m_\theta = 1$.

Given θ, Z_i is a (normal) random variable whose support is the entire real line. Hence, there is positive probability that the experiment continues at least to the m_θ^{th} trial. That is, denote $P_\theta(Y_i > c_i: i = 1, \ldots, m_\theta - 1)$ by $\mathbf{k}$, and we know that $\mathbf{k} > 0$. We use this fact as follows. The experimenter's stopping rule is to halt at the first $n = N$ such that $Y_n \leq c_n$. Thus, as $c_j < c_{m_\theta}$ for $j > m_\theta$, having reached the m_θ^{th} trial the experimenter halts no sooner than the first $n = M$ such that $Y_n \leq c_{m_\theta}$. Thus, if the m_θ^{th} trial is reached, $N \geq M$. Then

$$P_\theta(N < \infty | Y_i > c_i{:}\, i = 1, \ldots, m_\theta - 1)$$
$$\leq P_\theta(M < \infty | Y_i > c_i{:}\, i = 1, \ldots, m_\theta - 1).$$

But, $P_\theta(M < \infty | Y_i > c_i{:}\ i = 1, \ldots, m_\theta - 1) \leq P_\theta(M < \infty | M \geq m_\theta) < P_\theta(M < \infty) < 1$. The first of these inequalities follows because $c_i > c_{m_\theta}$: $i = 1, \ldots, m_\theta - 1$. The second inequality follows because $P_\theta(M < m_\theta) > 0$. And the third inequality is given by Lemma A. Let $\mathbf{k}'$ denote the conditional probability $P_\theta(N < \infty | Y_i > c_i{:}\ i = 1, \ldots, m_\theta - 1)$. Thus, max $\{\mathbf{k}, \mathbf{k}'\} < 1$. By the multiplication theorem, $P_\theta(N < \infty) = 1 - \mathbf{k} + \mathbf{kk}'$. Hence, $P_\theta(N < \infty) < 1$: there is positive probability that the experiment fails to terminate. □

Theorems 2.2 and 2.3 are established in a similar fashion, using Lemma A.

Lemmas 2.1 and 2.2 (Section II.2) are proven by cases: Use the *Strong Law of Large Numbers* whenever θ is not an endpoint of the hypothesis H, and use the *Law of the Iterated Logarithm* when θ is an endpoint of H.

REFERENCES

Anscombe, F. J. (1954), "Fixed-sample-size analysis of sequential observations", *Biometrics* 10: 89–100.

Ash, R. B. (1972), *Real Analysis and Probability*. New York: Academic Press.

Berger, J. O. (1985), *Statistical Decision Theory and Bayesian Analysis* (2nd ed.). New York: Springer-Verlag.

Berti, P., Regazzini, E., and Rigo, P. (1991), "Coherent Statistical Inference and Bayes' Theorem". *Annals of Statistics* 19: 366–381.

Chow, Y. S. and Teicher, H. (1988), *Probability Theory: Independence, Interchangeability, Martingales* (2nd ed.). New York: Springer-Verlag.

Cornfield, J. (1970), "The Frequency Theory of Probability, Bayes' Theorem, and Sequential Clinical Trials." In D. L. Meyer and R. O. Collier, Jr. (eds.), *Bayesian Statistics*. Itasca, IL: Peacock Publishing, pp. 1–28.

deFinetti, B. (1974), *The Theory of Probability* (2 volumes). New York: John Wiley.

Feller, W. (1940), "Statistical Aspects of E.S.P.", *J. Parapsychology*, 4, 271–298.

Jeffreys, H. (1939), *Theory of Probability* (3rd edition, 1961). New York: Oxford University Press.

Kadane, J. B., Schervish, M. J., and Seidenfeld, T. (1996), "Reasoning to a Foregone Conclusion", *J. American Statistical Association*, 91, 1228–1236. [Chapter 3.7, this volume]

Kerridge, D. (1963), "Bounds for the Frequency of Misleading Bayes Inferences", *Annals of Math. Statistics* 34: 1109–1110.

Regazzini, E. (1987), "deFinetti's Coherence and Statistical Inference", *Annals of Statistics* 15: 845–864.

Robbins, H. (1952), "Some aspects of the sequential design of experiments", *Bull. American Mathematical Society* 58: 527–535.

Savage, L. J. (1954), *The Foundations of Statistics*. New York: John Wiley.

Savage, L. J. et al. (1962), *The Foundations of Statistical Inference.* London: Methuen.

Index of Names

Subject Index

Printed in the United States
By Bookmasters